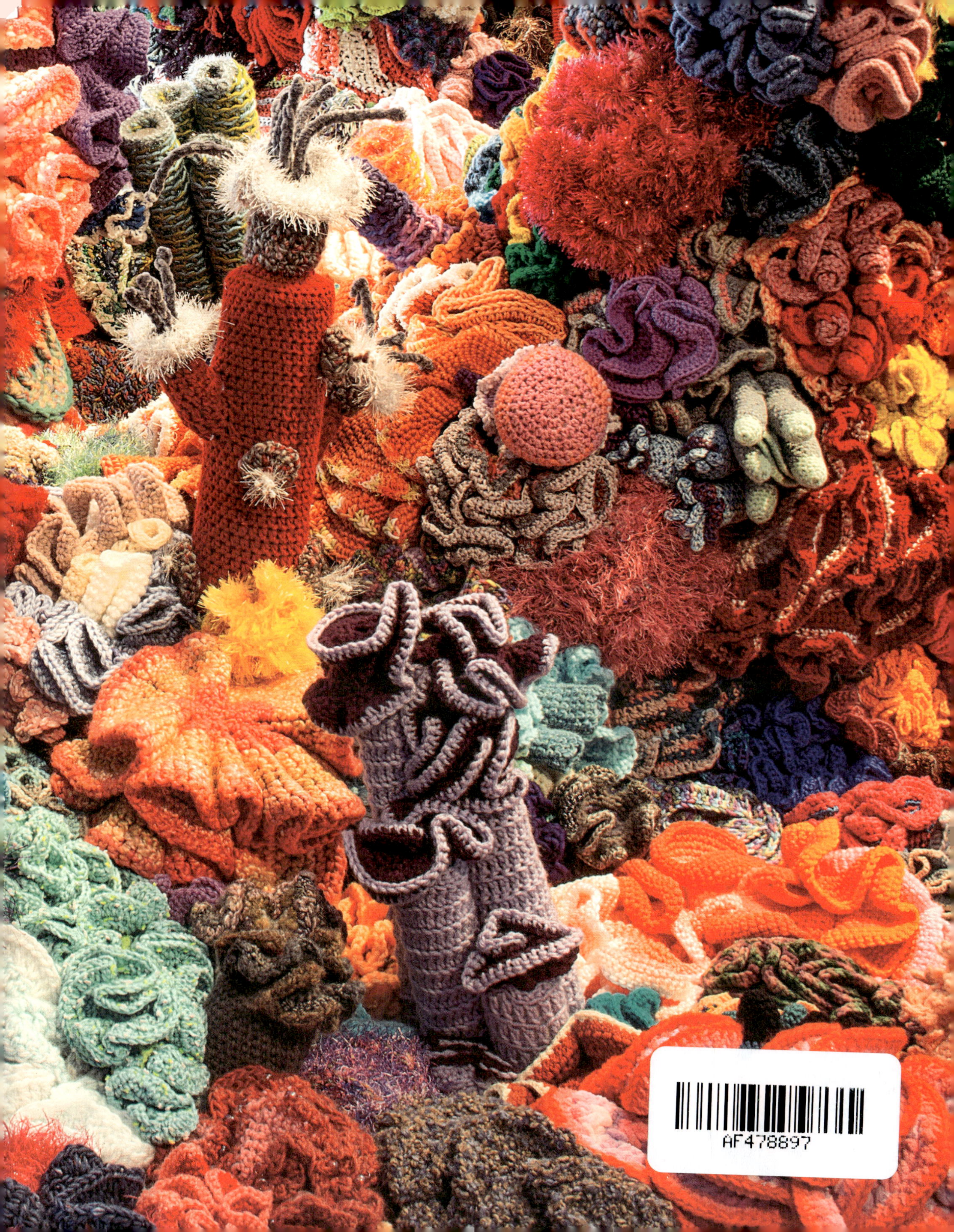

AF478897

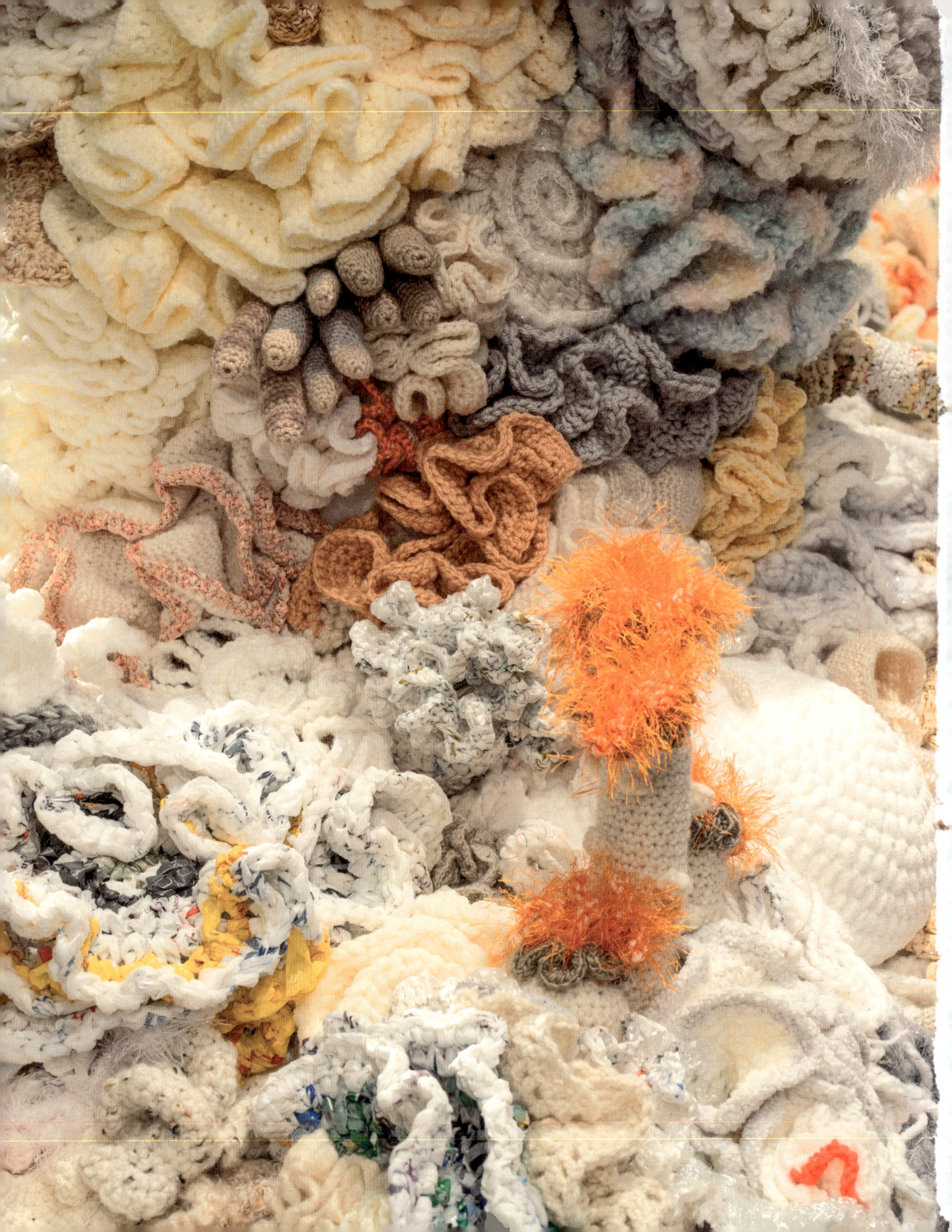

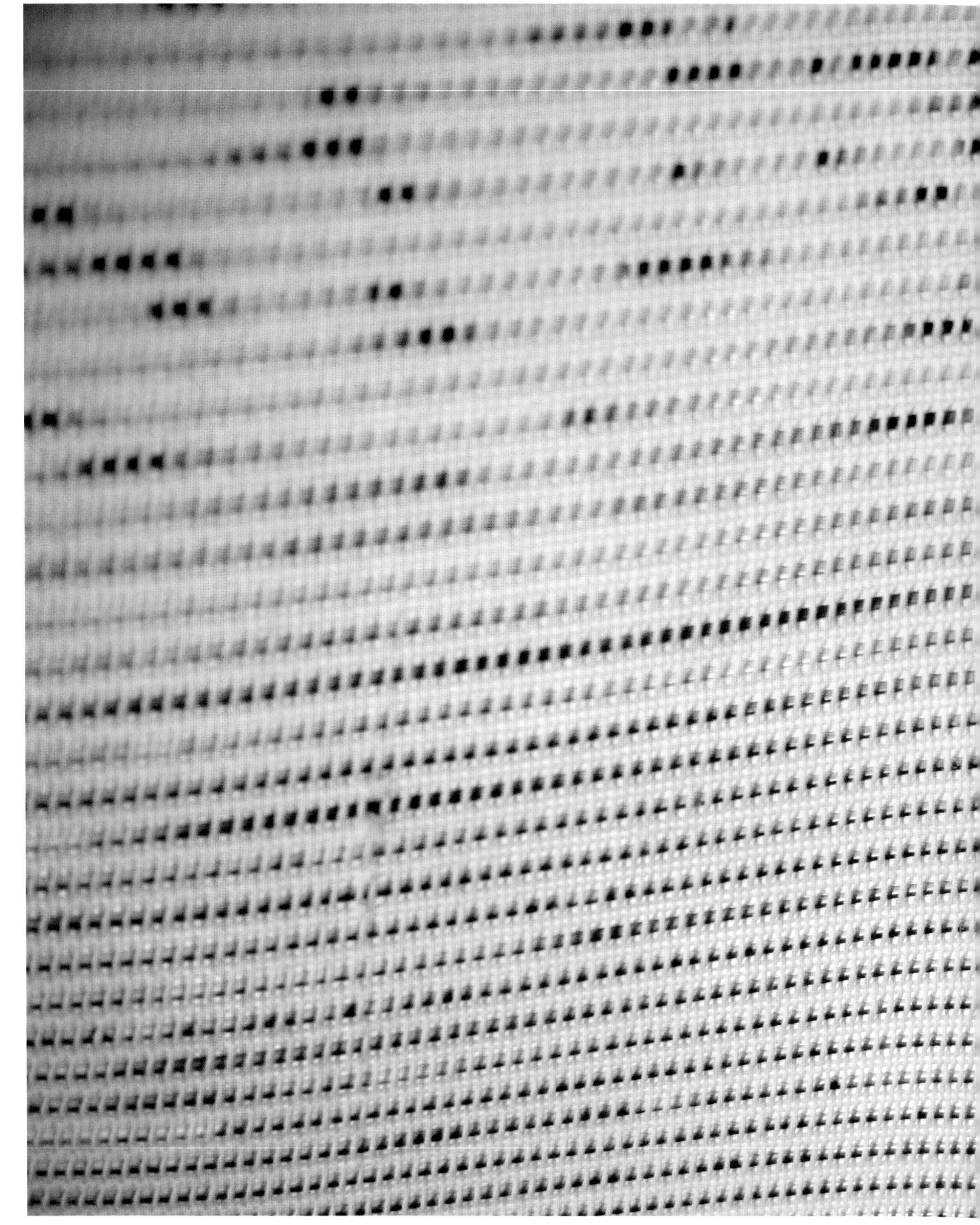

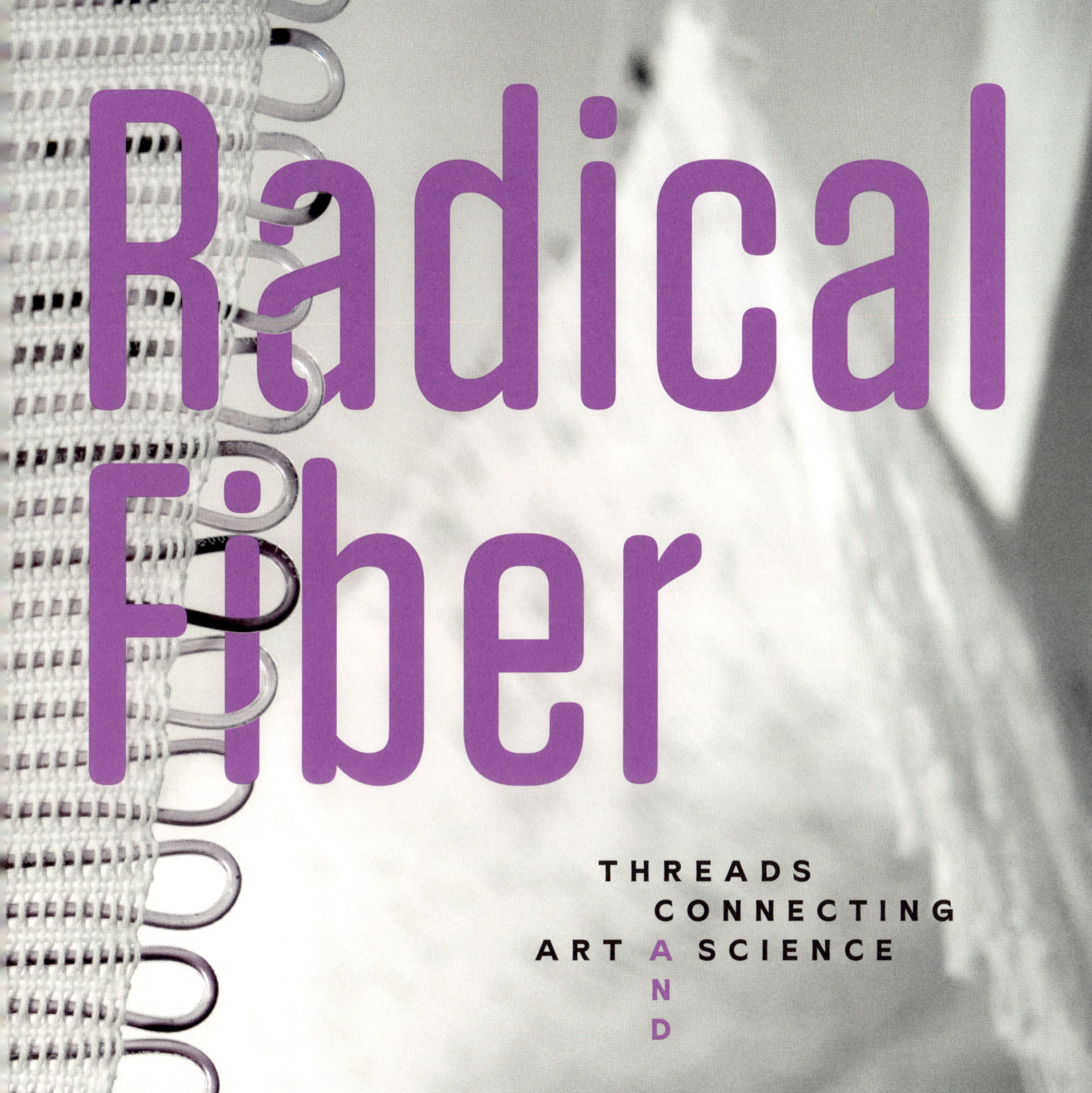

Radical Fiber

Edited by Rebecca McNamara

The Frances Young Tang
Teaching Museum and Art Gallery
at Skidmore College

DelMonico Books · D.A.P.
New York

This publication accompanies the exhibition **Radical Fiber: Threads Connecting Art and Science**
The Frances Young Tang Teaching Museum and Art Gallery
Skidmore College
January 29, 2022–June 12, 2022

Published in 2023 by The Frances Young Tang Teaching Museum and Art Gallery at Skidmore College and DelMonico/D.A.P.

The Frances Young Tang Teaching Museum and Art Gallery
Skidmore College
815 North Broadway
Saratoga Springs, NY 12866
518-580-8080
tang@skidmore.edu
tang.skidmore.edu

DelMonico Books
Available through ARTBOOK | D.A.P.
75 Broad Street, Suite 630
New York, NY 10004
artbook.com
delmonicobooks.com

Editor: Rebecca McNamara
Copy editor: Alison Hagge
Proofreader: Stephanie Cash
Design: Barbara Glauber/Heavy Meta
Color Separations: Echelon, Los Angeles
Printing and binding: SYL, Spain

Fonts: Sunset Gothic, Operator, Tungsten Rounded
Paper: Munken Polar 130 gsm

ISBN: 978-1-63681-040-9

Library of Congress Control Number: 2023937424

TABLE OF CONTENTS

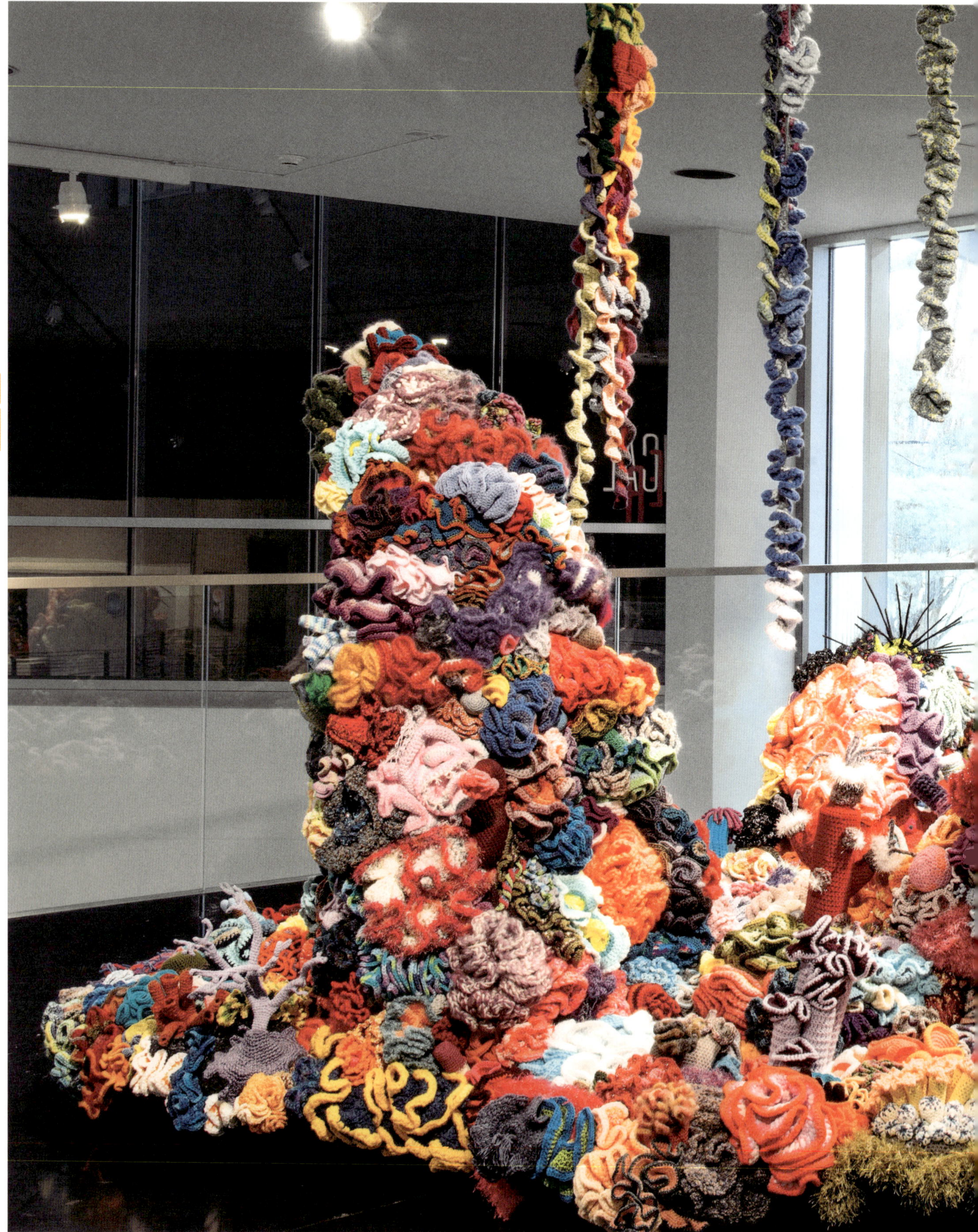

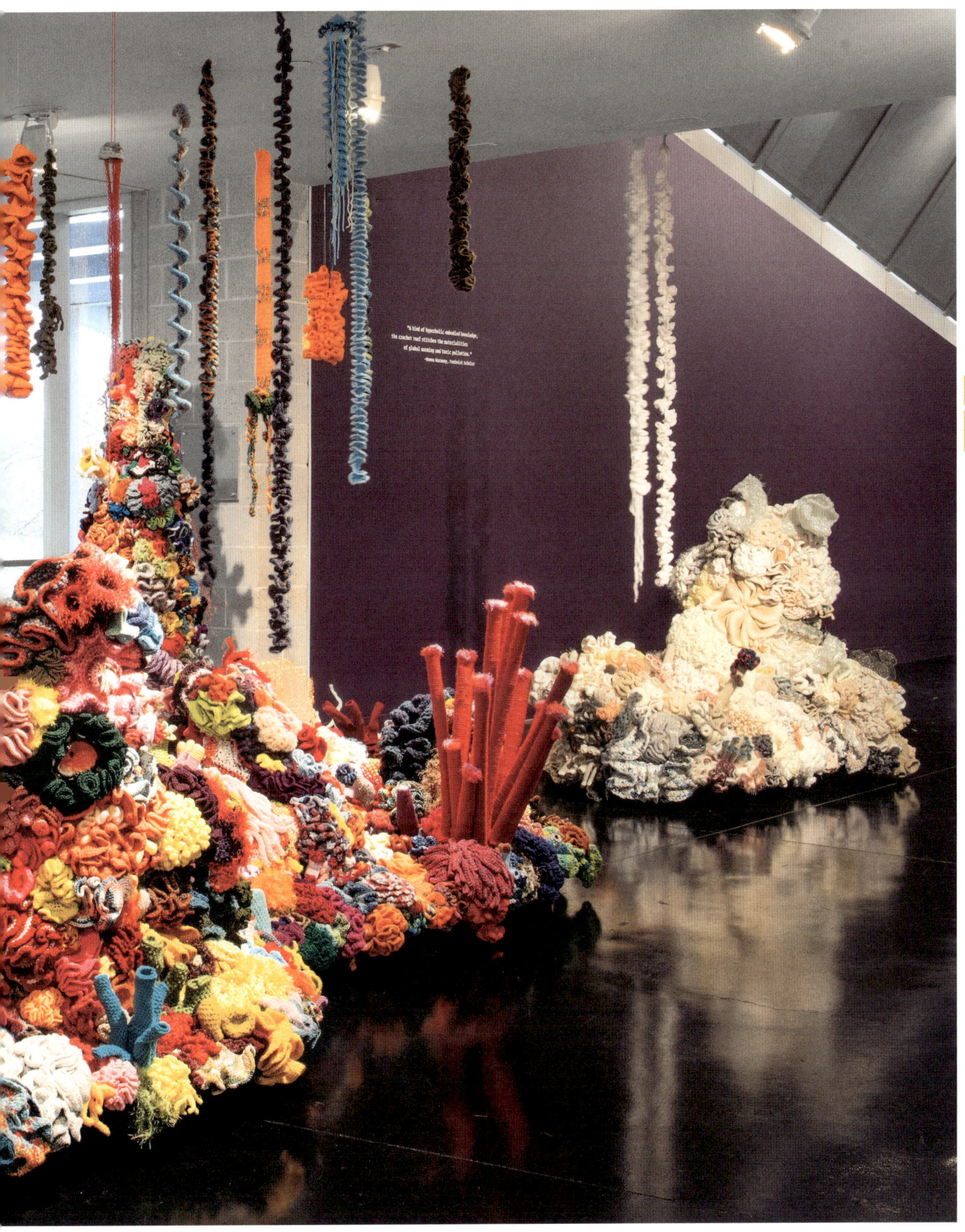
"A kind of hyperbolic embodied knowledge,
the crochet reef stitches the materialities
of global warming and toxic pollution."
—Donna Haraway, Feminist Scholar

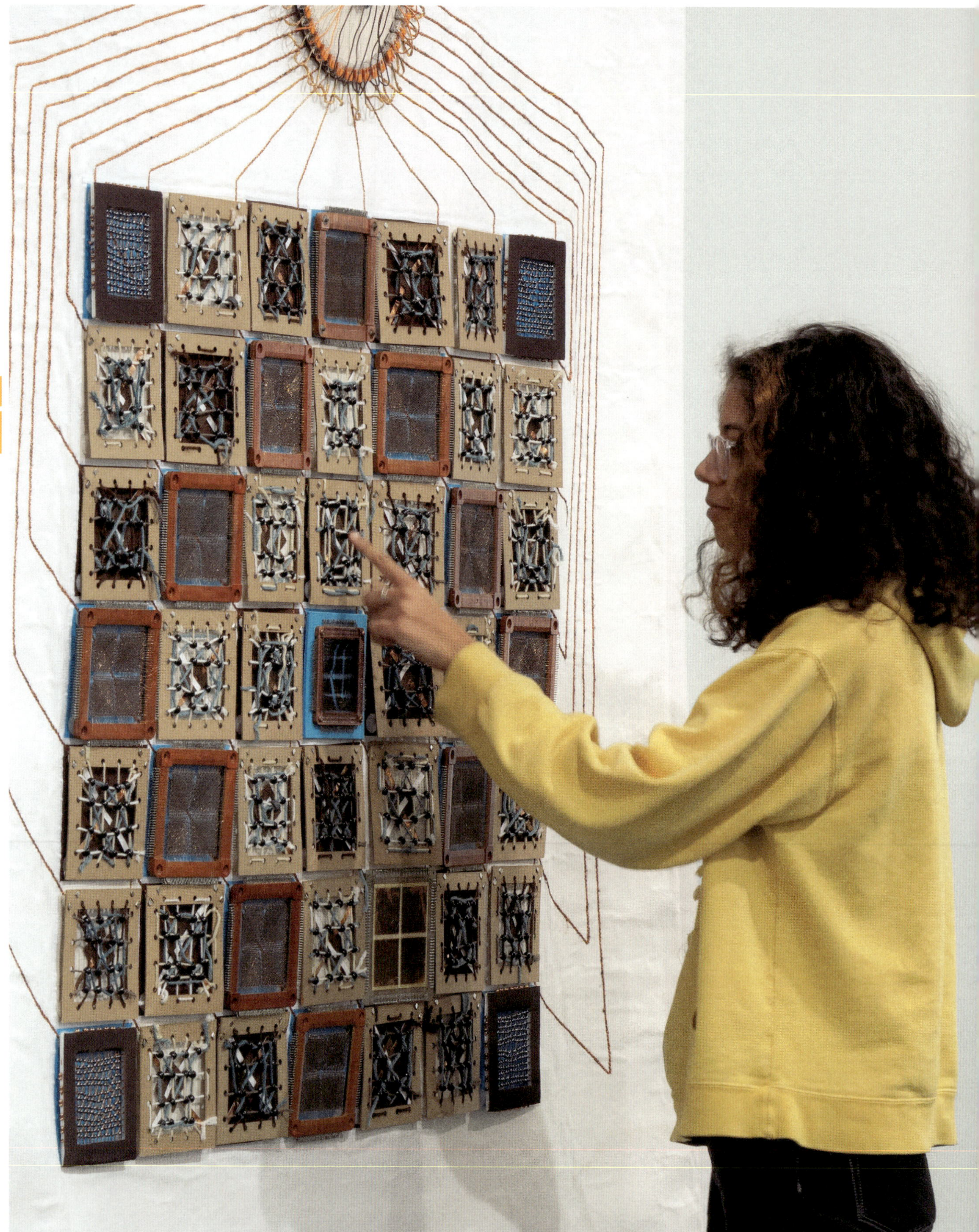

WELCOME

Art is key to all aspects of life. Beyond aesthetic engagement, it can also be a catalyst for new discoveries across the disciplines from scientific theory and humanistic research to new pedagogy. *Radical Fiber: Threads Connecting Art and Science* is a significant interdisciplinary undertaking that demonstrates the multifaceted connections between fiber arts, in particular, and the sciences, more broadly. This kind of creative gathering—inspired by ideas—is a hallmark of the Tang Teaching Museum's history of experimental exhibition making.

We are privileged to have the passion and excitement of Tang Associate Curator Rebecca McNamara, who championed and thoughtfully organized *Radical Fiber* and attended to all the details to make it happen. Congratulations, Rebecca! She benefited from an all-star advisory group of seven Skidmore College faculty including: Mark Huibregtse, Rachel Roe-Dale, and Rebecca Trousil from Mathematics and Statistics; Sarita Lagalwar from Neuroscience; Elaine Larsen from Biology; Aarathi Prasad from Computer Science; and Sang-Wook Lee from Art. We are grateful for their expertise and guidance. Their essays in this catalog—along with those of Denise Evert from Psychology and Neuroscience; and Rachel Seligman, Tang Malloy Curator and Lecturer of Mathematics and Statistics—offer robust examples of the expansive ways we can consider the intersections of art and science.

The exhibition opened with an ambitious two-day program: *Radical Fiber: A Symposium on Art and Science* was attended by over 300 participants for curator tours, panel discussions, and open conversations. Edited versions of those discussions can be found throughout this book.

Radical Fiber also prompted a major community art project, the *Saratoga Springs Satellite Reef*, part of the *Crochet Coral Reef* project by Christine and Margaret Wertheim and the Institute For Figuring. *Reef* programming brought new audiences to the Tang for fifty-eight programs throughout 2021. Thank you to Assistant Director for Engagement Tom Yoshikami and Special Events and Publications Manager Olivia Cammisa-Frost for making these and so many more successful programs for this project. The Tang supplied hundreds of Skidmore students and community members with crochet kits, and many became active crocheters, continuing to hone their craft even after the *Reef* project was complete.

Thanks to our fantastic installation crew, and to the entire Tang Museum staff for their dedication and enthusiasm. A sincere thank you to the many students who worked on this project, from early research and making

Left: Installation view, Radical Fiber, featuring the Core Memory Quilt
Previous spread: Installation view, Radical Fiber, featuring work by Christine and Margaret Wertheim with community makers

checklists to writing labels and giving public tours in the gallery. Your fresh insights continue to inspire us every day.

Thank you to Mary DelMonico and Karen Farquhar at DelMonico Books for partnering with us to publish this book, and thanks to the team that worked with Rebecca to produce it. Special thanks to Alison Hagge for her sharp editing skills, and to Barbara Glauber for her intertwined design.

We are fortunate to have the support of the Skidmore College administration, especially President Marc Conner and Vice President for Academic Affairs Michael Orr, as well as the Tang National Advisory Council, who champion innovation and experimentation in all areas of the museum. The exhibition and catalog were made possible by Friends of the Tang, and the symposium was made possible by The Alfred Z. Solomon Residency, which brings notable scholars, artists, and critics to Skidmore to address a wide range of issues in the visual arts. We thank these and all our supporters for helping us advance innovative explorations of art and ideas.

Ian Berry
Dayton Director

Karen Norberg, *Knitted Brain #2 (KB #2)* (detail), 2020–22, knitted cotton yarn, nylon zippers, metal zipper pulls, metal snaps, magnets, 3 ½ × 6 × 8 ¾ inches (element shown)

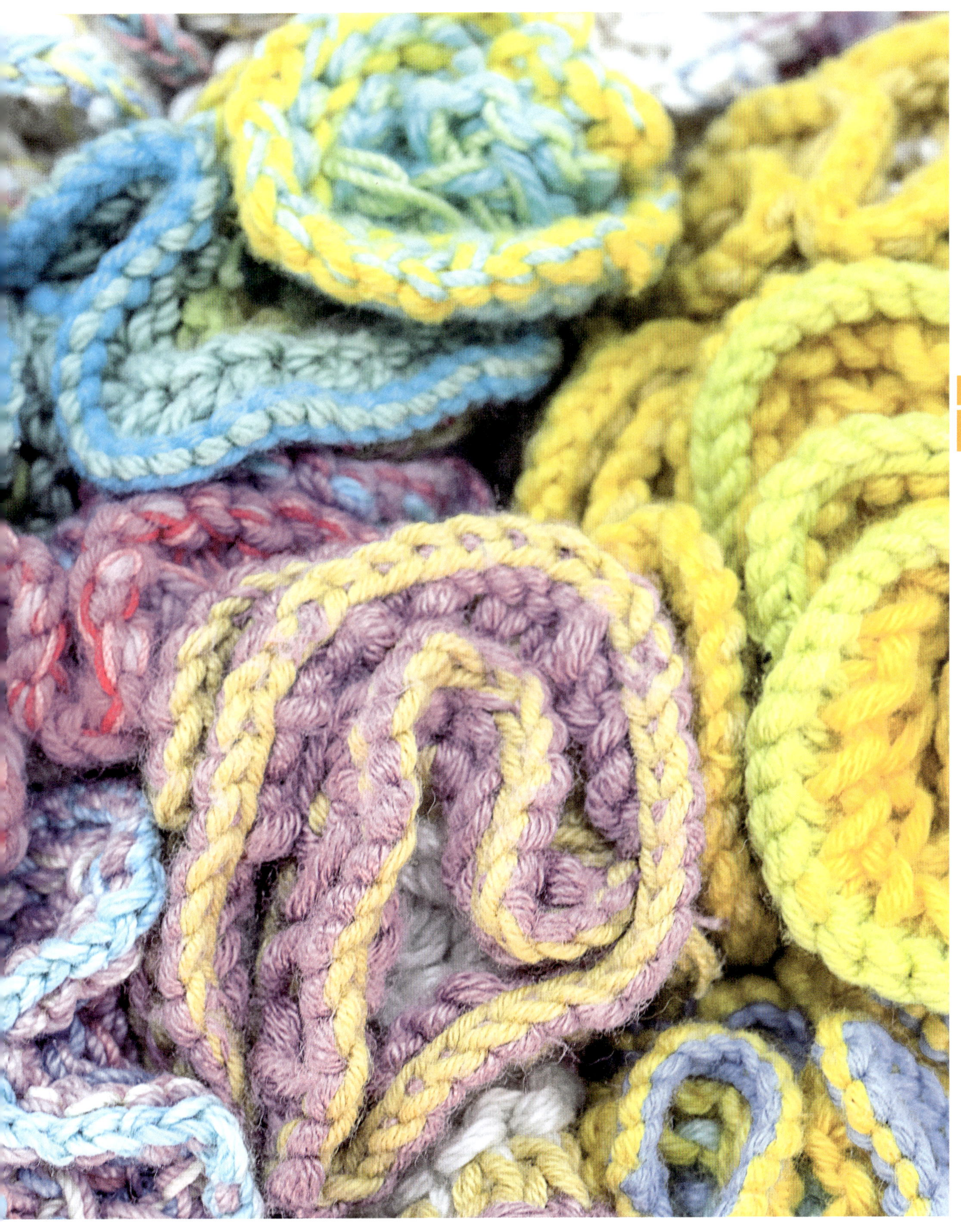

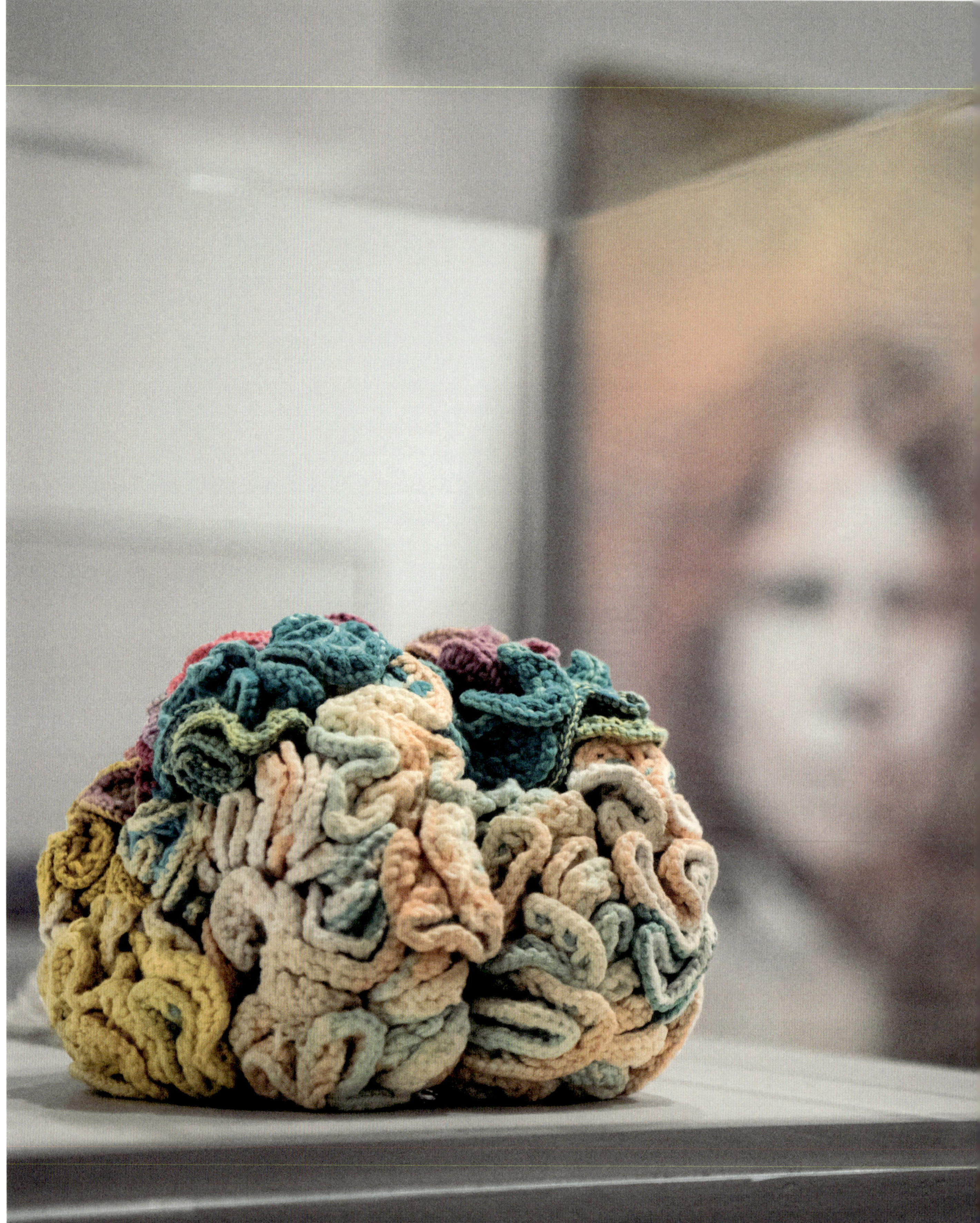

by Rebecca McNamara

SHAPE, MACHINE, BRAIN, BODY, COMMUNITY: AN INTRODUCTION

"It all began with string. Its invention changed the history of humanity." —Clare Hunter, *Threads of Life*[1]

Humans first developed string—drawn and twisted plant or animal fiber—more than 120,000 years ago. Since then, we have learned to knot, weave, knit, crochet, and stitch, developing societies that fish, hunt, build shelter, wear clothing, carry objects large and small, and complete all manner of tasks, thanks to this technology of string. Such world-changing innovations are so embedded in our everyday lives that we hardly recognize them—yet fiber as a material and its associated techniques continue to catalyze advancements across areas as diverse as digital technology, mathematics, neuroscience, and chemistry, among others, informing scientific theory, discovery, and pedagogy.

The exhibition *Radical Fiber: Threads Connecting Art and Science*, on view at The Frances Young Tang Teaching Museum and Art Gallery at Skidmore College from January 29 to June 12, 2022, presented historical objects and contemporary art, including mathematical models, artworks, patented innovations, and more, by makers such as artists, mathematicians, psychologists, hobby crafters, chemists, designers, and researchers. Their multifaceted, interdisciplinary work can be understood simultaneously as fine art, process-driven craft, and scientific tool, blurring established frameworks across fields. This accompanying publication expands on the artwork presented with new writing, edited transcripts from the two-day program *Radical Fiber: A Symposium on Art and Science*, testimonials from the crafters who created the social-practice sculpture the *Saratoga Springs Satellite Reef*, and full-color illustrations. Together, the exhibition, symposium, and publication demonstrate the value of interdisciplinary and collaborative thinking, emphasizing how artists change the way we experience and think about the world.

The exhibition was loosely organized into five thematic sections—shape, machine, brain, body, community—but one flowed into the other, with many works speaking to multiple themes. Lia Cook's *Connectome* (page 60), a weaving made on a modern Jacquard loom, represents the artist's own MRI scans and is named after a part of the *brain*, yet is equally evocative of *machine*. The biomedical implants by Ellis Developments (pages 46–47) are made with an embroidery *machine* and are literally inserted into a *body*. The *Core Memory Quilt* (page 100), which tells the story of connections among the bytes of core memory used in NASA's Apollo Guidance Computer that helped send US astronauts to the moon, was devised by four scholars from different disciplines, with the help of craft workshop participants—*machine* and *community*. The *Saratoga Springs Satellite Reef* (page 184) embraces *shape* (mathematics) and *community* in equal measure. Dario Robleto's *The Creative Potential of Disease* (page 22) connects artistic or functional sewing to surgical suturing (*body*) and craft to mental health (*brain*).

Fabrics comprise lines and grids and structures and algorithms, but, generally speaking, they are soft and flow and move and shift. This book also offers a shifting, flowing structure. And because textile metaphors are so irresistible, I offer an imperfect one here: consider the book as a weaving. Each conversation and essay is a warp (vertical) thread with artwork and ideas (representing the weft, or horizontal thread) weaving over and under this structure, disappearing behind the warp then reappearing somewhere unexpected. These behind-the-scenes threads—sometimes knotted-up and messy—are not irrelevant. Even when a thread (idea) isn't visible in a given moment, its presence remains, allowing it to reappear elsewhere. I encourage you to seek those interconnections between topics and moments of alternate perspectives. And remember: each thread, whether warp or weft, comprises smaller threads twisted together, many ideas forming and informing the whole.

In "Threads throughout History," artist Dario Robleto and scholar & curator Elissa Auther probe the value of artistic-scientific collaboration. Robleto challenges us to recognize the commonalities between artists and scientists, sharing his own experiences working directly with scientists and asking them questions that address the cultural and emotional ramifications of their fact- and data-driven work. Sarita Lagalwar, Skidmore Associate Professor of Neuroscience, looks at the history of how we use art to understand brain functionality, beginning with the work of Leonardo da Vinci. She then demonstrates how artist and psychiatrist Karen Norberg uses knitting to create brain models that can offer information in unprecedented ways. Denise Evert, Skidmore Associate Professor of Psychology and Neuroscience, worked with undergraduate students on an experiment to better understand emotional responses to photographic and woven portraits as presented by artist Lia Cook. Elaine Larsen, Skidmore Senior Instructor of Biology, presents the history of synthetic fabric dye leading to our modern pharmaceutical industry while detangling additional connections between fashion and medicine.

From wellness of the body in the past, the book moves to wellness of the body, mind, and spirit in the future—while remaining grounded in key historical moments. The conversation "Textiles, Technology, and Social Good" with materials scientist, inventor, and entrepreneur Trisha Andrew; researcher, artist, and educator Emilie Giles; computer-science educator and textile crafter Ursula Wolz; and moderated by Skidmore Assistant Professor of Computer Science Aarathi Prasad explores the ways embroidery, stitching, weaving, and other fabric techniques intersect with digital technologies, and how the combination of new digital technologies with fiber arts can generate opportunities that welcome more diverse audiences, such as blind and visually impaired users. They also

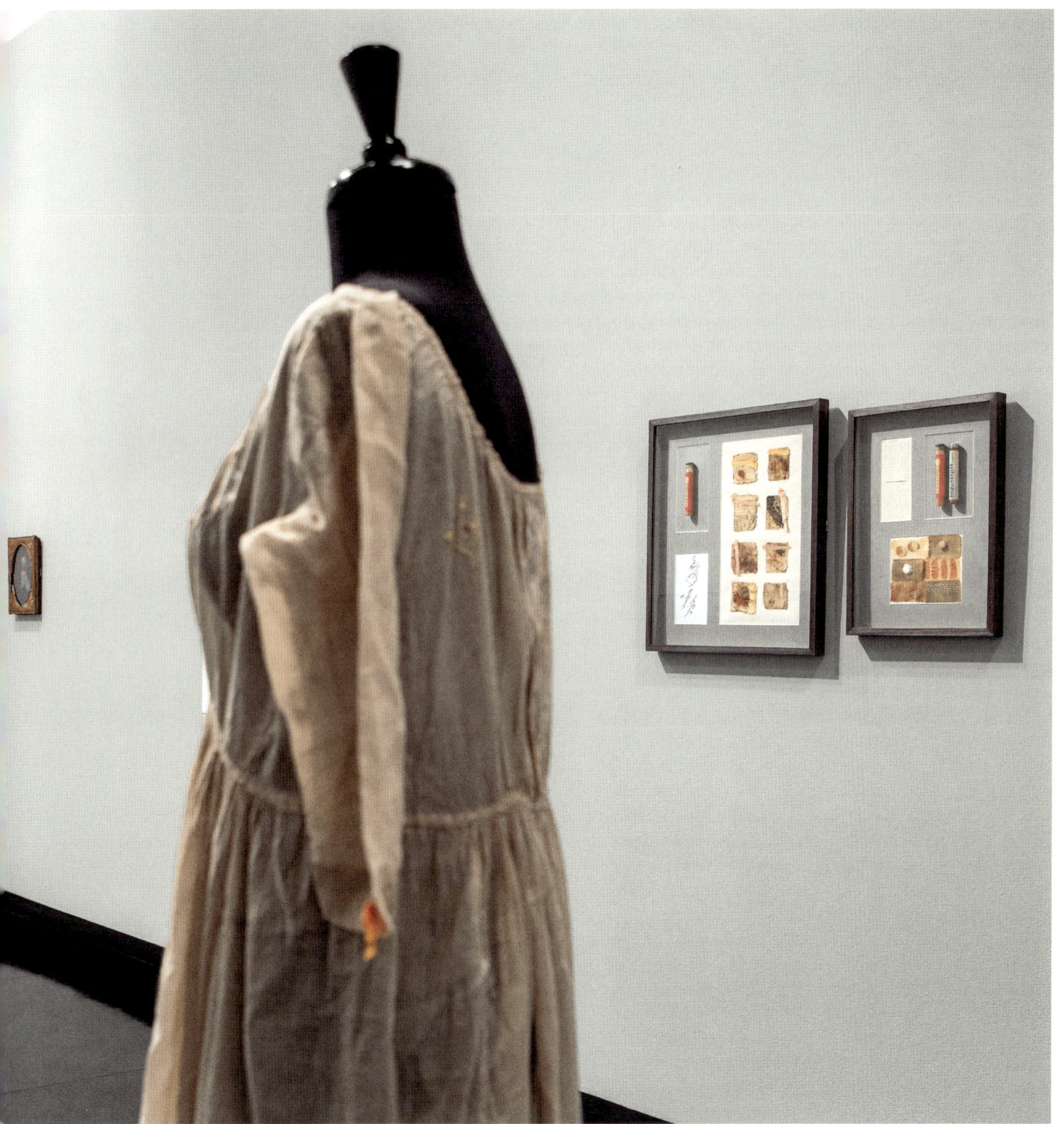
Radical Fiber

discuss the microelectronics that have the capacity to generate a more sustainable textile industry as well as the imperfections inherent in both fiber craft and coding—key concepts that arise in other conversations throughout the book. Two essays on the historical connections between weaving and computer technology, by Skidmore Professor of Art Sang-Wook Lee and by Prasad follow. In her essay, Prasad further challenges us to innovate a more inclusive, accessible software industry that embraces more women and people of color. Rebecca Trousil, in examining the elastic properties of specific fibers versus knit fabrics, considers the new fabrics that may result from this clearer understanding.

"Making Visible: Math, Craft, Culture" features math-artist, writer, and curator John Sims; anthropologist and quipu maker Jeffrey Splitstoser; and mathematician & artist Daina Taimina; and is moderated by science writer Stephen Ornes. They discuss the ways in which fiber-based mathematical models can impact classroom pedagogy and the embodied knowledge gained in crafting an object with distinct mathematical or numerical properties—and the importance of sensing and recognizing (im)perfection in both mathematics and craft. Skidmore Professor of Mathematics Mark Huibregtse offers an overview on the history and mathematics behind hyperbolic geometry, which Taimina manifested as a usable, physical model through crochet. Skidmore Professor of Mathematics and Statistics Rachel Roe-Dale similarly discusses modeling the Lorenz Manifold with crochet. Roe-Dale and Skidmore Lecturer in Mathematics & Tang Malloy Curator Rachel Seligman then discuss more broadly the uses of models for pedagogy, visualized with various fiber techniques, including quilting, knitting, and crochet.

The final conversation, "The Future of Textiles and Sustainability," features textile professional and consultant Preeti Arya, fashion industry entrepreneur Alissa Baier-Lentz, nanotechnologist and inventor Juan Hinestroza, and is moderated by Skidmore Associate Professor of Environmental Studies & Sciences and food studies scholar Nurcan Atalan-Helicke. This group presents the stark impact the textiles industry has wrought on our natural resources through pollution and its drastic carbon footprint in materials production and manufacturing; they offer solutions for more environmentally sound management of the textile industry, to be enacted by individuals, corporations, and governments.

Interspersed are additional artworks beyond the scope of any one essay or conversation, but which further expand our understanding of the depths of these connections. The book concludes with a section devoted to the *Saratoga Springs Satellite Reef*, part of the worldwide *Crochet Coral Reef* project by Christine and Margaret Wertheim and the Institute For Figuring. My essay describes this

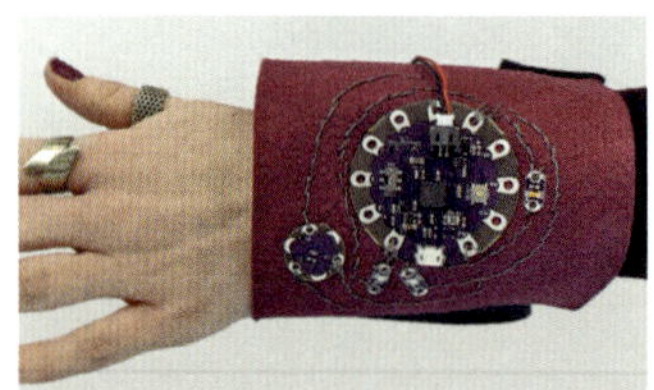

The Tang held robust interdisciplinary public programming throughout the run of the exhibition. For "Wearable E-Textile Programming with Arduino Lilypad," Assistant Professor of Computer Science Aarathi Prasad and I worked with Skidmore's IdeaLab as well as scholar/artist Emilie Giles to lead participants in a three-hour workshop to create wearable light-up e-textile devices. This program, which took place in Skidmore's new Billie Tisch Center for Integrated Sciences, brought the museum to the other side of campus and brought together Skidmore community members and the public to teach both fiber craft and coding and demonstrate the benefits of working together across fields.

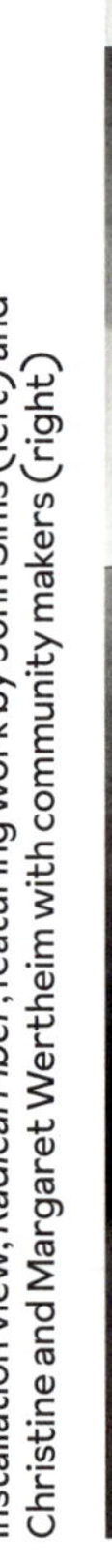

Installation view, *Radical Fiber*, featuring work by John Sims (left) and Christine and Margaret Wertheim with community makers (right)

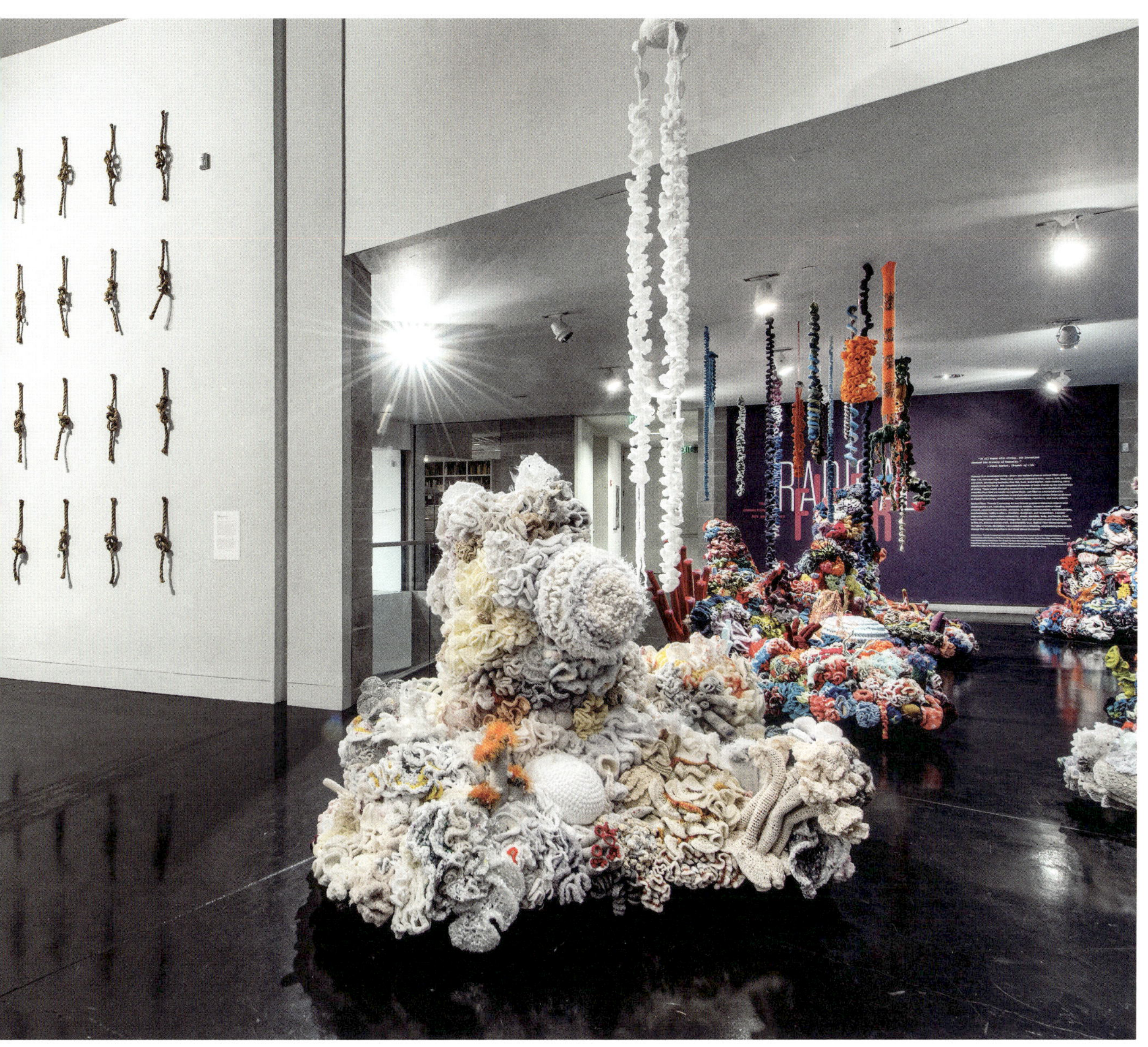

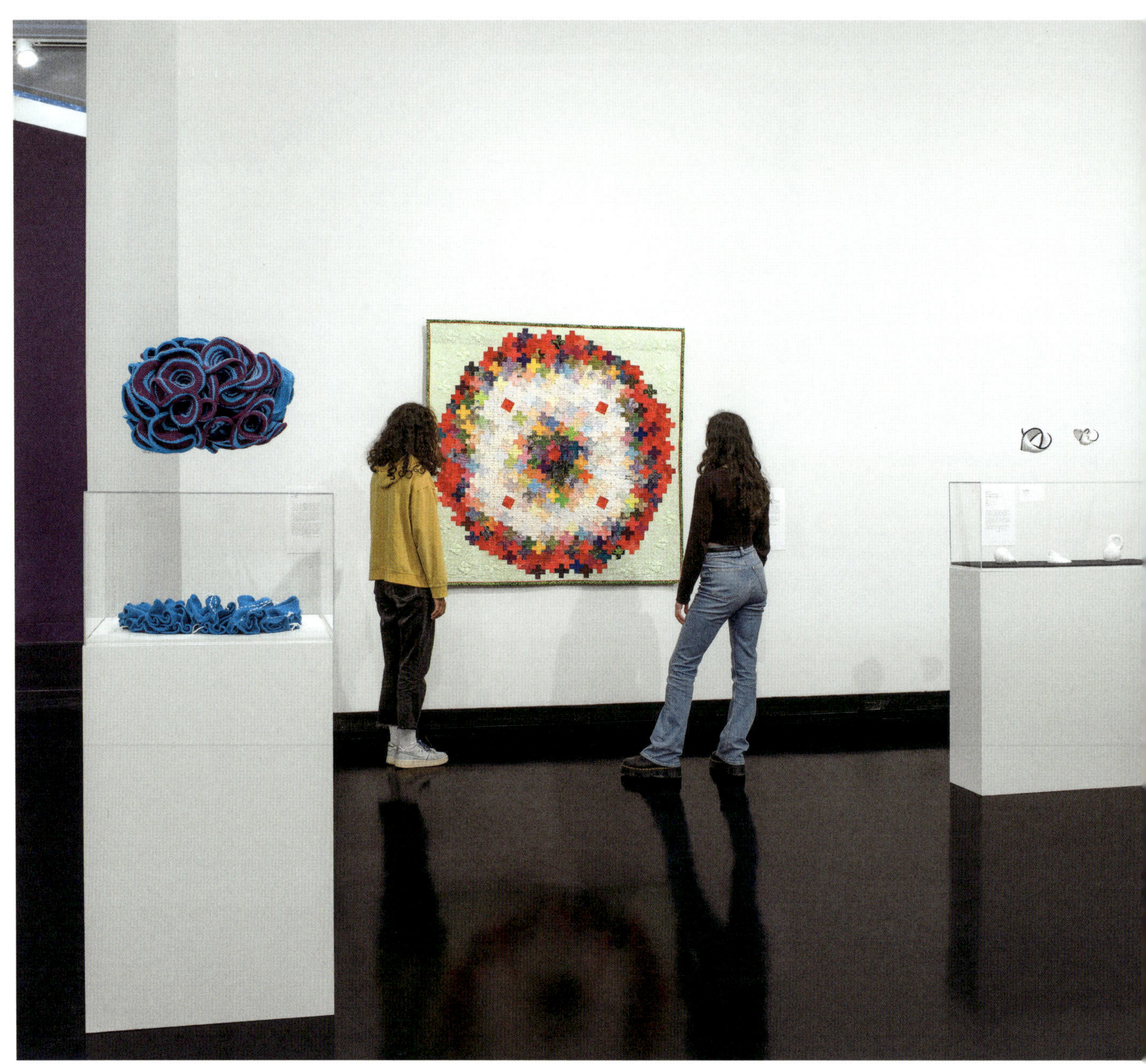

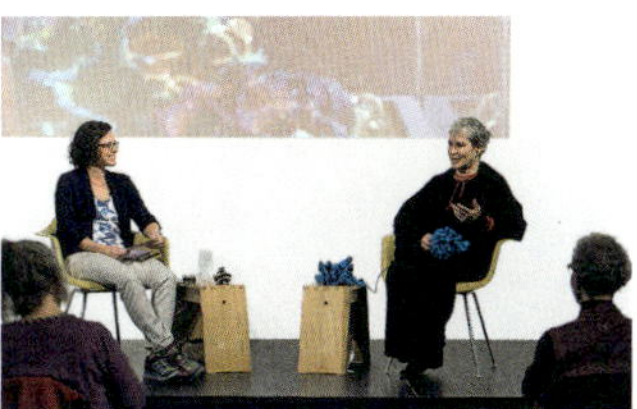

Lectures included a Dunkerley Dialogue with Skidmore Associate Professor of Geosciences Amy Frappier (left) and artist Margaret Wertheim (right), shown here, and the spring 2022 Distinguished Scientist Lecture, cosponsored by the Geosciences Department and The Charles Lubin Family Chair for Women in Science, featuring Logan Brenner: "The Coral Time Machine: Using Coral Geochemistry to Reconstruct Our Oceans."

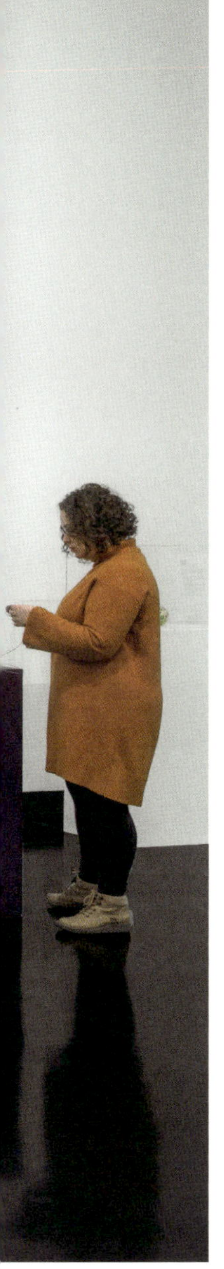

community-collaborative art-making project in greater detail, and the names and personal stories of makers follow. Participating in the project by crocheting corals—or glitched hyperbolic plane models—offered a sense of community during the early years of the COVID-19 pandemic; taught non-Euclidean geometry; and brought much-needed attention, for both participants and exhibition visitors, to the negative impact climate change has wrought on our Earth's oceans. The *Reef* event series and other programs held in conjunction with *Radical Fiber*—in crocheting, spinning, weaving, and sewing e-textiles—encouraged embodied knowledge. Participants in all programs gained new fiber skills or strengthened existing skills, both of which led to clearer understandings of the fiber that surrounds us.

Many types of art can and do intersect with, inform, or transform the sciences. So why fiber? Clare Hunter's quote at the beginning of this essay is not over-wrought. String is one of humanity's earliest technologies. Elizabeth Wayland Barber has redubbed the period beginning around 40,000 years ago "the String Revolution"—giving it equal weight as the Stone Age, the Bronze Age, the Industrial Revolution, and all the other hard-edged ages and revolutions we more regularly learn about.[2] Look around. What would you see without string? This book is not intended as a comprehensive compendium of fiber art–science interconnections. Instead, I hope it forms a starting point to think about where, how, and why fiber exists in our world. How do we perceive fiber at different moments or in different spaces? What has fiber offered us, and what will it hold for us tomorrow?

1   Clare Hunter, *Threads of Life: A History of the World through the Eye of a Needle* (New York: Abrams, 2019), 205.

2   Elizabeth Wayland Barber, *Women's Work: The First 20,000 Years: Women, Cloth, and Society in Early Times* (New York: W. W. Norton, 1994), 70.

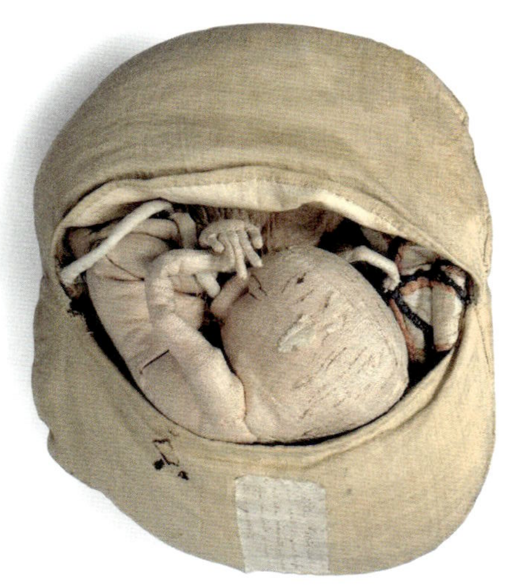

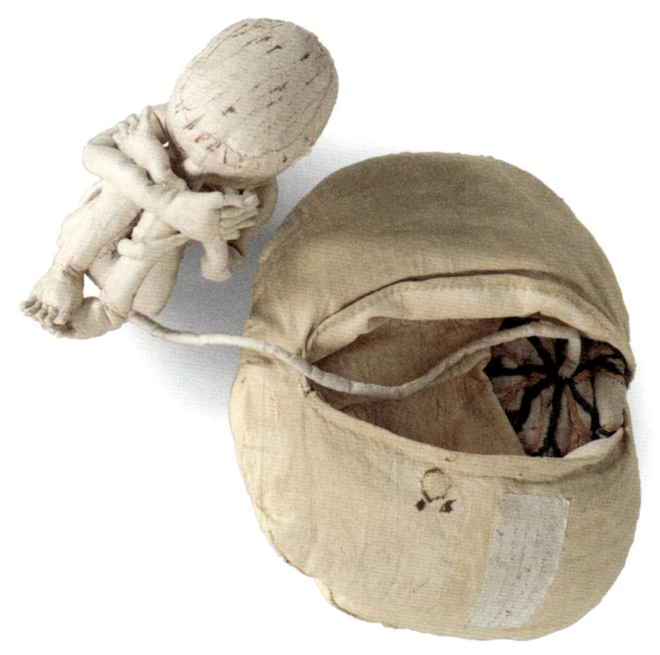

Angélique Marguerite du Coudray (1712–1790), *mannequin d'accouchement* (birthing mannequin), before 1778, fabric, collection of the Musée Flaubert et d'Histoire de la Médecine. Coudray, France's premier midwife in the second half of the eighteenth century, invented and used "machines" (or mannequins) like this one to train poor and illiterate women—and eventually two-thirds of French midwives—throughout the country on female anatomy and the birthing process.

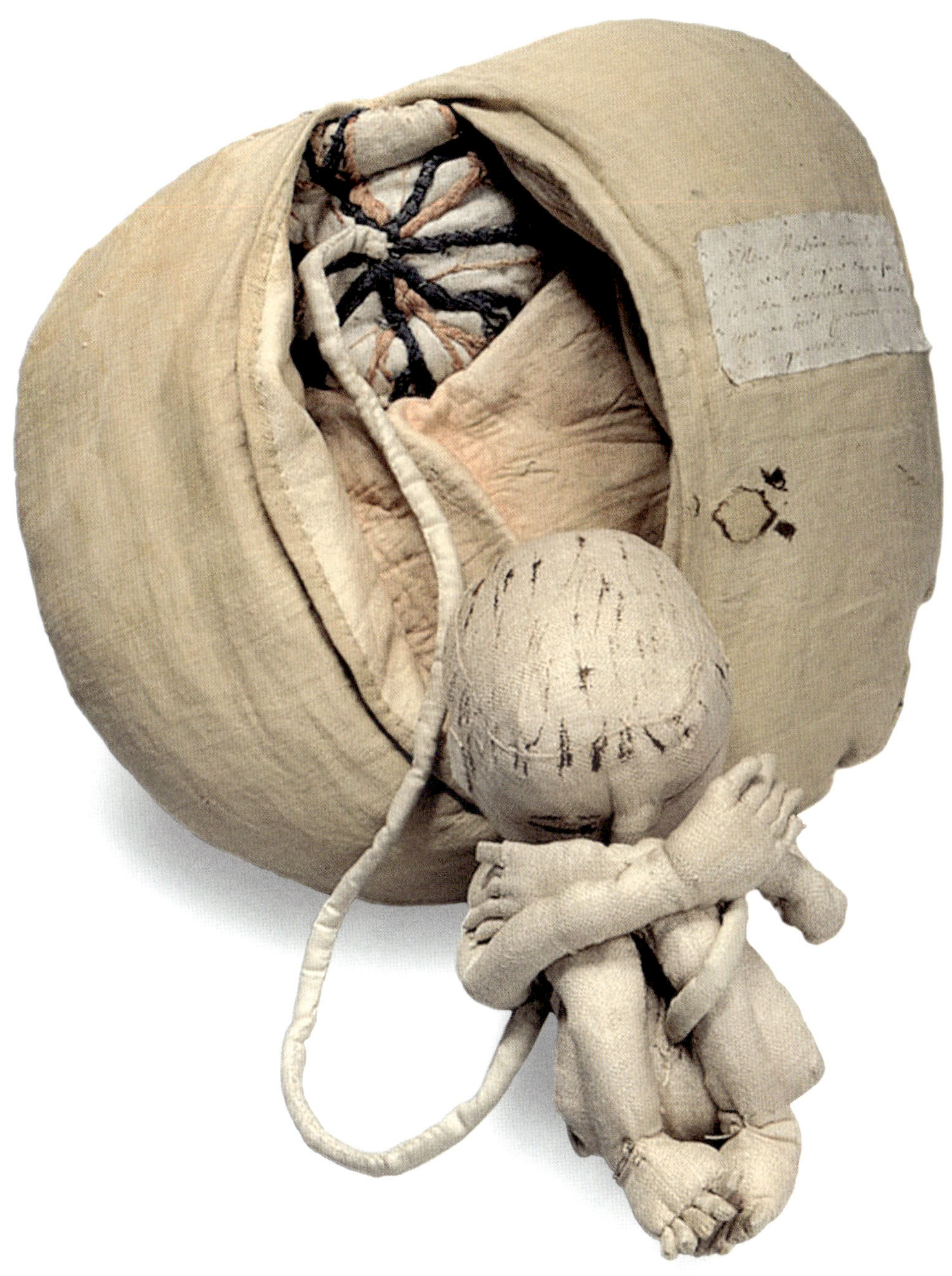

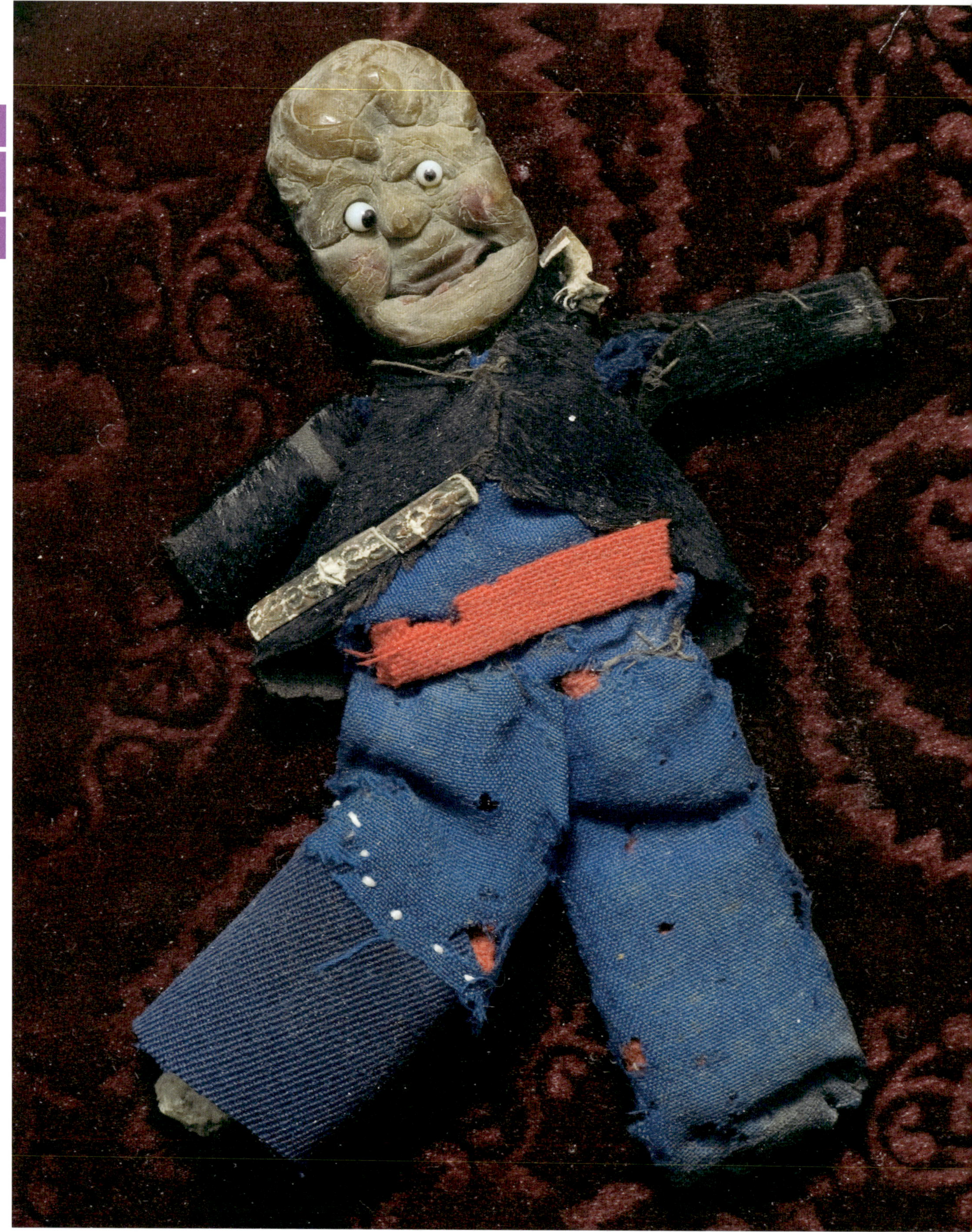

THREADS THROUGHOUT HISTORY

In Conversation: Dario Robleto
and Elissa Auther

Dario Robleto, *The Creative Potential of Disease* (detail), 2004, a self-portrait doll made by a Union Civil War soldier amputee while recovering in the hospital, mended and repaired with a modern-day surgeon's surgical needle and thread, new pant leg material made from a modern-day soldier's uniform, cast leg made from femur bone dust, and prosthetic alginate treated with *Balm of a 1000 Foreign Fields*, vegetable ivory, collagen, melted shrapnel and bullet lead, cold cast steel and zinc, polyester resin, rust, 12 ⅛ × 10 × 1 ⅝ inches (overall)

DARIO ROBLETO: What's so exciting about this symposium is that the arts and humanities and sciences—we all get to share an intellectual space together. I've spent much of my career trying to cultivate such encounters, and people don't always want to do it, or don't expect or desire it. This lingering—and, I think, entirely unnecessary—suspicion that our fields sometimes hold against each other should continually be challenged.

Elissa and I want to talk today about some big-picture topics related to collaboration between the arts and sciences, and then we'll discuss craft, fiber, and textiles and selected examples of my work. For me, the act of stitching carries so much poetic weight. And, in the spirit of this art-science connection, I also want to stress the medical variation on that idea, suturing, which is important to the way I personally conceptualize how I try to bond things together. I spend a lot of time thinking about the language artists and scientists use when they are describing what they're really after in their practices, because, embedded in that language, there are often residual clues about what we once shared in common and what, I think, we still share in common.

Stitching and *suturing* are good examples; each word describes a similar physical act but, each, in its own way, suggests something different disciplinarily and a unique type of sensibility. But they're more aligned than we realize. To get to that, I want to explain what we mean when we use the word *sensitivity* from both an artistic and a scientific point of view. Like the distinction that I'm making between *stitching* and *suturing,* there is a similar dynamic going on between, on the one hand, the poetic use of a word like *tenderness* and, on the other hand, the scientific use of a word like *resolution.* These words have different tones for different contexts, but they both identify a particular stance to the world around us. They both, in their own ways, measure the thoroughness of our observations. And this is why I find a word like *sensitivity* is a perfect meeting ground between the poetic tradition of *tenderness* and the scientific tradition of *resolution.*

I want to recharge this word sensitivity as this meeting ground between art and science because, at some core, base level, both disciplines are trying to increase the *sensitivity* of their observations. Whether that means we're staring longer and more deeply into each other's eyes or turning the magnification lens up on our microscopes, practitioners in both fields are trying to see what hasn't quite been seen before—in each other and in the universe.

I have spent many years working with scientists in a variety of fields, in particular, synthetic biology, astrobiology, neuroscience, and glaciology. In the spirit of fiber and textiles, I'd like to share a particular story that deeply impacted me as an artist and that I continue to engage with to this day. When I was a little boy, in the comic section of the newspaper, there was a cartoon called *Ripley's Believe It or Not!* I stumbled on this incredible story about the first time an artificial heart was implanted in a human, and it completely stunned me. I didn't think such a thing was possible. I had only recently seen *The Wizard of Oz,* and I just thought this story was a complete fairy tale. We've lost touch with this moment in history—which I've spent many years studying— and I still can't believe it. One day I even got to hold this heart, now housed in the Smithsonian's National Museum of American History. This procedure was an Apollo-like event; there was twenty-four-hour press coverage, headlines around the world. And the reason why, to state something pretty obvious, is because the heart matters. For most of history, across

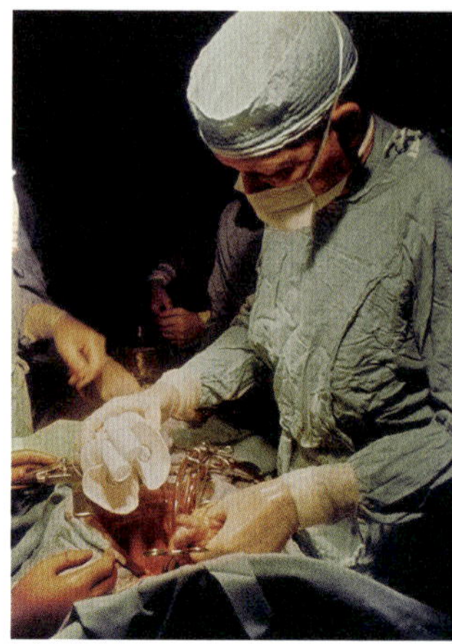

Denton Cooley, chief of cardiovascular surgery at Houston's St. Luke's Episcopal Hospital, performs the first artificial heart transplant on Haskell Karp, age 47, at the Texas Heart Institute, April 4, 1969.

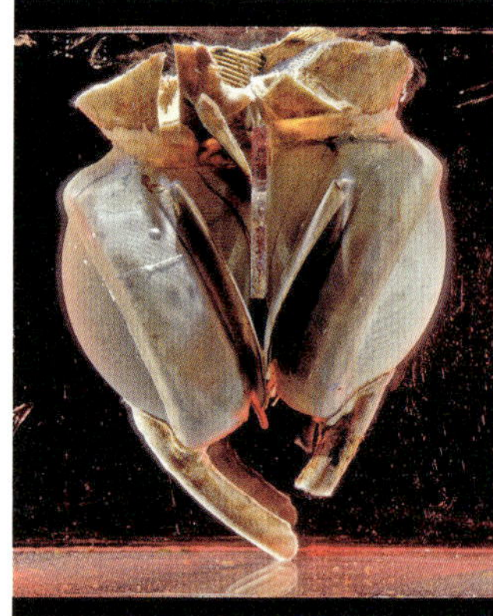

Liotta-Cooley artificial heart, 1969, Dacron, polyurethane, Silastic, 7½ × 9 × 6½ inches (overall, as stored); Domingo S. Liotta delivered the experimental artificial heart he was developing with Michael E. DeBakey to Cooley during Haskell Karp's heart transplant surgery when the patient's heart was not pumping enough blood to survive. It became the first completely artificial heart to be implanted into a human. Karp lived three days with the artificial heart before a donor heart was implanted.

time and cultures, the heart wasn't simply a metaphor. It was considered the literal vessel of the soul, this conduit between the immaterial and material domains. Our great poets and philosophers and priests have all weighed deep questions about identity and emotional authenticity, fate and faith, around the heart.

When modern science wants to pull that vessel out and replace it with a machine, inevitably, important questions arise. The image of the day the surgery occurred is as important as an image of Neil Armstrong landing on the Moon because similar things were at stake. Dr. Denton Cooley performed the surgery, and in his right hand, he's holding the artificial heart. And in his left hand, he's holding Mr. Haskell Karp's now-dead heart. I'm so glad a photographer caught this transitional moment because there's a lot going on—the human coming out and the machine going in. Nothing like this had ever happened before. You could argue that the past is in one hand and the future is in the other. In one hand is the poetic heart, with millennia of religious, philosophical, and emotional meaning. In the other hand is the scientific heart, which could be labeled simply as a complicated pump or a muscle, no more or less important than our kidneys—an organ that we can upgrade through better design.

There is a little-known fact about what happened just moments after this picture was taken; it's not something you'll find in the official narratives. I want to argue, as an artist, that something pretty incredible happened—and it creates an opportunity to think about this divide between science and art. When Dr. Cooley began the process of stitching this new technology back into Mr. Karp's body, to his arteries and valves, he was confronted with a problem that, incredibly, no one had accounted for. His suturing needle wasn't sharp enough. This new experimental material they were using, called Dacron, was dipped into a substance called Silastic, creating a density entirely different from the human body. Dr. Cooley struggled to pierce through this space-age fabric, requiring him to increase his force of movement to do so, then immediately reduce his speed and pressure for the more delicate human tissue. If you know sewing firsthand, imagine making a shift in tactile sensitivity while someone's life is at stake and while you're ushering in a new era of mechanical humans. It slowed Dr. Cooley down tremendously, jeopardizing the surgery, and he eventually had to send someone to get a sharper needle. So, after all these advancements—millions of dollars in his hand, years of research by dozens of people—the success of the surgery in that moment came down to the sharpness of a needle. But, of course, sharpness is a form of sensitivity. And in the history of such sensitivities, no one had ever been tasked to find a sharpness that could suture a machine to a human where their heart had once been. And, to Dr. Cooley's credit, no one had ever been tasked, impromptu, to regulate their own sensitivity of touch in order to connect a machine to a human. This sharpness immediately turned into a metaphor and an opportunity for the arts and sciences to think about the necessary ways we can collaborate and complement our sensitivities in such moments.

There's a point I often like to make: sometimes an advancement in science is so profound that it transcends its field of origin and it becomes impossible—and, I think, even irresponsible—for other disciplines not to weigh in. **It's one thing to make a machine heart, but it's an entirely different matter to stitch that machine into a human being and ask them to live and love and find meaning with it.** By now, we have culturally absorbed this development and have come to some acceptance.

I'll share two other advancements that I'm deeply invested in and that I would argue are presenting a new round of questions that the arts and humanities should be paying attention to. The first advancement is called *the beatless artificial heart*, invented by Dr. Cooley's protégé, Dr. O. H. "Bud" Frazier. After decades of pretty much failing in various ways to solve the artificial heart problem, Dr. Frazier proposed something maybe more radical than the mechanical heart. He suggested that we could design a heart and remove the need for it to beat, remove the pulsatility of the device. He did make this heart and he implanted it in Mr. Craig Lewis about a decade ago. The ramifications of this device are that the patient has no heartbeat, has no pulse, and has no EKG.

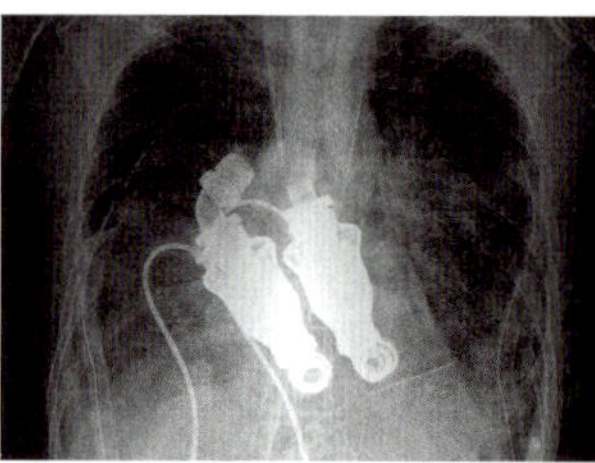

Beatless, or pulseless, artificial heart, developed by O. H. "Bud" Frazier, as seen in an X-ray of Craig Lewis, age 55, who, in March 2011, became the first human to be implanted with the device, two continuous-flow centrifugal pumps. He lived for six weeks before dying due to amyloidosis, an underlying disease.

Mr. Lewis was the first human in history to register as a flat-line on every monitor in the room, and still he lived for several weeks with this device. This is groundbreaking stuff. In the need to save lives, are we willing to let go of some of our defining features, the pulsatility of our hearts?

The second advancement comes from the brilliant scientist Dr. Doris Taylor. Instead of pursuing a mechanical solution to the heart problem, she's proposing that we could

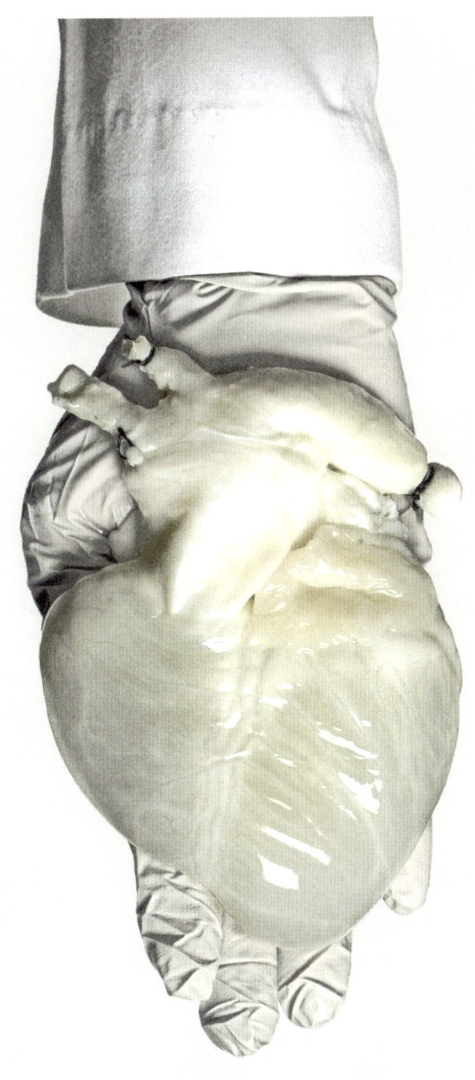

Doris Taylor, ghost heart, c. 2012

continually implant our own hearts, over and over. We haven't been able to solve the problem of rejection: every time we put another person's organ into our body, even if given and sustained with great care, our immune system immediately attacks it. Dr. Taylor is instead suggesting: what if we could give our own hearts back to ourselves? It's such a beautiful idea. And she has figured out a way to literally—and I mean this in both the technical and all the beautiful, metaphorical ways you want to imagine—*wash a heart clean* by using certain detergents and water. She can take a heart and wash away every cell—fat cells, protein cells—out of it. What's left behind is what's called an extracellular matrix, or, more evocatively, what she calls a *ghost heart*. And it's one of the most beautiful things I've ever seen. I like to tell her that I think she has accidentally found the fundamental form of the heart that poetry has been searching for for millennia. The question is whether you can reinject the washed heart with your own heart cells, repopulate it, get it to come back online, so to speak, and implant it back in a human. It's a stunning development.

These technologies need their equivalent advancements in practical issues, such as a properly sharp needle or a frictionless thread. Ponder the material sensitivities needed to stitch a ghost heart into a living body. Think about that for a moment. **But these advancements also need to be considered within poetry and art and philosophy so that we can comprehend and find meaning within such significant changes to the way we define ourselves.** For me, the most important thing is that we might expand the ways we care for and empathize with one another.

ELISSA AUTHER: That was a wonderful framing of your engagement with science and the questions that have driven your practice. I'd like to further set the stage by summarizing two key points related to the perception of the divide between fields of knowledge, especially the divide between the arts and humanities and the sciences.

My first example is a humorous one and comes from a novel I recently finished—*Double Blind* by Edward St. Aubyn. He explores what the jacket copy describes as "the profundity and obscurity of the divide between the nature of electrochemical activity in the brain and the experience of being conscious." There's a character by the name of Saul. He's an AI researcher who has created a cap that contains recorded scans of electromagnetic activity of famous people. He monetized this contraption, so you can wear this cap and have the same experience. For instance, a famous monk who contemplates the Stations of the Cross—you can have direct access to this monk's emotional experience.

I'd like to quote from the book. Here Saul is talking to the venture capitalist who has acquired a controlling interest in his company, Brainwaves, about how our concepts of the brain and the mind don't always overlap in the ways he would like. He says, *Everything happens for a reason…but unfortunately, when it really matters, we don't know what the reason is…I mean, experience accuses science of being reductionist and authoritarian, while science dismisses experience as subjective, anecdotal, and self-deceived. We have an absurd situation where the first-person narrative of* experience *and the third-person narrative of* experiment *shout insults at each other from either side of an explanatory gap, that huge, huge explanatory gap.*[1] That's one way of describing this apparently unbridgeable divide.

My second example is based on my own experience. Prior to joining the museum world, I spent decades on a college campus where it was not infrequent for people in the arts and humanities to express how they felt dismissed, like the stepchildren of the liberal university, or that they had lost out to the sciences in some way. It never resonated with me personally, but I know we're all familiar with that position. It can devolve into a situation where the arts and humanities are positioned as owning the moral high ground—with arts and humanities faculty being the only people who really care about making the world a better place or fostering their students' development as responsible, empathetic adults in a world that they see being ruled by science that is overspecialized or quantifies and reduces life to data or abstractions. Of course, that perspective is flawed. The humanities and arts don't have a monopoly on moral insight. People in all fields care very much about making our world a better place and they certainly care about their students. And, of course, the arts and humanities are also characterized by specialization—some would even argue overspecialization—expertise, and a commitment to the production of new knowledge.

Can you talk about your interest in trying to bridge this divide?

DR: There is some necessary humility in what you just said. I personally am very sensitive about not framing the humanities and arts as taking the moral high ground, but I also acknowledge the humanities have a lot to say about things that sometimes the sciences aren't equipped to speak on or interested in discussing. I watched Dr. Frazier perform a heart transplant, and afterward I asked him what, I soon realized, were to him terribly overromanticized questions about the heart. He got a bit frustrated, and he said something to me that I'll never forget: "Dario, do you want to live or do you want to die?" When he's working on the operating table, he doesn't care about the issues I was raising. If keeping his patient alive means eliminating a defining feature of our humanity, the beating heart, that's a price he's willing to pay.

Dr. Frazier holds the world record for the most heart transplants. He has saved thousands of lives. Nobody can dispute his moral commitment to the practice of care. Nonetheless, I would argue—without it being some critical comment on who he is as a person—that if his desire to care comes with a redefinition of a feature of our humanity, the *beating* heart, then it's not really up to just him anymore. It's up to others to, at minimum, have a conversation about it. Just as I love and value the history of philosophy and its grappling with mind-body problems, when I look into the history of artificial hearts, those problems arise in a very different and practical way.

There's much to learn in relation to these deep human questions about where we—our subjective selves—are located in our bodies, where love resides, the transition from the heart to the brain. We take for granted that we've moved all the problems of consciousness to the brain. Nonetheless, the metaphor of the heart just won't break. I find it fascinating that when you want to nonverbally communicate to someone that you love them, you don't put your hand on your head, you put your hand on your heart. Why? What lingering values do we carry in these metaphors?

I wonder, how do we engage each other without taking on a self-righteousness? As much as I value Dr. Cooley and Dr. Frazier, once those patients left the operating table, they were onto the next patient. Nobody discussed what it meant to live with those hearts. For example, after the surgery, Mr. Karp's wife asked her husband if he still loved her. There's no room to entertain that question in the clinical setting because that would be quite silly to someone like Dr. Cooley, but I want to really ponder Mrs. Karp's question. What does it mean? We don't know. No one ever had a spouse with an artificial heart before she did. In so many ways, Mrs. Karp was at the forefront of asking important questions that aren't recognized the same way as the technical, medical questions of the heart. And we can do that type of work in the humanities.

EA: That is a great segue to the more practical aspects of these kinds of collaborations. Perhaps you could talk about your on-the-ground observations? For instance, what was your role within the research team that worked on the ghost heart? Were you an observer? An ethicist? An interlocutor? How did they perceive you? And what did you try to do with them?

DR: Those are such good questions because in school there wasn't anyone to consult with about this model of the artist I'm trying to create. Thinking of that picture of Dr. Cooley, I wish a poet had been in the operating room. I think it is a weakness in our culture that this idea is weird to say out loud, because, to me, it's an obvious thing to have said out loud.

I put a lot of value in the act of observation—both in the sense of tenderness and in the sense of resolution. **The most challenging part of this work is observing.** In our collaborative efforts, how can we get past the back-patting stage, when we're just happy that we noticed each other, and actually make a convincing argument that working together might change our practices in ways that matter to one another? What in the world does an artist have to contribute to artificial heart design that might matter to the patient? I have tried to take that step with Dr. Frazier. He's driven by engineering purposes; to him, the beatless heart is a better system because it solves some physiological problems. But he's not asking whether an emotional price might be paid for having a beatless heart. I don't want to say he's not interested. As I said earlier, he's focused on saving the life. And I get it. We all want our heart surgeons to be focused on saving our life when we're on the table. But why not have someone else pondering, "What does it mean one, five, ten years later, when you've lost any internal sense of your own beating heart?" I like to say that no one has fallen in love with a beatless heart. No one has made art with a beatless heart. No one has danced or worshipped their God with a beatless heart. In the moment, on the operating table, this seems irrelevant, but it's not later on. At minimum, it is an interesting question to ask. And culturally, is the price of having a beatless heart visible in your emotional maturity, your own sense of self, or how others treat you? I don't know. These are questions the humanities are equipped to ask. So, in the design stage of a beatless heart, why not fold in poetic questions along with engineering ones? Could the beatless heart still function in the way Dr. Frazier hopes, but could the beat be worked back in to serve these psychological and emotional needs?

I've had these conversations with Dr. Frazier. That's the place I try to get to with these projects—to deep observations, conversations, and research by embedding myself with the team. Then I make my own work in response. And I have this third goal of uniting the poetic and the scientific questions for the same end. In this case, what success looks like for Dr. Frazier is very particular and what success looks like for me is a little different. Can we unite those visions and meet each other's goals? That's what I'm always striving for.

EA: Recalling those ideas about the tenderness and the technical that you mentioned earlier: How do you unite those two aspects in your work, in reaction to, or influenced by, your experience as part of these teams?

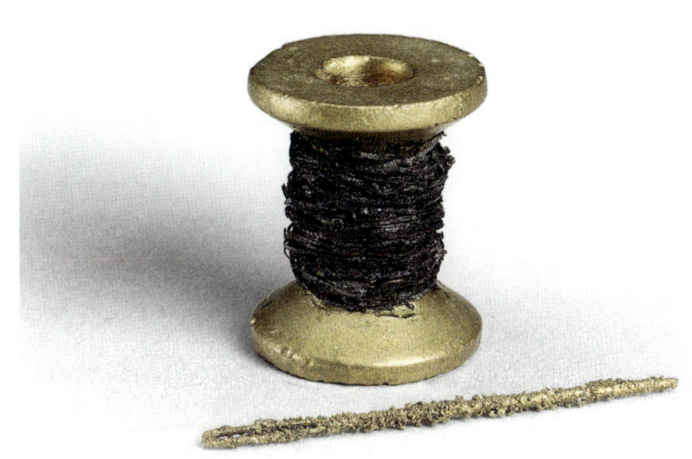

Dario Robleto, *Untitled (Patsy Spool)*, 1998, vinyl record, iron pyrite (fool's gold), glue, 1½ × 1 × 1 inches; Patsy Cline's "Fall To Pieces" 45 rpm vinyl record was slowly sliced along the outer rim until reaching the center, then connected into one long thread and spooled.

DR: I've been talking about what influence I might have with the scientist or the surgeon, but being embedded in the thought process and being sensitive to those topics definitely changes my work in particular ways. I also am a huge music person with a DJ background. *Untitled (Patsy Spool)* came out of some of those ideas. I took Patsy Cline's 45-RPM record featuring "I Fall to Pieces." Using an X-Acto knife, I slowly, slowly carved along the outer edge, almost unwinding it, as if I was peeling an apple. And I unraveled her voice. Then I spooled the "thread" around this spool I had made of fool's gold.

It was a way of thinking about these types of sensitivities. The thread has one very practical function, which is to hold something together. But how could I charge it differently? In this case, through Patsy's voice and what she's singing about. She's literally singing about falling to pieces. I think she's one of the strongest, most powerful humans who has ever lived—and yet, she's singing about something incredibly fragile. These are the types of sensitivities that I attribute to studying these points in science that require a new way of looking at the world. In other words, I don't think I could have made this piece without years of pondering the artificial heart and the differing ways we can "give" our hearts to one another.

With *You Make My World a Better Place to Find,* for two years, I secretly picked a little piece of lint, or thread, sometimes a hair off of various people's shoulders—a little action we all have probably done to ourselves or others, a small gesture of kindness or affection. I gathered it all together and spooled it, like with the record, onto a wooden spool of my grandmother's.

Deeper Into Movies (Buttons, Socks, Teddy Bears, and Mittens) is similar in spirit. I took the thread from my very first baby blanket, which symbolizes my first interactions with a fiber in the world, and I completely unraveled it. I took snippets of that thread, and I seamlessly inserted it into spools of thread from various places, like thrift stores and fabric stores. I would unspool the thread quite a lot, insert a little fragment of my blanket, and slowly respool the thread back into its original form. Then I would take those spools back to wherever I found them.

These early works were influenced by these scientific questions—even though the science part is not obvious right

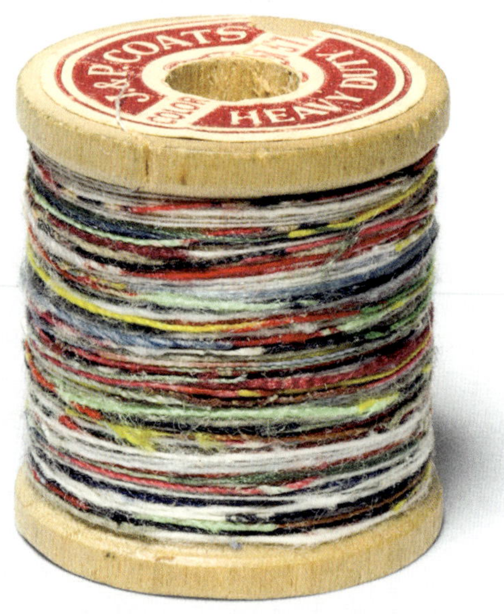

Dario Robleto, *You Make My World a Better Place to Find,* 1996–98, lint, thread, various debris and particles, the artist's grandmother's antique wooden spool, dimensions variable; For the past two years I have been secretly collecting lint, thread, etc., from friends, acquaintances, and strangers (for example, a piece of lint on someone's shoulder, a hair hanging on a forearm). I have connected all this debris into one long thread, which I then spooled. From this spool of debris I repaired a blanket, a tiny pair of mittens, sewed buttons back on, repaired tears in clothes, and various other things.

away. Thinking about sensitivity, how can art frame these actions in a way that elevates them to monumental moments, even though they may be something as trivial as gathering some lint off of someone's shoulder?

EA: These examples connect to your work with the pulse wave and its translation into prints and sculptures. You've always been very interested in making connections through time—connections to human experience from an earlier point for the viewer in the present. For instance, through recordings, like the Patsy Cline record you sliced and transformed into an actual connective thread, you metaphorically highlighted the way in which sound recordings allow one to connect to the past. It brings up another aspect of your work that I think is distinctive: your labor-intensive handwork. This aspect has developed to a really high intensity. Could you discuss your commitment to craft and the handmade?

DR: *The Creative Potential of Disease* is from the body of work that looks at craft on the battlefield, which was largely dictated by battlefield nurses. For most of the history of war, nobody had any frame of reference to understand that psychological trauma has physical ramifications—what we now call PTSD. But before that knowledge, during the Civil War, soldiers could officially be categorized with "cowardice" rather than PTSD. What I found so moving and fascinating about this period—and I made *The Creative Potential of Disease* in response—is that the nurses of course knew about psychological and physical trauma. In some way, they knew that the science had not yet caught up to the soldiers' experiences. And so, they began giving these recuperating soldiers craft projects, often while they were still in bed. I became a collector and historian of these objects. This one, in particular, really hit home. It's a little doll that was made by a Union soldier while he was recovering in the hospital from a leg amputation. And, as people probably know, the Civil War was unprecedented in the number of amputations. The nurse gave him his uniform, his bullet-ridden uniform, to craft a self-portrait. And he made this little doll. Here is a soldier, likely suffering from PTSD, certainly in shock from losing a

Dario
Robleto

Elissa
Auther

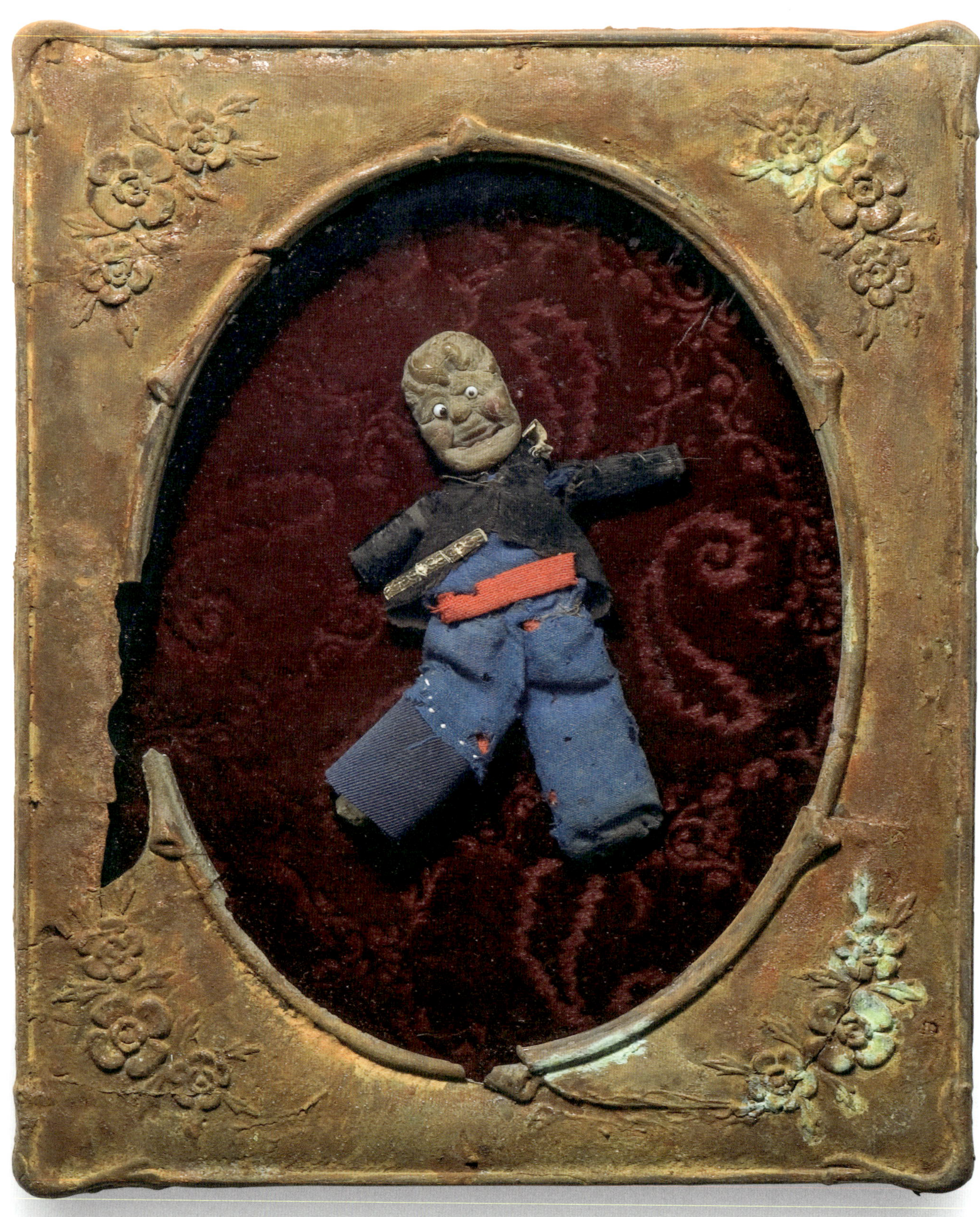

limb, who is trying to come to terms with this new reality of himself through the fabric of his own uniform.

The point you made about time is important with this piece because, in this body of work, I challenged myself to ask, **is there a way to retroactively mend and care?** I very much want to help, today and in the future, but I'm also interested in whether you can extend care to the past. And so, the leg on the left is made with a modern soldier's uniform and stitched on with a modern-day surgical thread. It's me, through time, trying to finish something that he couldn't finish. Underneath the pant is a little prosthetic leg that I made for him, which connects to a tradition I learned about that really struck me: soldiers took it upon themselves to literally carve their own new limbs, often from wood—it was before anyone expected the government to supply such things.

I was venturing into what an artist can do to retroactively heal the past. I hold a lot of hope that metaphors can serve as an act of healing, but I wanted to know if there were

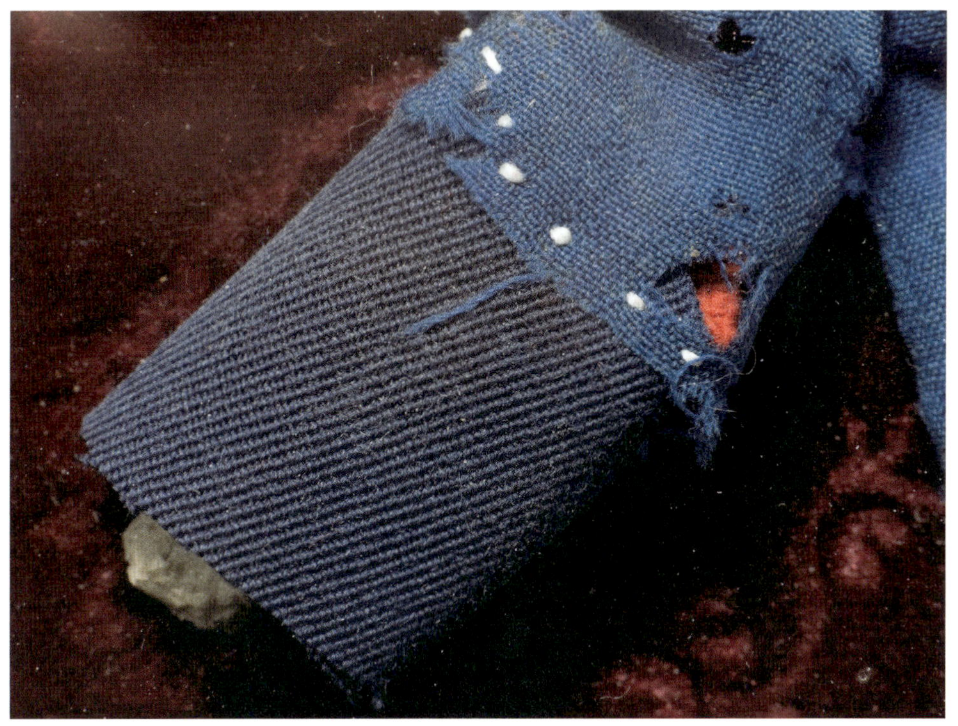

other ways to do it. At the point where the prosthetic limb that I made for him is touching the cotton under the body, I gave him medicine. I created a homemade medicine that I applied to the joint, which turned into its own sculpture, *Balm of a 1000 Foreign Fields*, which speaks to the home remedies the soldiers were using at the time. I wanted to see if I could merge this premedical, home-remedy logic—which was often firsthand knowledge of plants and roots and herbs, made, of course, with the love of wanting to care for a wounded husband or son—with my artist logic to make this medicine.

It was practical but also ingenious of the nurses to use the materials of war as the materials of healing. I find it so radical that those nurses upended those materials and allowed the soldiers to find some way to live again with them.

EA: I love the connections between the small handmade doll and the psychology of trauma and early forms of art therapy, which, of course, developed over time. By World War II, certainly, there were programs in the United States for returning soldiers to engage in that kind of activity in an organized way.

With *Elegies of Proxima b*, I'm struck thinking, "Wow, here's someone who's doing some serious research into science." But at the same time, it's a fabulous *fabula* that can be appreciated on the level of composition and color alone. I can lose myself in the science backstory or through free association and fantasy facilitated by the object. It almost appears contradictory.

One thing that stands out for me in terms of art's relationship to science is the precise, exquisite handcrafting of the object. I can't help but be reminded of early examples of scientific and medical equipment—all of which was handmade for very specific purposes. Those machines are absolutely gorgeous. It's also a good example of how the Industrial Revolution didn't entirely kill craft. In some areas there was flourishing of object making during the Industrial Revolution—one of them being the making of precision machines with precision tools that had to be invented. Could you talk about where precision or virtuosity fits into your scientific project?

Dario Robleto, *Balm of a 1000 Foreign Fields* (detail), 2004, homemade balm (almond oil, beeswax, home cultured antibiotics, pollen from artillery plant, Death's herb, rupturewort, winter's bark, bugleweed, honeysuckle, sweet chestnut, Tree of Heaven root, life everlasting root, life root, bittersweet, antler velvet, ground vinyl dust from Marlene Dietrich's "This World Of Ours"), cast and carved bone dust from every bone in the body, dirt from various battlefields, bovine cartilage, WWI US military medical blanket, iron salvaged from the sea, melted bullet lead, zinc, nickel,

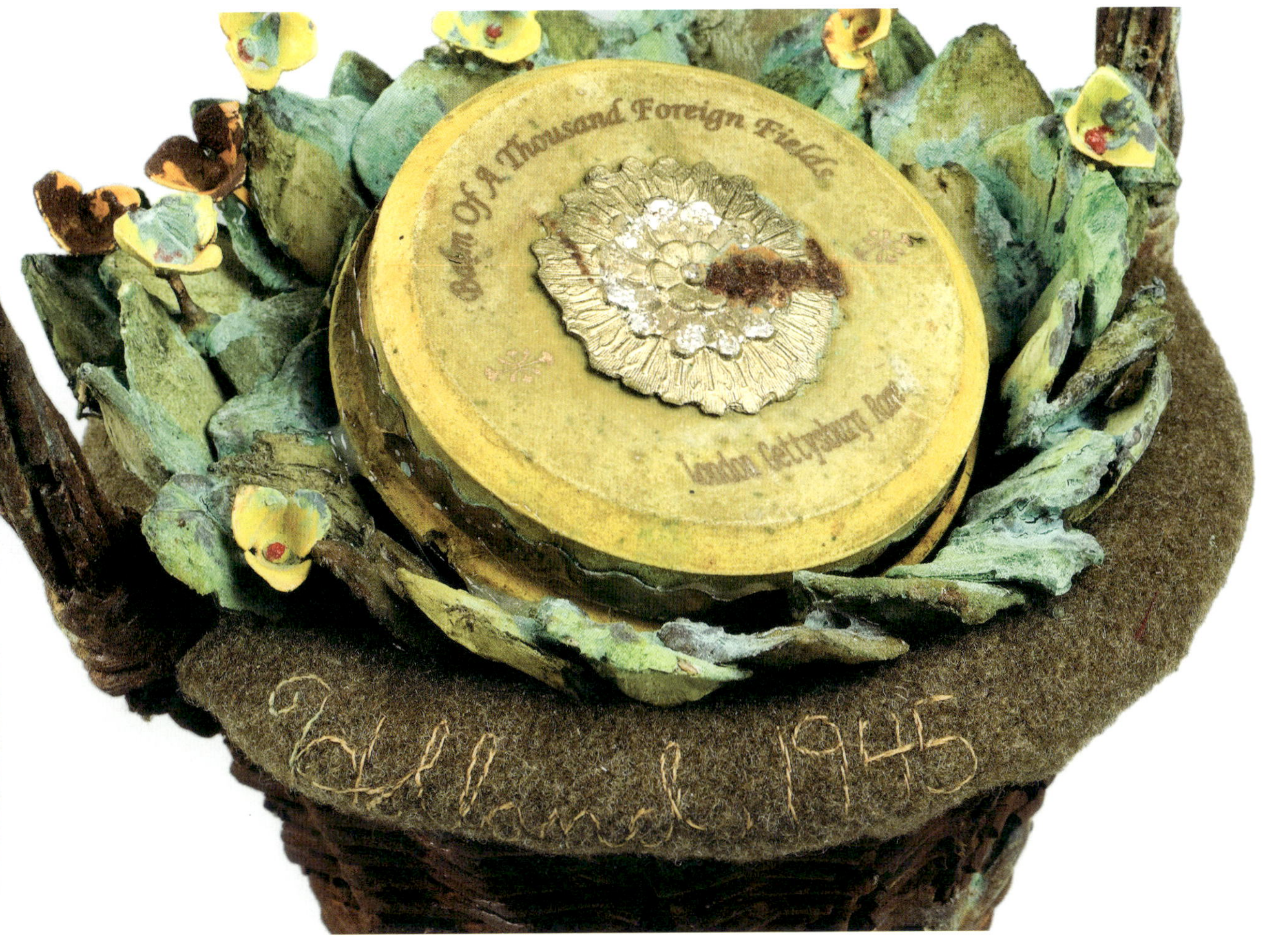

DR: Another area that I'm deeply invested in is astrobiology, the SETI Institute, and the search for extraterrestrial intelligence. I would argue that SETI—by asking what may be the most profound question we can ask: are we alone in the universe?—requires, like the beatless heart, like the ghost heart, a new era of sensitivities. And not just technical sensitivities; if we make contact with another intelligence tomorrow, think about the types of poetic, philosophical, cultural, and, for many, spiritual sensitivities we would need to quickly evolve as a species.

With SETI, I've been spending a lot of time thinking about this question: if we got a message tomorrow, what would we say back? This is a profound problem. And there are all kinds of proposals. If you ask almost any scientist, they will usually say math or physics—that should be our first response. It's a great response and makes a lot of sense, especially if physics is the one universal language of the cosmos. But I ask, what if an artist could choose which equation to send? That's a weird and uncomfortable thing to ponder. Can the answer be motivated by what we have in common? In other words, can an equation or physical law express both a gasp of science *and* a poetic stance, measurement *and* meaning, the brain *and* the heart? That's where I would start, rather than the military position of what's different about us, which often dictates this conversation.

One of my favorite science discoveries in the past decade is that the first color life produced on our planet was likely pink. Instead of the blue-green hue of today's Earth, it's very likely the planet was quite pink and rosy because this is the color that the earliest cyanobacteria produced.

I wanted to make a series of sculptures that, if found by another life form, would ask: "Did you also begin life in a pinkish tone? Maybe your early oceans were also rosy." We don't know if all life tends to start this way, but it did here, and maybe we can nod to each other that we all got off the ground in this hue.

My sculpture *Elegies of Proxima b* was directly related to this discovery of the color of early life on our planet and the recent discovery of an Earth-like planet outside our system: Proxima Centauri b. I made a series of sculptures as proposals, as initial gift-giving acts. I'm building off of craft traditions—in particular, paper crafts and seashells, butterfly wings, all kinds of things that have a longer history. This is an updated example of how a deep engagement with a scientific field is provoking new ways of thinking about not just increasing my sensitivity but also how I make things. The motivation for this work was an initial act of recognizing we were also once pink. And it was an initial gift to signal that bit of knowledge as common ground.

AUDIENCE: Thinking specifically about college and university campuses, how do we get more people from outside of the

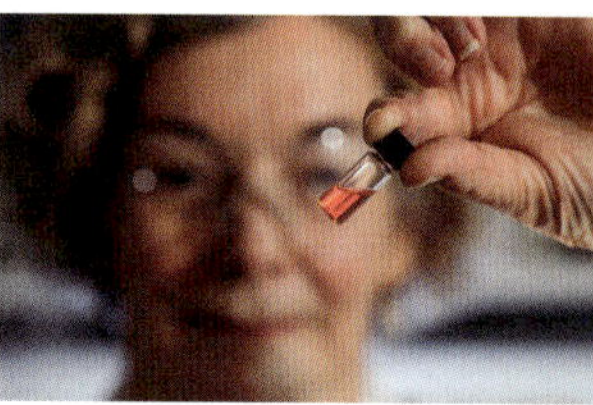

This vial of porphyrins represents the oldest intact pigments in the world, extracted from rocks in the Saharan desert, with the pink color dating back 1.1 billion years. Research was led by a team from the Australian National University Research School of Earth Sciences.[2]

arts to be open to the value of art, both for its own sake and for the sake of what art can bring to different fields? Elissa, I'm thinking also about your work investigating the distinctions between art and craft and considering how we define things and how we appreciate things differently based on their definitions.

EA: **The way we categorize objects can be an obstacle.** A lot of pretensions float around, terms like *high art* or *fine art*, which can turn people off because they feel like they have to be experts to engage. On the college campus it's very important to be doing the kinds of things the Tang is doing. First of all, connect art to a curriculum. That way, students can engage from the perspective of being in the classroom, where they're led through discussions. This involvement with interpretation gives people access to what's going on. Contemporary art is never easy, even for art students. They often feel they don't have a comfortable entryway into the field.

There has been such a huge resurgence of people making things at home, and it spans a lot of different kinds of materials and practices. You have all kinds of sewing that you might do for the home, or crocheting and the *Saratoga Springs Satellite Reef*. But then it expands out dramatically—to woodworking, to boatmaking, and beyond. If someone has access to community resources, like a kiln or a glassworks, that's another example. But I think those are harder. Sticking to practices and materials that are very close to home and that engage people creates a great entry point.

There are many contemporary artists who use materials traditional to craft for this reason. They politicize craft and use it as a tool to reduce the pretensions of the contemporary art world, offering people access and a connection to the creativity they have already developed in their own lives to a broader field or discipline, whether they want to call themselves artists or not.

For more than a decade now, there has also been an explosion of artisanal culture in the areas of food, wine and spirits, and even cannabis. There are all kinds of forms of creative expertise outside of what we think of as "objects" for the gallery and "object making." One idea for college campuses—and, to a certain extent, museums, if they have the social space for it—is to create an event with someone who is an expert in an artisanal product who can share it with others. Focus on the social interactions and the ways you can introduce creativity that exist outside of object making.

DR: That's a great point about moving away from the object. Something I'm hyperaware of is that most people in the public have very few encounters with living artists or scientists. And even more rare, people tend not to see artists and scientists talk to one another publicly. In my experience, the way to build trust is to spend time together. I've learned so much about the misconceptions we might hold about each other—but then we are able to get over those misconceptions by spending time together.

High on my list of misunderstandings that scientists often hold about artists is what rigor looks like in our field. We have different methodologies, although I think they're closer than we realize. It's hard to overcome the stereotype of the flighty, splashing-paint-on-the-canvas artist who is moved only by emotions and not intellect. That's just a terrible cliché. When I'm around scientists, I try not to overdo the fact that I'm thinking and observing like they are. It just needs to happen. And so, I value universities that build the infrastructure for artists to be in residence at research institutes. I've had great opportunities to do that, including one now at Northwestern University. And one of the successes in that program for me was when I was invited to speak at a synthetic biology conference. When we started our collaboration, it was hard for them to imagine doing this—even to let me talk about my art and be there as a thinker. **I want us to see what we have in common so that we can combine our expertise on the very real problems that we have today.** We don't spend enough time together. And we don't hear each other think out loud and get a problem wrong and work through the problem; we don't observe each other's rigor. Back to stereotypes: This idea of the cold, calculated, emotionless scientist is also a ridiculous exaggeration. One of my favorite things that Carl Sagan said was, Why should science cede the spiritual to religion? The questions that science asks should move your heart as much as your mind. I find this to be very true when I spend time with scientists.

> **AUDIENCE:** Dario, how do you source your materials, specifically that doll and the contemporary uniform in *The Creative Potential of Disease*?

DR: A big part of my practice is being a collector. I collect lots of things—war memorabilia, meteorites, dinosaur bones, all kinds of stuff. When I come up with an idea, I can get very specific about what I'm looking for. And at this point, I have a really great network of people to ask. For war memorabilia, I can get highly specific when I'm on my search—all the way to the particular bullet lead from a particular battlefield.

Part of *The Creative Potential of Disease* was engaging with contemporary veterans. Like those nurses who were taking that material of war and finding new possibilities for it through an act of healing, I tried to do the same with contemporary veterans. The pant leg was from a uniform donated by one of those veterans who I worked with. It was really important that he felt like he was part of that process, too—that he was nodding through time to a relatable event that, unfortunately, is never going to go away.

> **AUDIENCE:** Are there any texts you would recommend for someone without a scientific background to begin engaging in these types of discoveries? Are there any specific writings that fall either on the poetic side or the scientific side, or anywhere in between?

> **EA:** From the arts side, I recommend Glenn Adamson's recent publication *Craft: An American History*,[3] which examines artisans and their contribution to American history. I wouldn't say that it bridges the divide between art and science, but it offers important connections to technology, design, and entrepreneurship that are not necessarily part of the story of fine art.

DR: You just can't answer this question without mentioning Carl Sagan. He did so much for me, and for many generations, because his books aren't some technical jungles that you have no chance of ever understanding. He initiated the era of popularizing science at a time when it was not cool. The Neil deGrasse Tysons of our time would not exist without Carl Sagan's bravery, which democratized scientific knowledge in a way that inspired people. His career suffered because he appeared on Johnny Carson and other shows. His credibility was called into question, which is ludicrous. I don't think I'd be the artist I am today without having engaged with Carl Sagan.

The Demon-Haunted World by Carl Sagan and Ann Druyan[4] is a book about science, but it's also about how you can question the world and be curious about it in a systematic way that inches you closer to a type of truth. Sagan also opened the door to the spiritual aspect. To hear a scientist say that spirituality is not something to shy away from, that it's a human experience of awe . . . Science can deliver on that over and over and over. It certainly has for me.

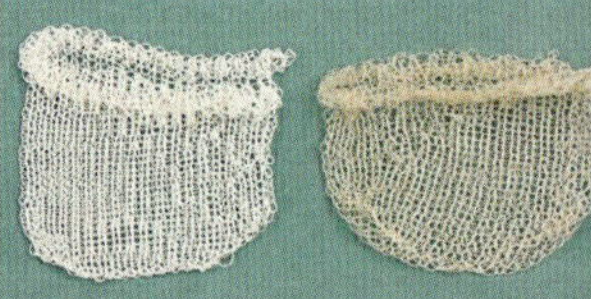

Knitted cardiac support devices have been used for canine heart failure.[5]

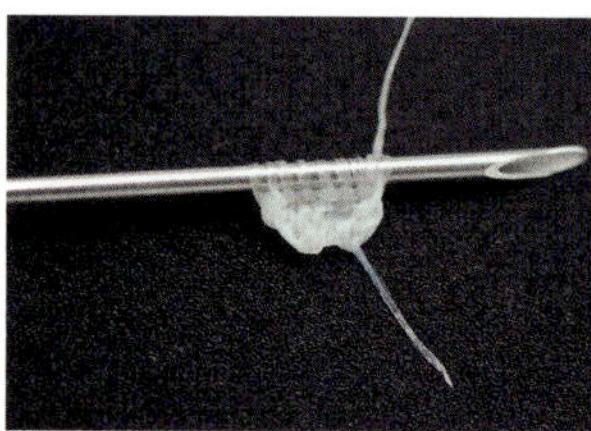

Yarn made from human skin cells can be knit for tissue grafts or organ repair.[6]

1   Edward St. Aubyn, *Double Blind* (New York: Farrar, Straus and Giroux, 2021), 27–28.

2   N. Gunnel, A. M. McKenna, N. Ohkouchi, et al., "1.1-Billion-Year-Old Porphyrins Establish a Marine Ecosystem Dominated by Bacterial Primary Producers," *PNAS* 115, no. 30 (2018), www.pnas.org/cgi/doi/10.1073/pnas.1803866115.

3   Glenn Adamson, *Craft: An American History* (New York: Bloomsbury, 2021).

4   Ann Druyan and Carl Sagan, *The Demon-Haunted World: Science as a Candle in the Dark* (New York: Ballantine Books, 1996).

5   Martin W. King, et al., "Structural Design, Fabrication and Evaluation of Resorbable Fiber-Based Tissue Engineering Scaffolds," in *Biotechnology and Bioengineering*, ed. Eduardo Jacob-Lopes and Leila Queiroz Zepka (London: IntechOpen, 2019), 61–90, doi.org/10.5772/intechopen.84643.

6   Laure Magnan, et al., "Human Textiles: A Cell-Synthesized Yarn as a Truly 'Bio'" Material for Tissue Engineering Applications," *Acta Biomater* (March 15, 2020), DOI: 10.1016/j.actbio.2020.01.037.

Dario Robleto, *Sparrows Sing to an Indifferent Sea* (detail, top; overall, bottom), 2019, earliest waveform recordings of inhalation and blood flowing from the heart during various auditory experiences (1876–96), rendered and 3-D printed in brass-plated stainless steel; lacquered maple,

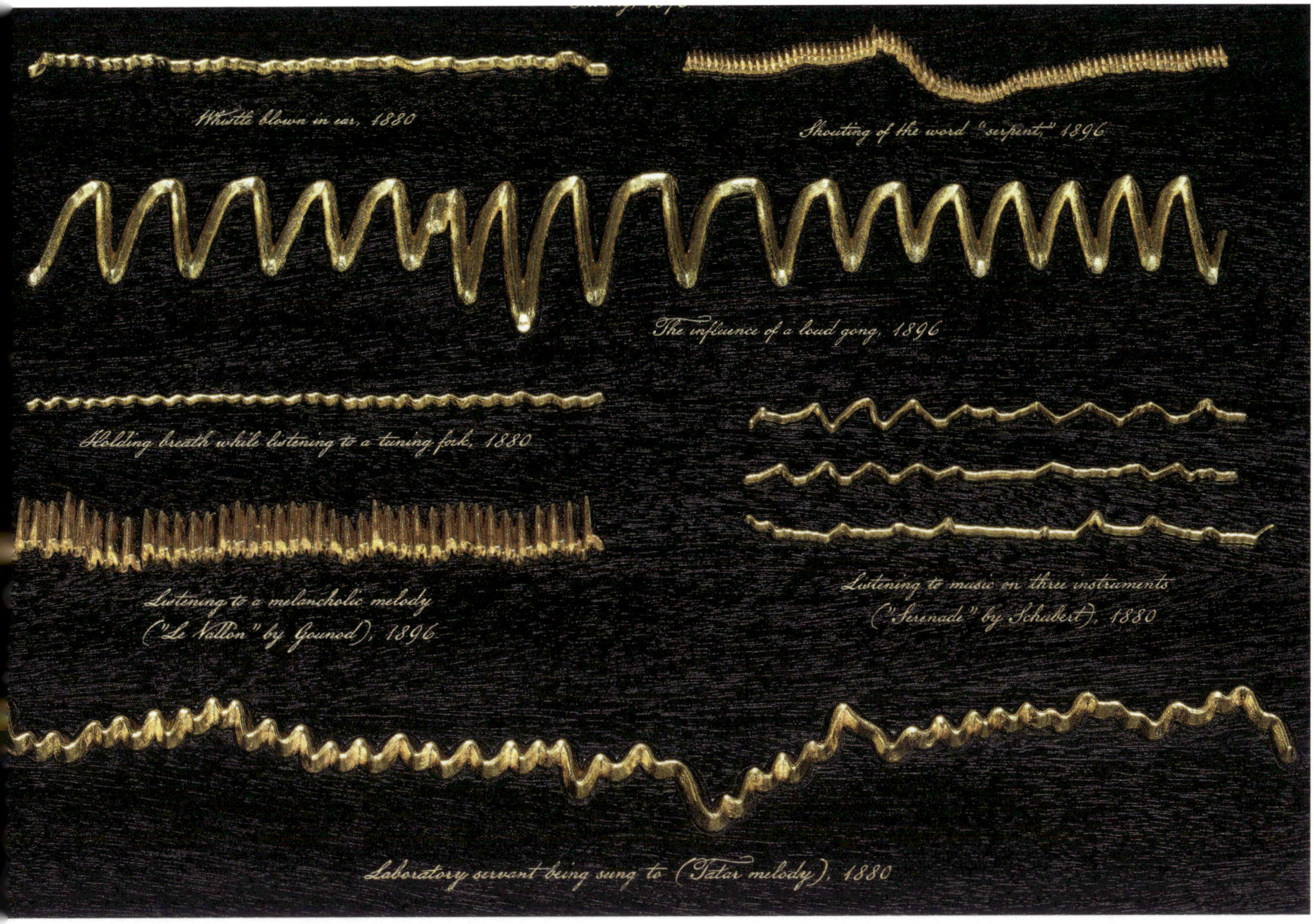

Whistle blown in ear, 1880

Shouting of the word "serpent," 1896

The influence of a loud gong, 1896

Holding breath while listening to a tuning fork, 1880

Listening to a melancholic melody ("Le Vallon" by Gounod), 1896

Listening to music on three instruments ("Serenade" by Schubert), 1880

Laboratory servant being sung to (Tatar melody), 1880

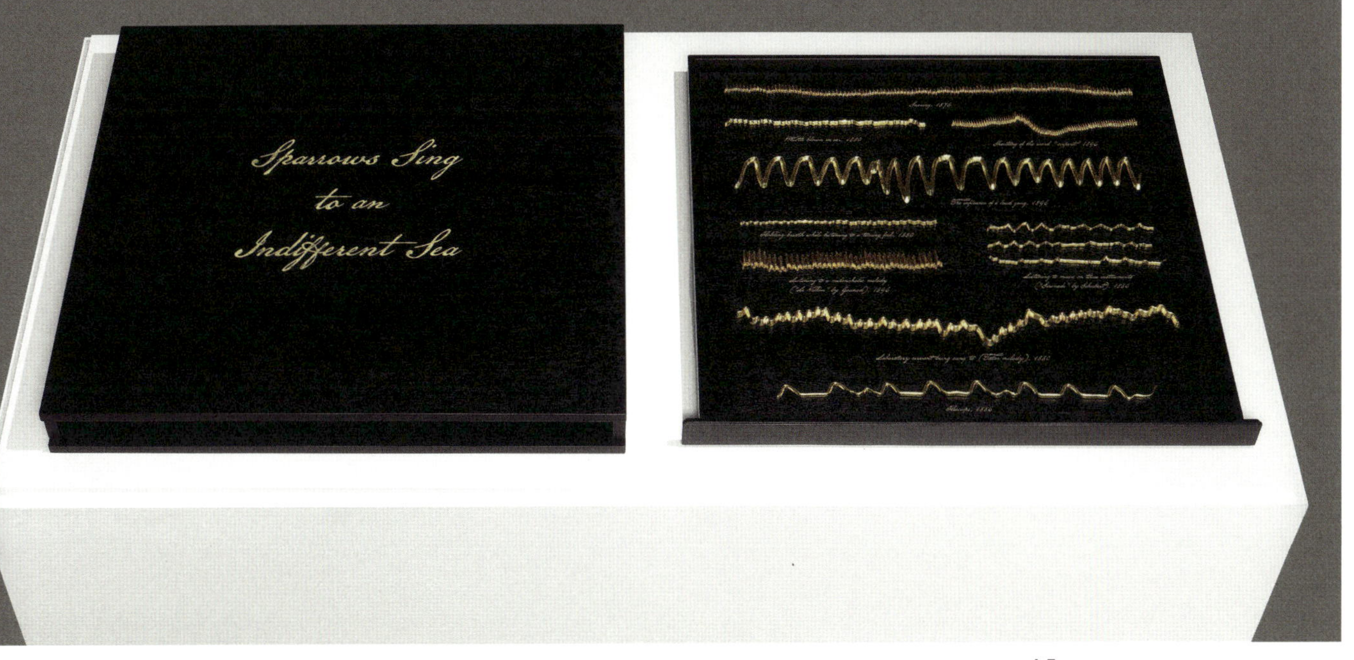

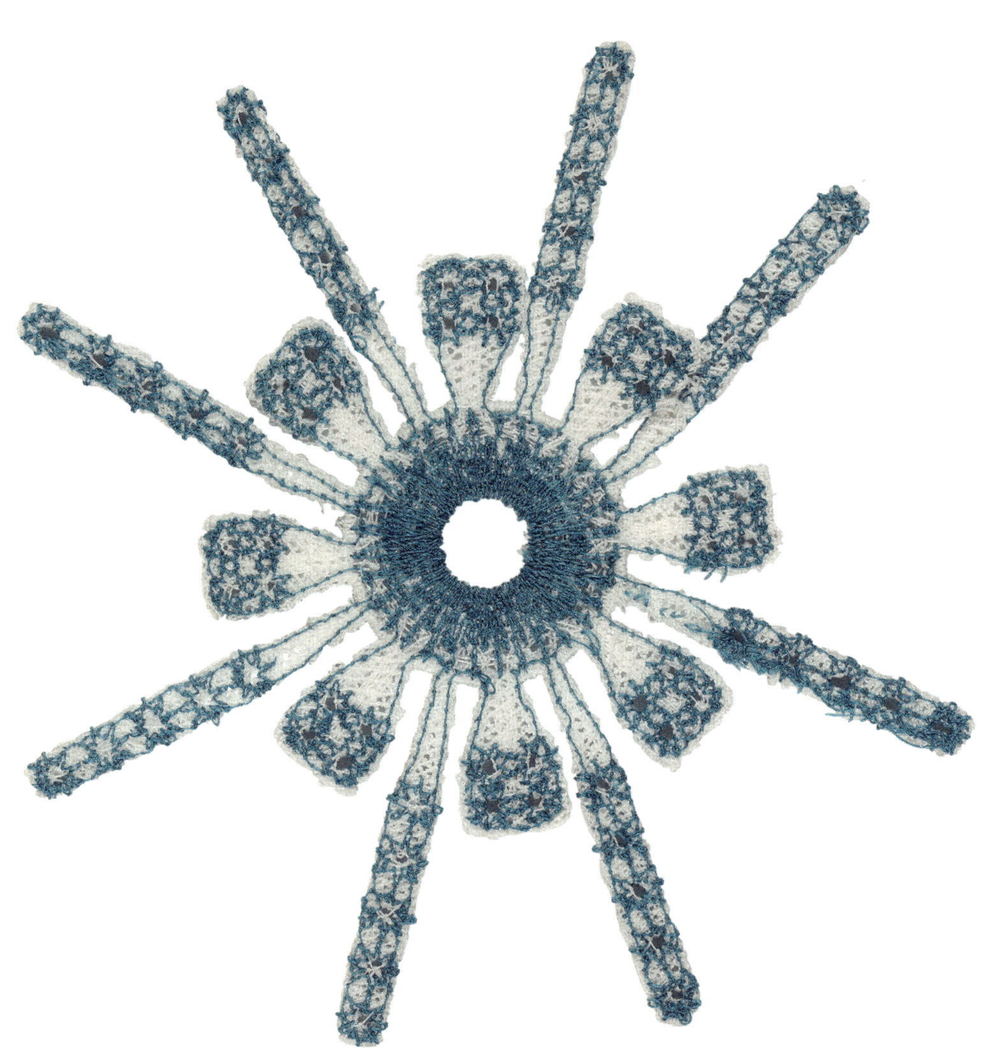

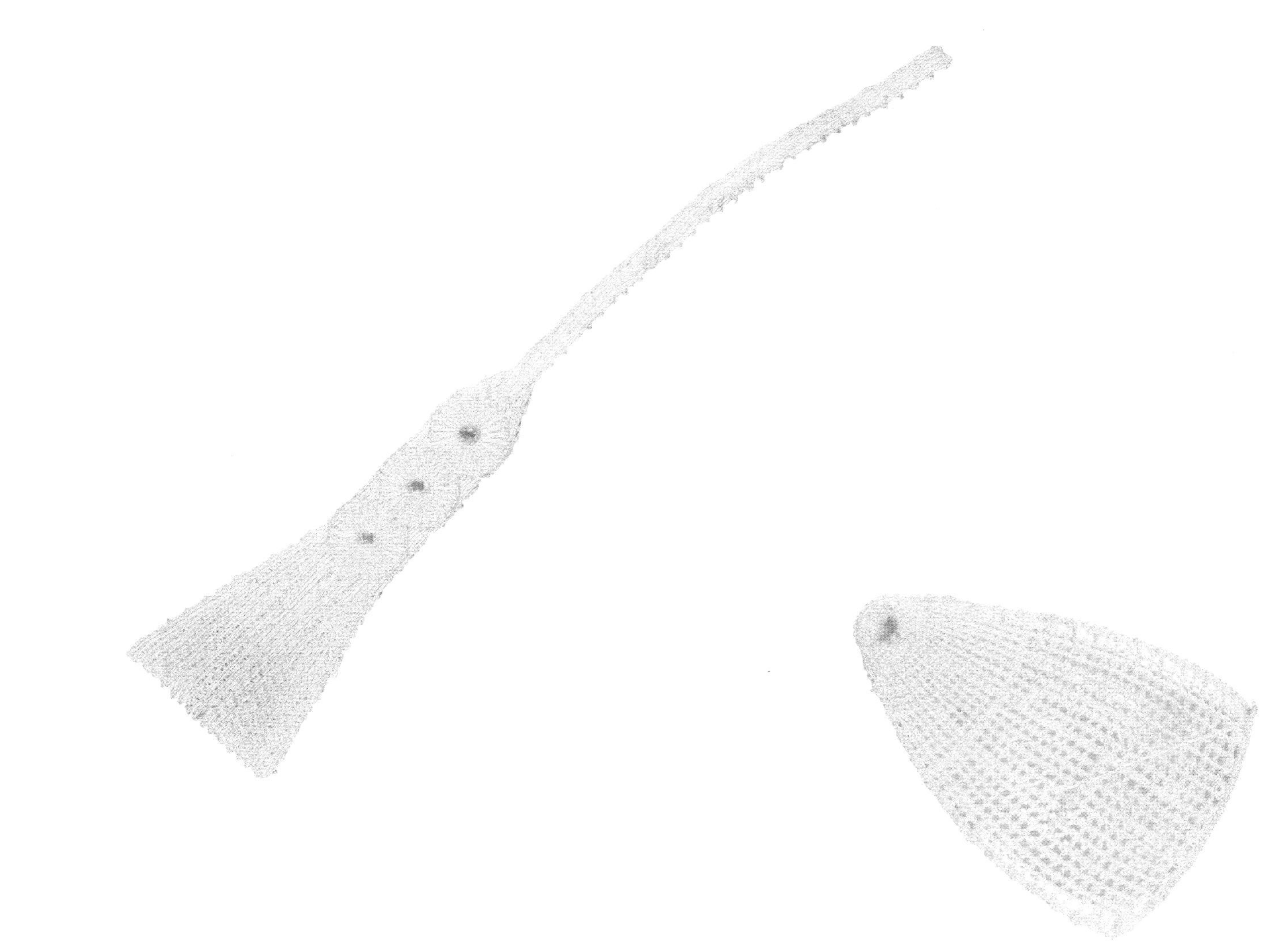

Ellis Developments, Peter Butcher (textile designer), bioimplantable devices, developed 1997–2003, embroidered polyester; device for reconstructive shoulder surgery, 2021, 5¾ × 5¾ inches (left); device for torn rotator cuff, 2021, 6¼ × 1¼ inches (center); device for torn rotator cuff, 1997, 2⅜ × 1½ inches (right)

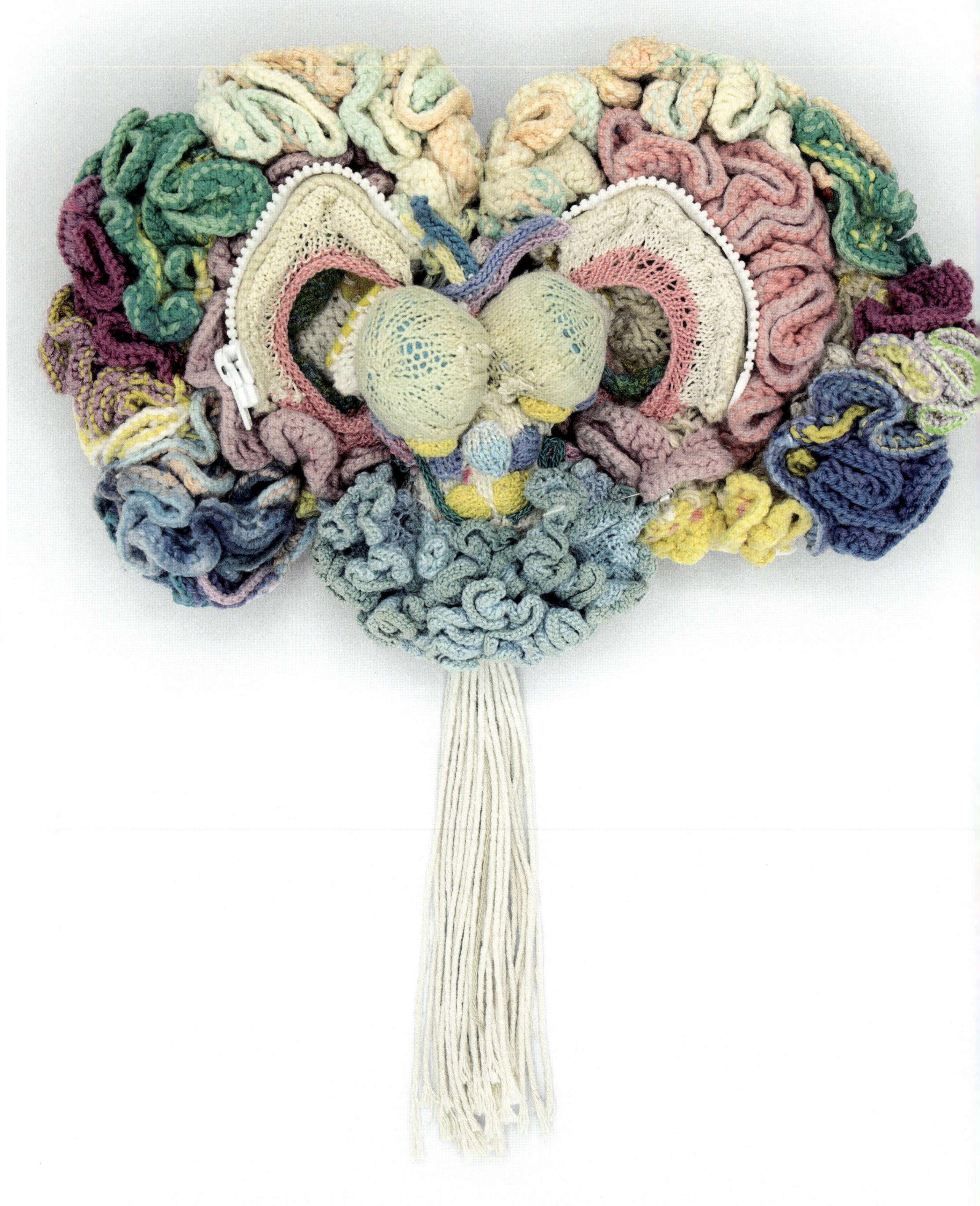

by Sarita Lagalwar

THE NEURAL ARTISTRY OF THE BRAIN

Our understanding of brain function is rooted in art. Leonardo da Vinci—engineer, artist, and neuroanatomist—rendered drawings and paintings of post-mortem human and animal brains to help him deduce neuroanatomical function.[1] Using fifteenth-century technology, he injected whole brain tissue with hot wax, allowing the wax to embed in the brain's three-dimensional structure. Through this technique, Leonardo was able to see and then accurately depict the brain's fluid-filled cavernous ventricular system in his drawings. Recognizing the ventricular system allowed him to further position brain structures surrounding the ventricles. Through this process, Leonardo formed hypotheses on how the brain interprets sensory information and carries out motor commands.

While Leonardo's drawings depict whole brain structures, the brain's inherent beauty at the cellular level was made visible by the advent of the light microscope, which came into widespread usage in the late nineteenth century. Santiago Ramón y Cajal, an artist and scientist, stumbled upon the beauty of the neuron when assembling a book on histology, the study of tissues at the cellular level. Serendipity came into play, for he took the novel approach of impregnating the brain tissues of animals with a silver stain that was invented and quickly dismissed by the scientist Camillo Golgi.[2] Unlike the inventor, Ramón y Cajal realized that the Golgi staining technique, which notoriously only penetrated portions of tissue, was selective in the neurons it recognized yet comprehensive in its ability to "paint" the entirety of the neuron. For the first time, the full cell body, dendritic branches, and in some cases the axon were visible. Ramón y Cajal depicted his microscope images through his artistic drawings and painstakingly stained adjacent slices of tissue numerous times in order to piece together the cytoarchitectural layers of functional systems (for example, the three layers and multiple cell types that constitute the retina of the eye).[3] His stains and depictions led to his formation of the neuron doctrine, which outlines the basic principles of nervous system organization that are

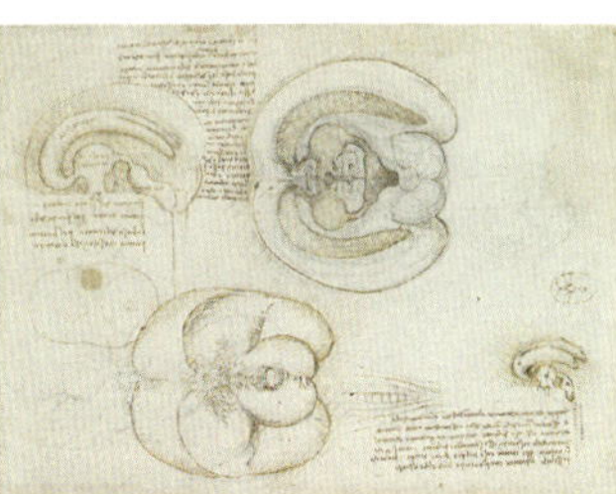

Leonardo da Vinci, *The brain*, c. 1508–9, black chalk, pen, ink, 7⅞ × 10⅜ inches

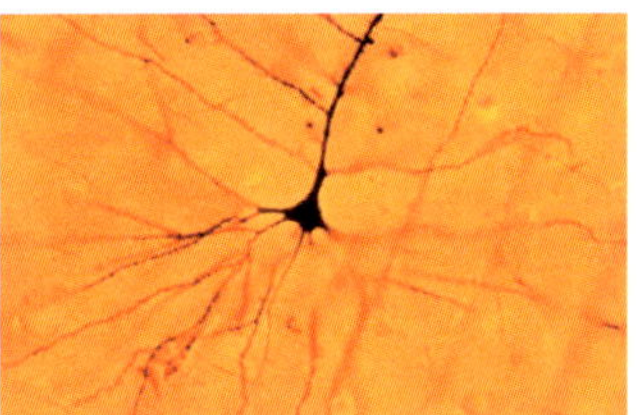

Bob Jacobs (Laboratory of Quantitative Neuromorphology, Department of Psychology, Colorado College), pyramidal cell of the prefrontal cortex stained using Camillo Golgi's technique.

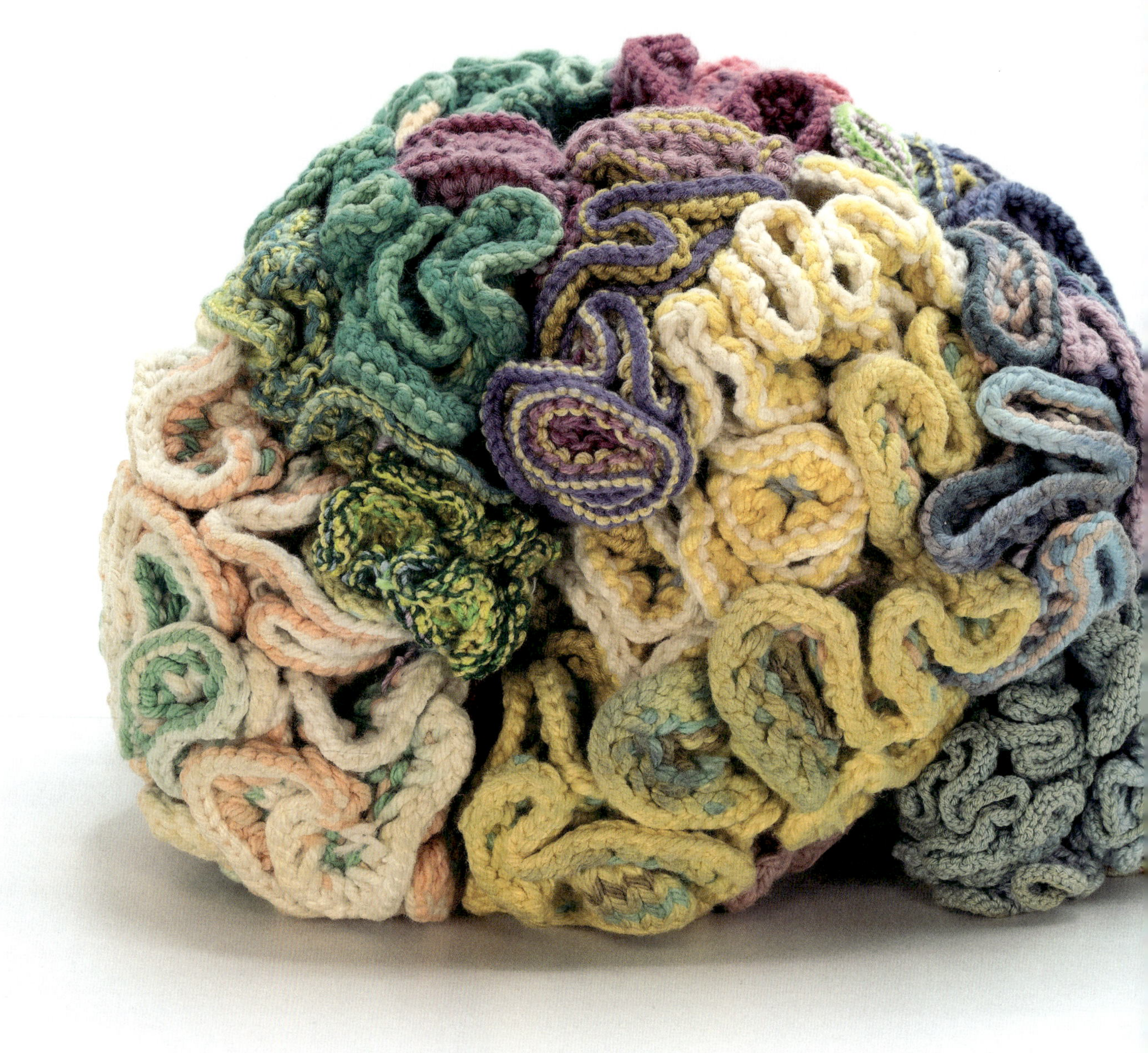

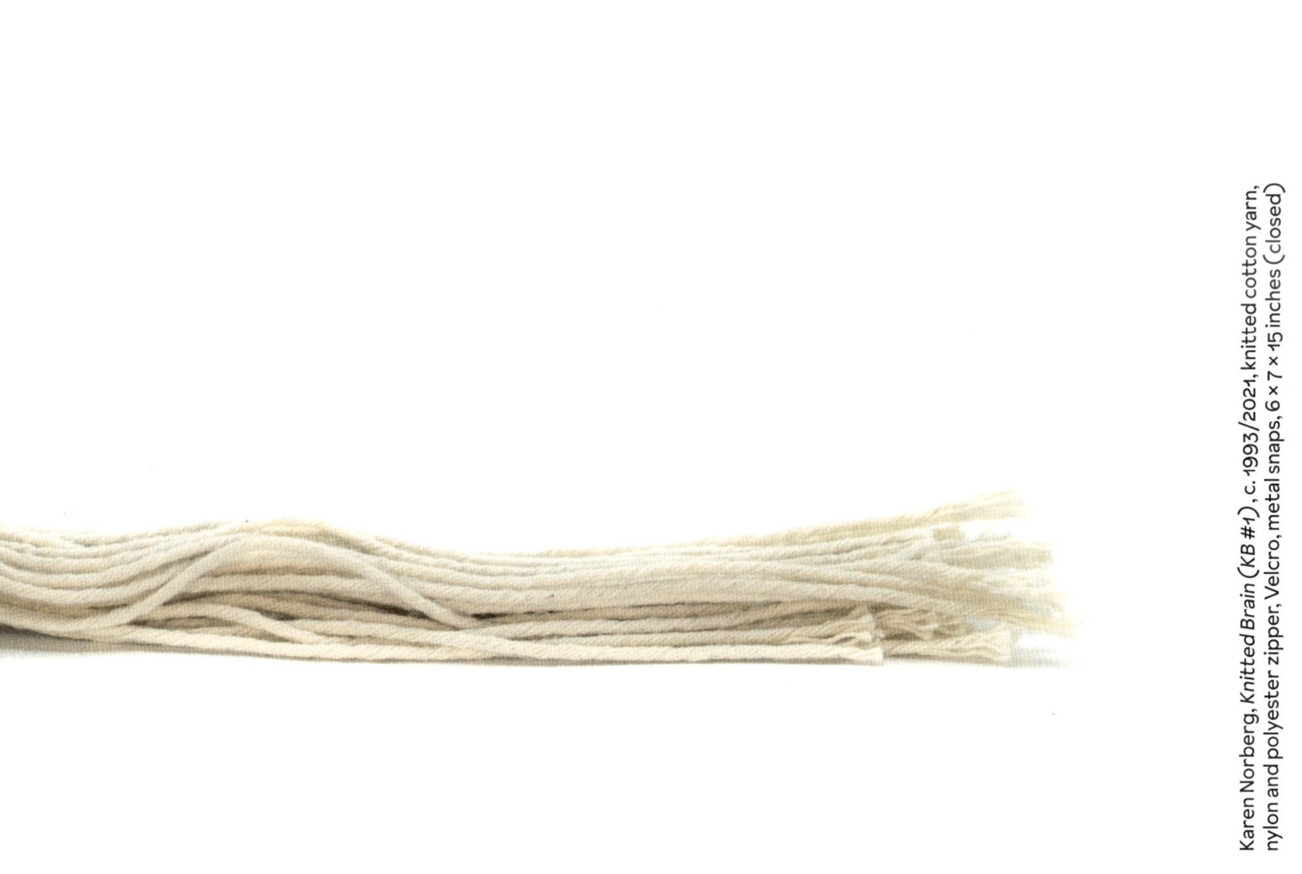

Karen Norberg, *Knitted Brain (KB #1)*, c. 1993/2021, knitted cotton yarn, nylon and polyester zipper, Velcro, metal snaps, 6 × 7 × 15 inches (closed)

still recognized today. This achievement crowned Ramón y Cajal as the father
of modern neuroscience.

Karen Norberg's *Knitted Brain* follows in the neuroartistic history of
these giants. As with the works of Leonardo and Ramón y Cajal, the technology
of the time inspired the art. By the mid-1990s, neuroanatomical tools—including
numerous stains, the advent of novel advanced imaging techniques, and
electrophysiological recordings—provided scientists with a more accurate
understanding of the brain's structure and neural pathways. Three-dimensional
brain models, typically made out of hard plastic parts, were created to help
students grapple with the relative locations of structures and microstructures.
But the plastic brains are limited. To access the internal regions of the brain,
the outer pieces must be disconnected and removed, which makes it difficult
for students to comprehend proximity among regions as they navigate through
three-dimensional planes. *Knitted Brain* overcomes those limitations, as
the knitting maintains connections among regions when the model is opened up.
Moreover, the tightness or looseness of the knots allows Norberg to depict the

brain's cellular density, with the cerebellum yarn
wound tight, realistically depicting the fact that
half of our brain's 86 billion neurons are packed into
an area that is only 10 percent of our full brain volume. Notably, the Internet was in its infancy when
Knitted Brain was initially created, and magnetic
resonance imaging (MRI) scans were not widely
available. Norberg relied on two-dimensional photographs from anatomical atlases in order to knit
and stitch together the three-dimensional model.
Knitted Brain #2, created in 2022 for the *Radical
Fiber* exhibition, was crafted in a new technological

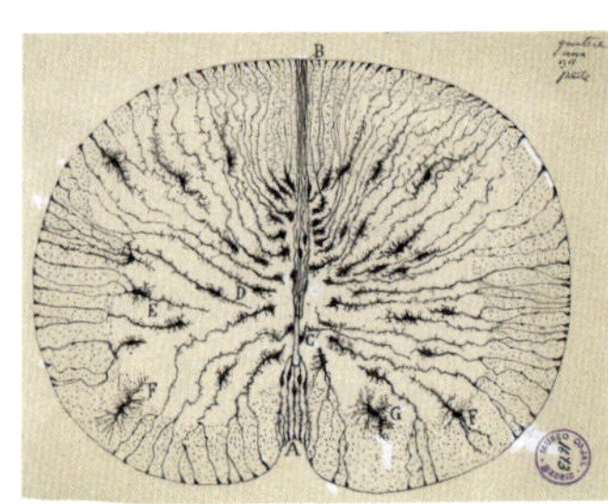

Santiago Ramón y Cajal, glial cells of
the mouse spinal cord, 1899, ink and pencil
on paper

era—one in which three-dimensional models and reconstructions of the brain,
garnered from advanced capabilities in clinical imaging, are widely available
on the Internet. As a result, *Knitted Brain #2* features greater anatomical complexity and enhanced accuracy. Moreover, functional studies have progressed
significantly during the last thirty years, and this advance is notable in the
functional color-coding of *Knitted Brain #2*. For example, when looking at
the oval-shaped thalamus in the center of the brain, one sees not a single color
of yarn, but a rainbow spectrum of stitches reflecting the variety of sensory
and motor functions regulated by individual thalamic nuclei. By tracing the colors
throughout *Knitted Brain #2*, one can start to gain insight into the multimodal
workings of the brain, the integration of widespread brain regions in regulating
function, and the complexity and diversity of individual regions.

Knitted Brain and *Knitted Brain #2* were labors of love for Norberg.
The artist created the first *KB* at a time in her life that reflected significant
career and life changes. The project was both an intellectual challenge and
a personal hobby that brought tremendous joy to Norberg and her colleagues,
who observed the progress firsthand. She knitted the second cerebrum with

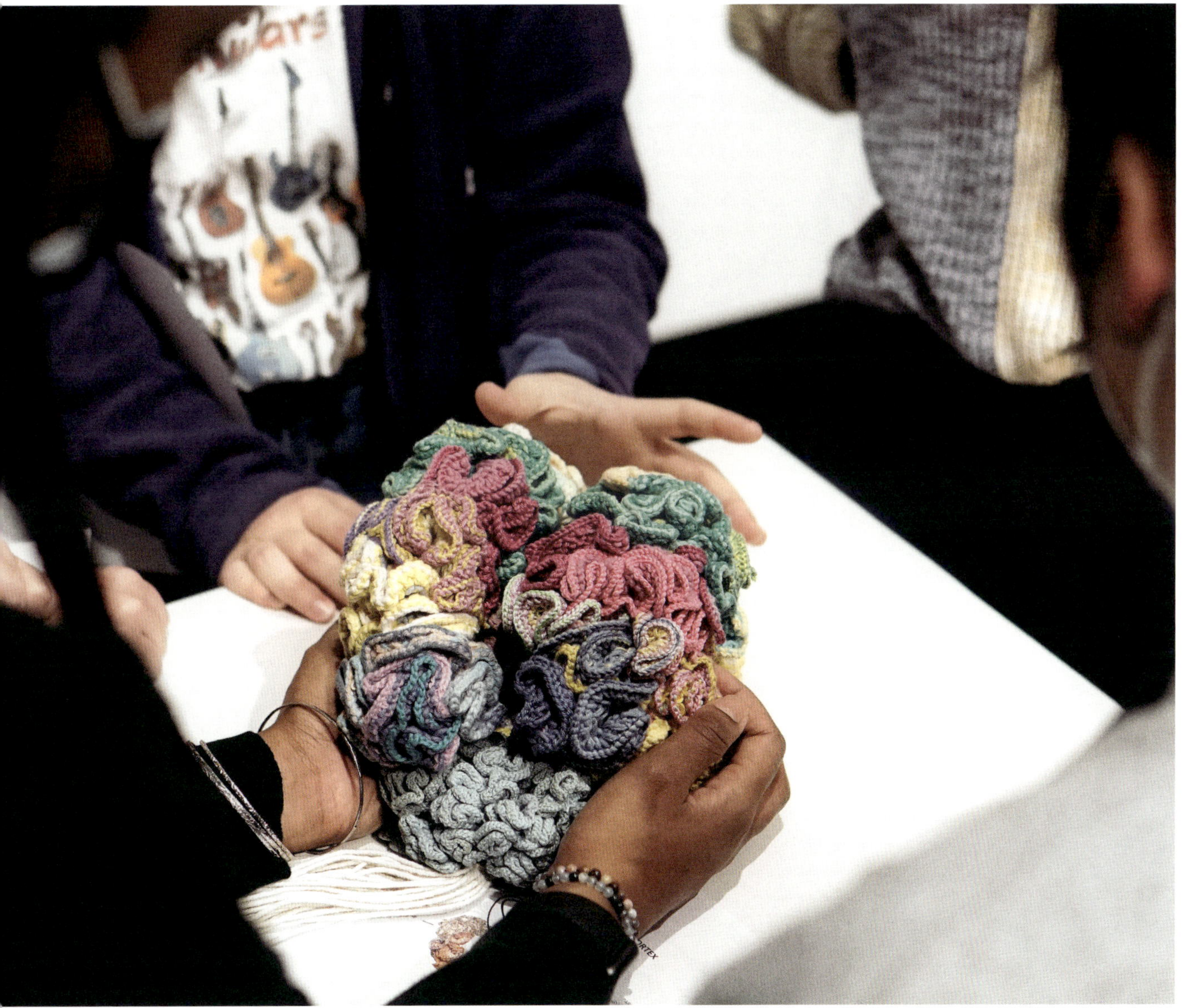

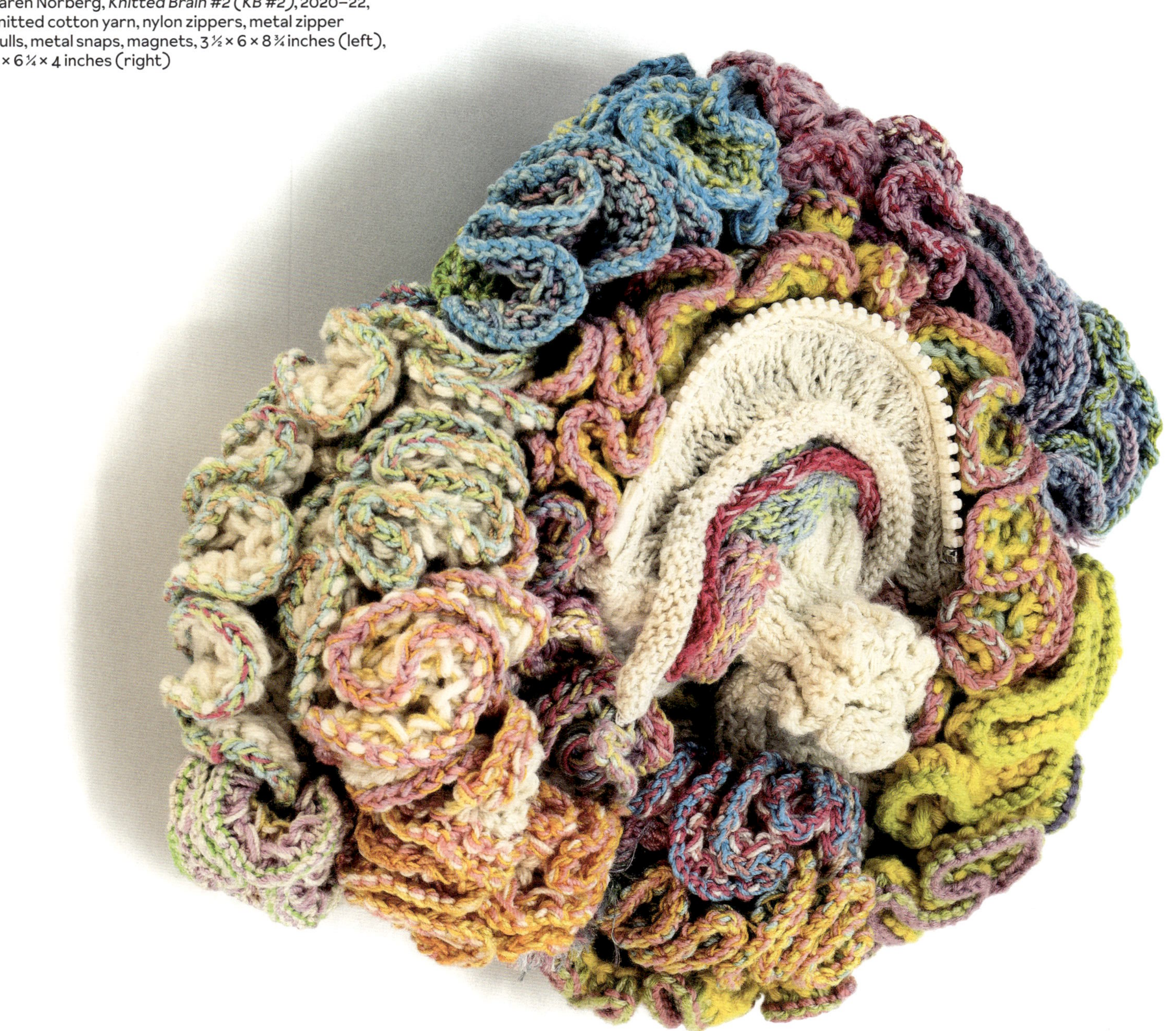

Karen Norberg, *Knitted Brain #2 (KB #2)*, 2020–22, knitted cotton yarn, nylon zippers, metal zipper pulls, metal snaps, magnets, 3 ½ × 6 × 8 ¾ inches (left), 5 × 6 ¼ × 4 inches (right)

the goal of making a teaching tool, one that numerous undergraduate students of neuroanatomy and visitors to the Tang Teaching Museum used as an entry point into the discipline of neuroscience. It was not constructed as a final piece, but as a work in progress. All viewers are urged to visualize the brain from different vantage points, provoking canonical neuroscientific and humanistic questions: How do different regions of the brain work together to drive behavior? Where are our memories stored? How do our neural fibers house our identity?

1   Sophie Fessl, "The Hidden Neuroscience of Leonardo da Vinci," Dana Foundation, September 23, 2019; dana.org/article/the-hidden-neuroscience-of-leonardo-da-vinci.

2   Marina Bentivoglio, "Life and Discoveries of Santiago Ramón y Cajal," the Nobel Prize, April 20, 1998; www.nobelprize.org/prizes/medicine/1906/cajal/article.

3   "The Beautiful Brain: The Drawings of Ramón y Cajal," Grey Art Gallery, New York University, January 9, 2018; greyartgallery.nyu.edu/exhibition/beautiful-brainthe-drawings-santiago-ramon-y-cajal.

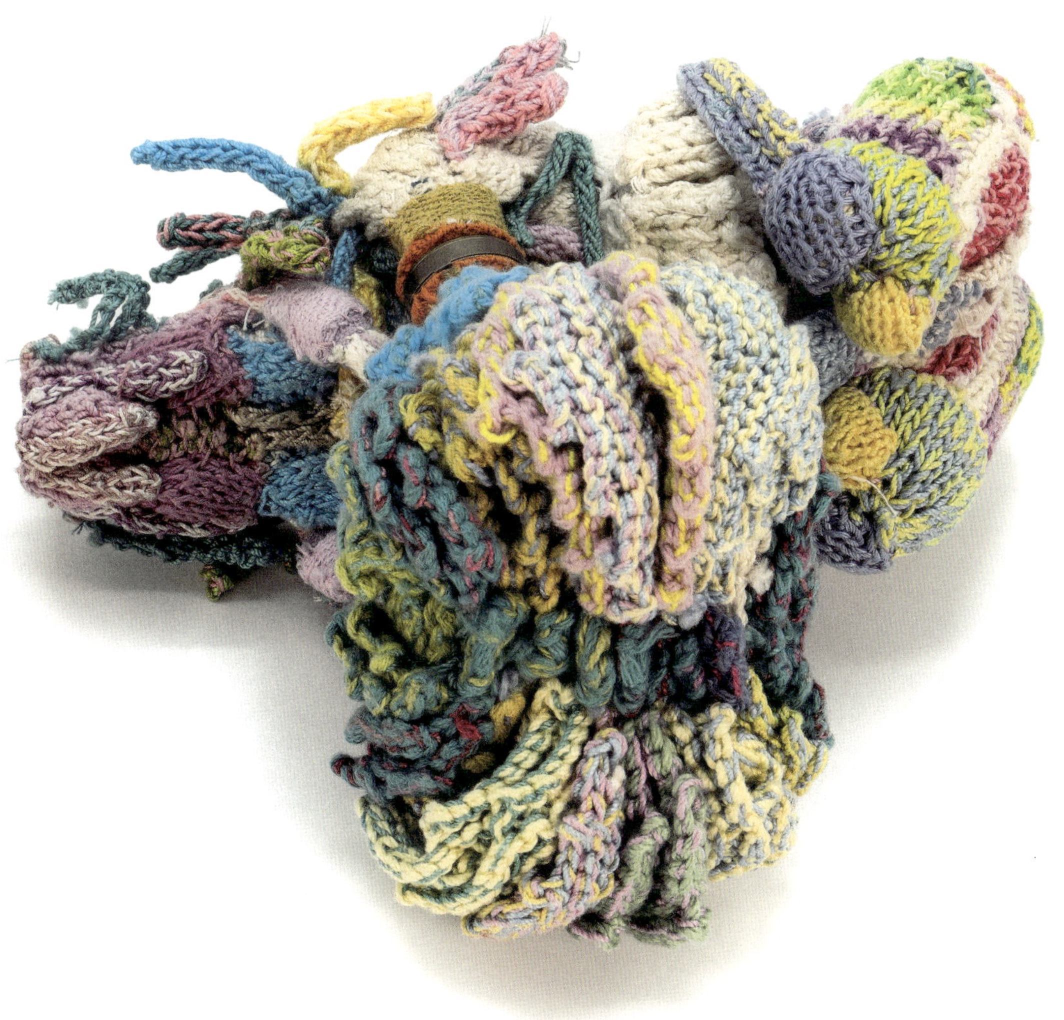

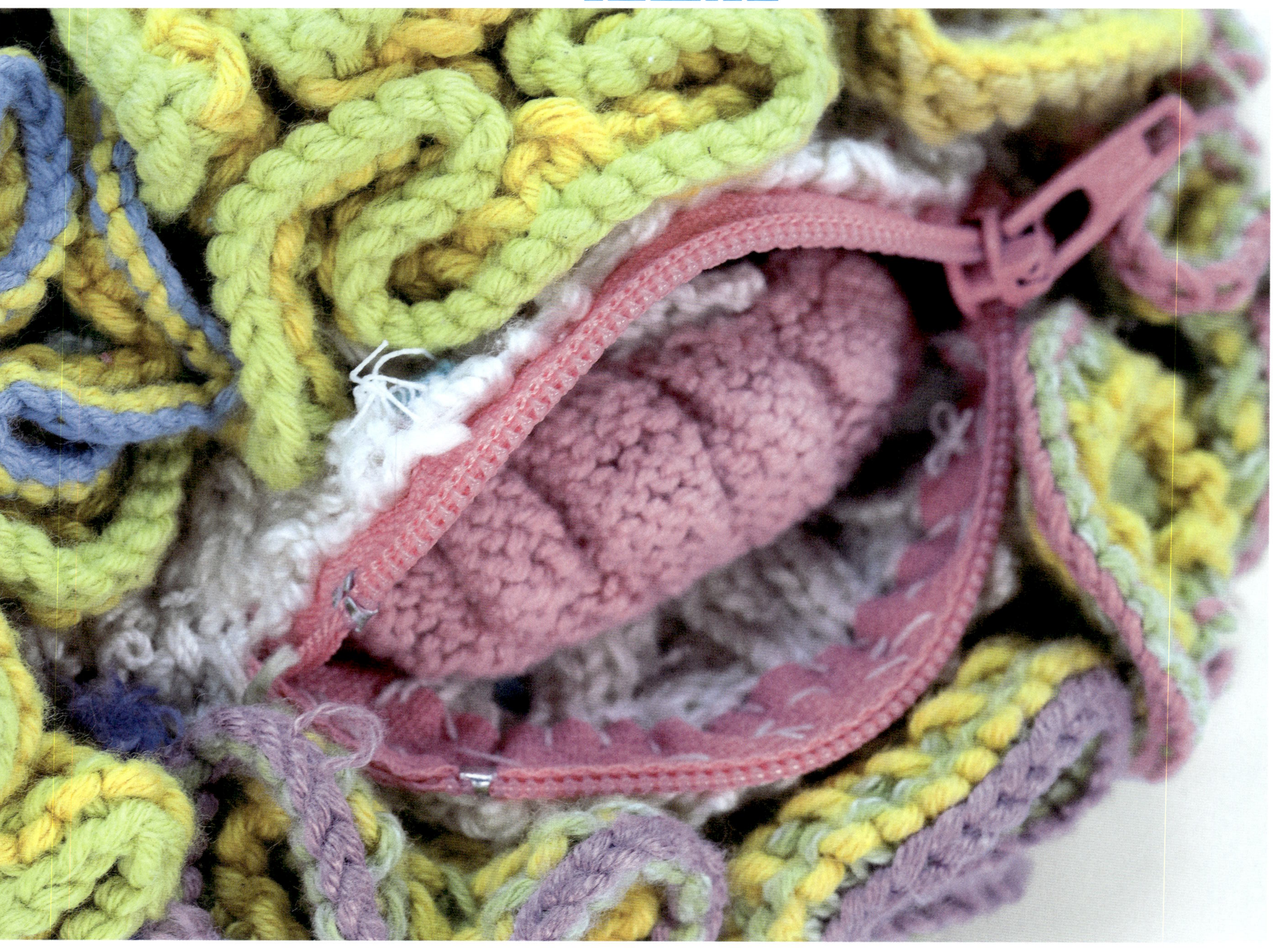

Karen Norberg, *Knitted Brain #2 (KB #2)* (details), 2020–22, knitted cotton yarn, nylon zippers, metal zipper pulls, metal snaps, magnets, 3 ½ × 6 × 8 ¾ inches (element shown)

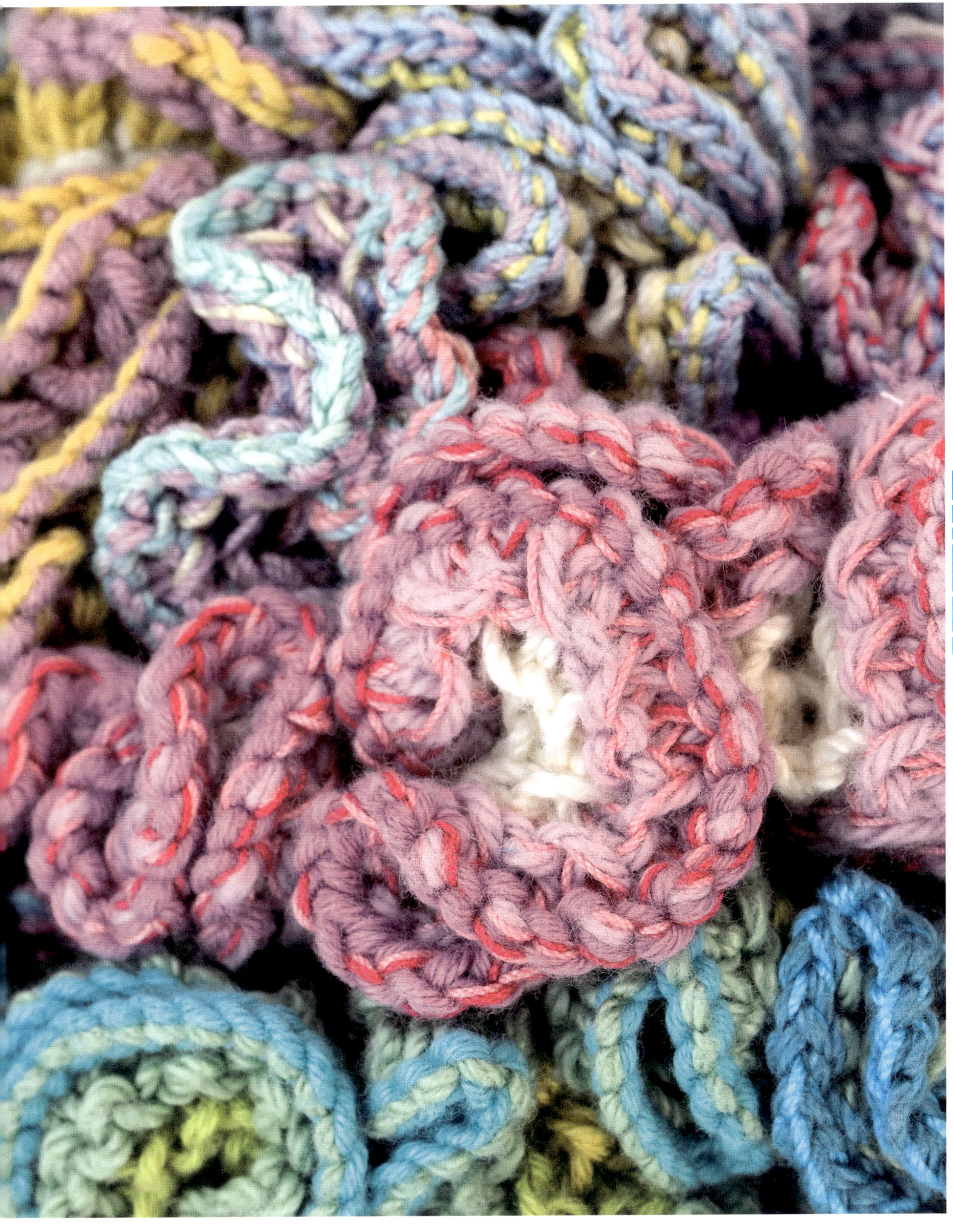

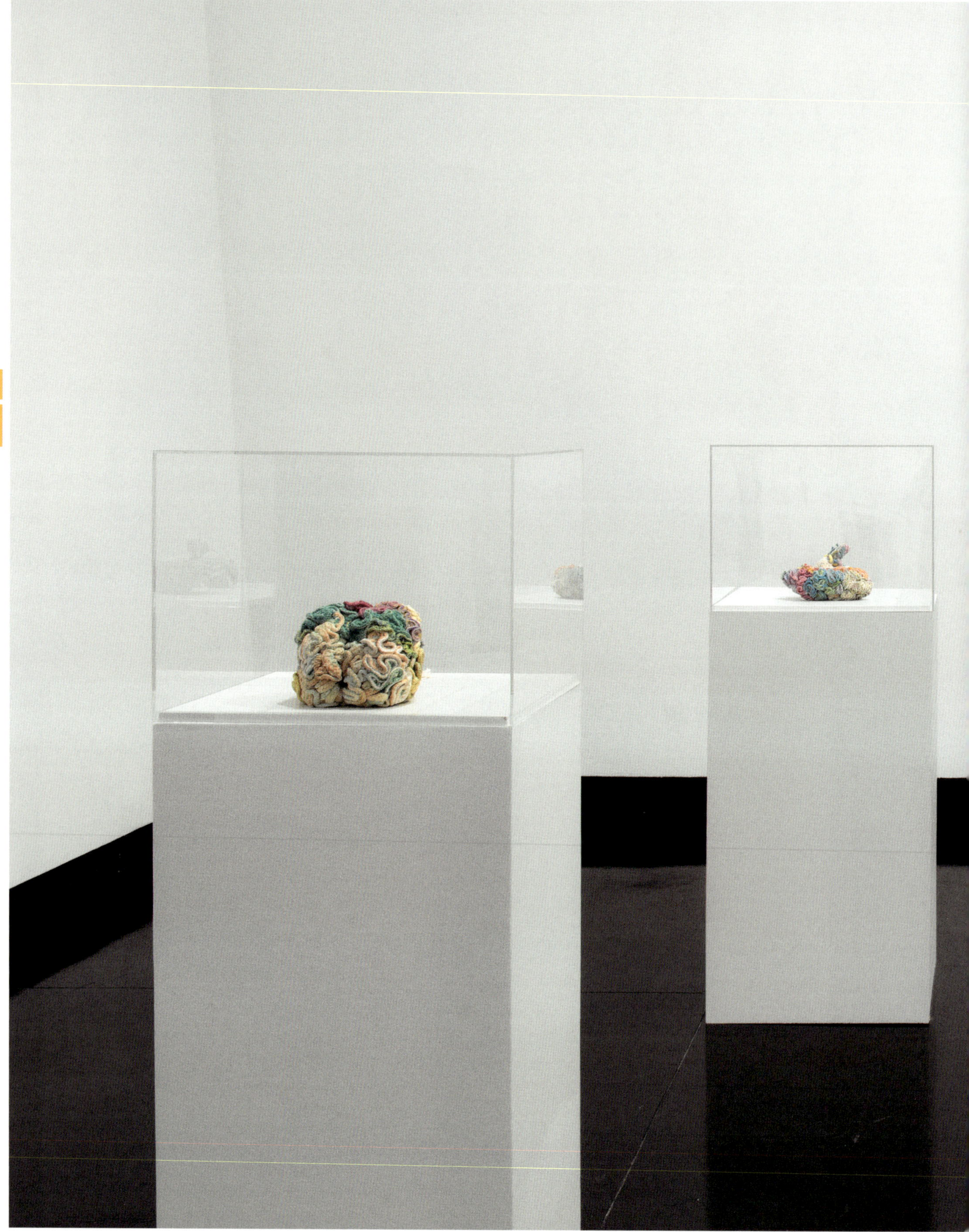

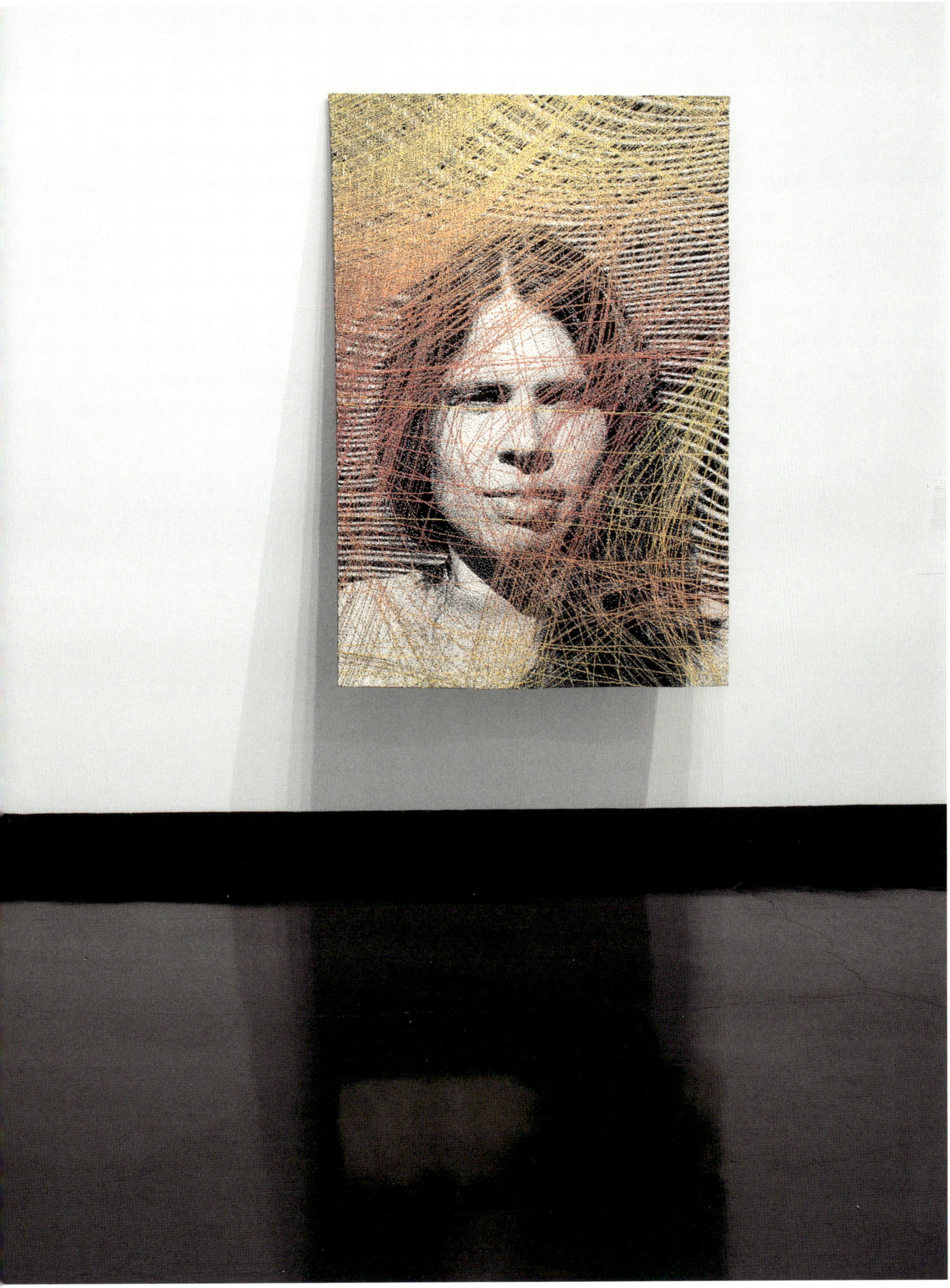

Installation view, *Radical Fiber*, featuring work by Karen Norberg (left) and Lia Cook (right)

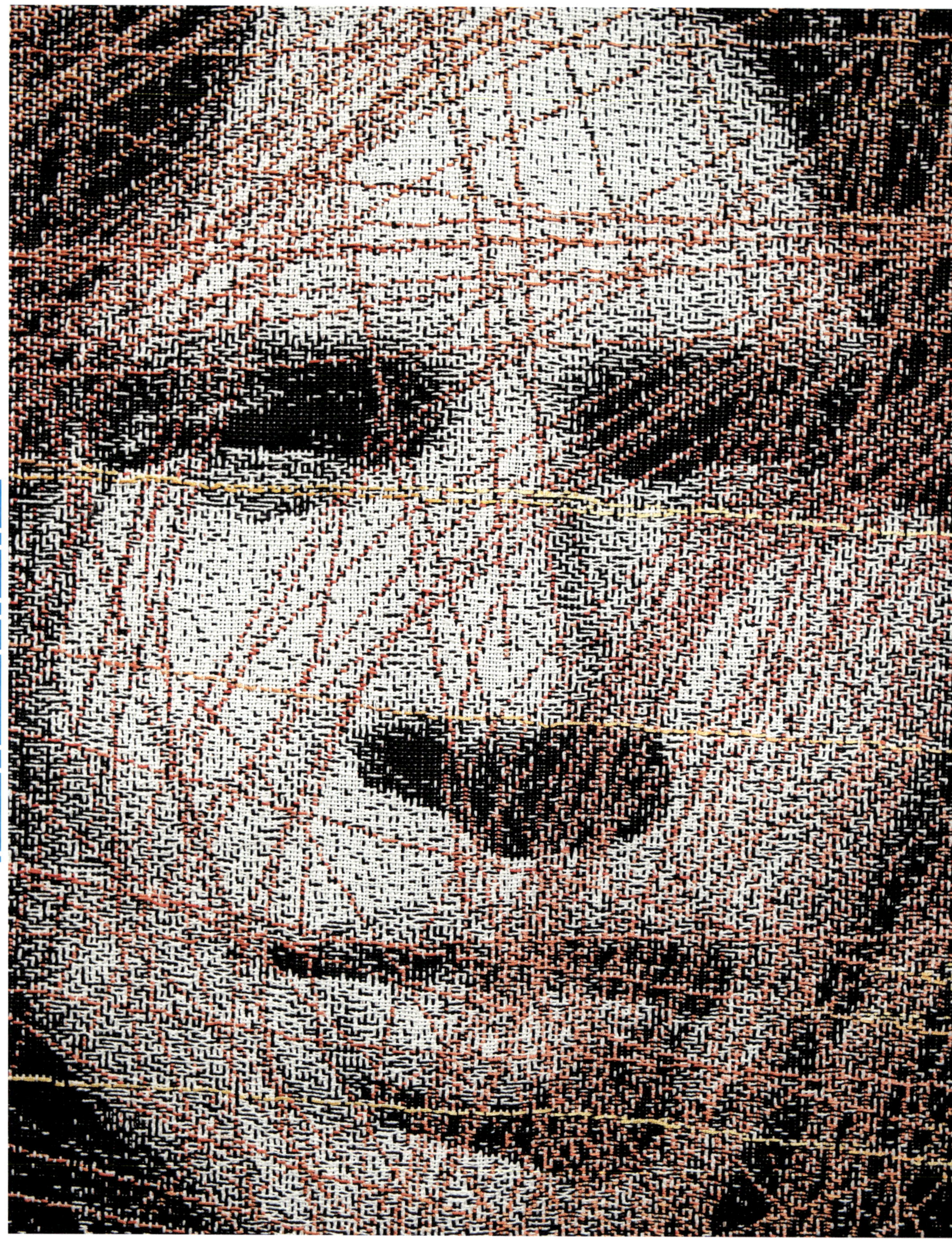

by Denise Evert

A PSYCHOLOGICAL AND NEUROSCIENTIFIC STUDY OF EMOTIONS IN TEXTILES

As Associate Professor of Psychology and Neuroscience at Skidmore College, my research attempts to understand how the organization of the brain mediates emotional and attentional processing. Coupled with my interests in understanding brain mechanisms associated with the creation and perception of art, I explore these ideas in the Skidmore courses I teach, including "Divided Brain/Integrated Mind," "Creativity and the Brain," and "Creating Minds," and through a research laboratory I run with undergraduate student researchers. The confluence of this work with the Tang Teaching Museum's *Radical Fiber* exhibition provided a unique opportunity for my independent research students to engage with the emerging field of *neuroaesthetics*. Through an interdisciplinary, cross-campus collaboration, we partnered with artist Lia Cook, who has long been interested in the innate emotional quality of her textiles and the emotional connections that individuals forge with these works.

In my Cognitive Neuroscience laboratory, three undergraduate student researchers from the Skidmore class of 2023—Lilia Sattler (majoring in Neuroscience and Psychology), Victoria Thorpe (Neuroscience), and Kelby Wittenberg (Psychology and Anthropology)—examined viewer responses to close-up portraits in a set of Cook's original cotton and rayon woven tapestries and in printed photograph versions of the same images. The research study was designed to address three main questions:

> Do viewers rate the emotionality of portraits depicted in textiles higher than corresponding photographic representations?

> Are portraits rated as more emotional when processed preferentially by the right hemisphere of the brain given its overall specialization for emotional processing?

> Does viewer emotional state impact emotionality ratings of the portraits?

Seventy-two Skidmore undergraduates served as participants in the study and were asked to rate the emotionality of depicted facial expressions (for happiness, surprise, fear, anger, sadness, and disgust) in each of ten textiles and ten photographs. Each image (whether textile or photograph) was presented one at a time in a randomized order to either the left or the right of the seated participant. Participants also completed two inventories to assess their current mood state.

Results indicated that for the textile/photograph pairs in which emotionality ratings differed, textiles were rated higher in emotionality than corresponding photographic representations. Further, textile images processed preferentially by the right hemisphere of the brain were rated higher in emotionality than those processed preferentially by the left hemisphere, consistent with right-hemisphere specialization for emotional processing. Lastly, preliminary findings suggested that participants with higher anxiety scores on the mood inventories exhibited different patterns of brain hemisphere differences than those with lower scores, but more participants need to be tested for conclusive results. Students in my research lab are continuing to study hemispheric differences in emotional processing to better understand the functional organization of the brain as well as factors that may influence patterns of hemispheric differences.

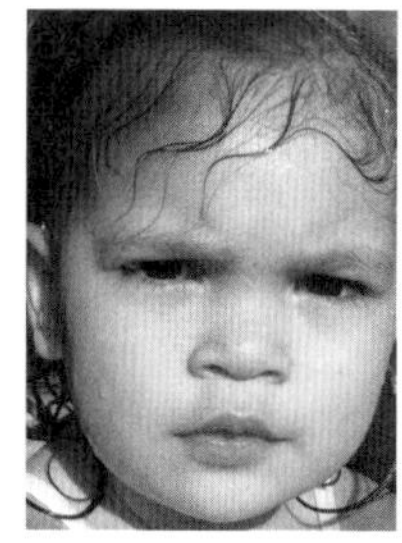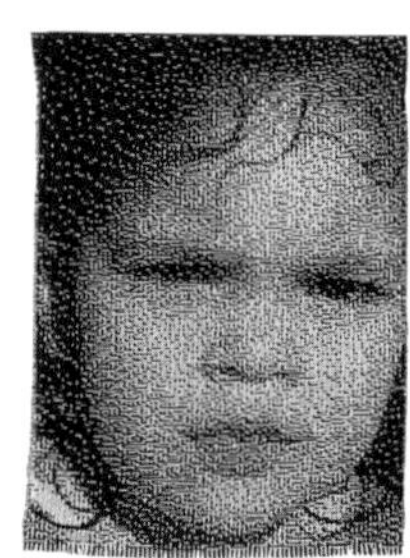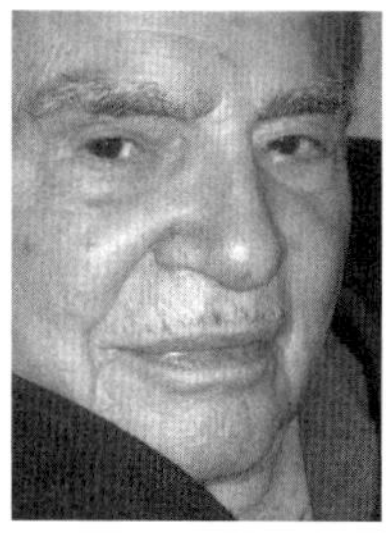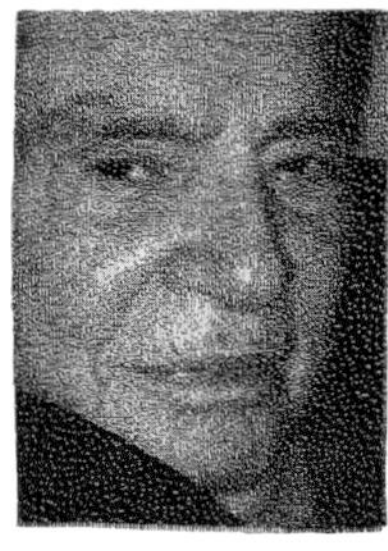

The research study in Denise Evert's Cognitive Neuroscience laboratory used ten pairs of images of faces; for each pair, one version was printed (left) and one version was woven (right).

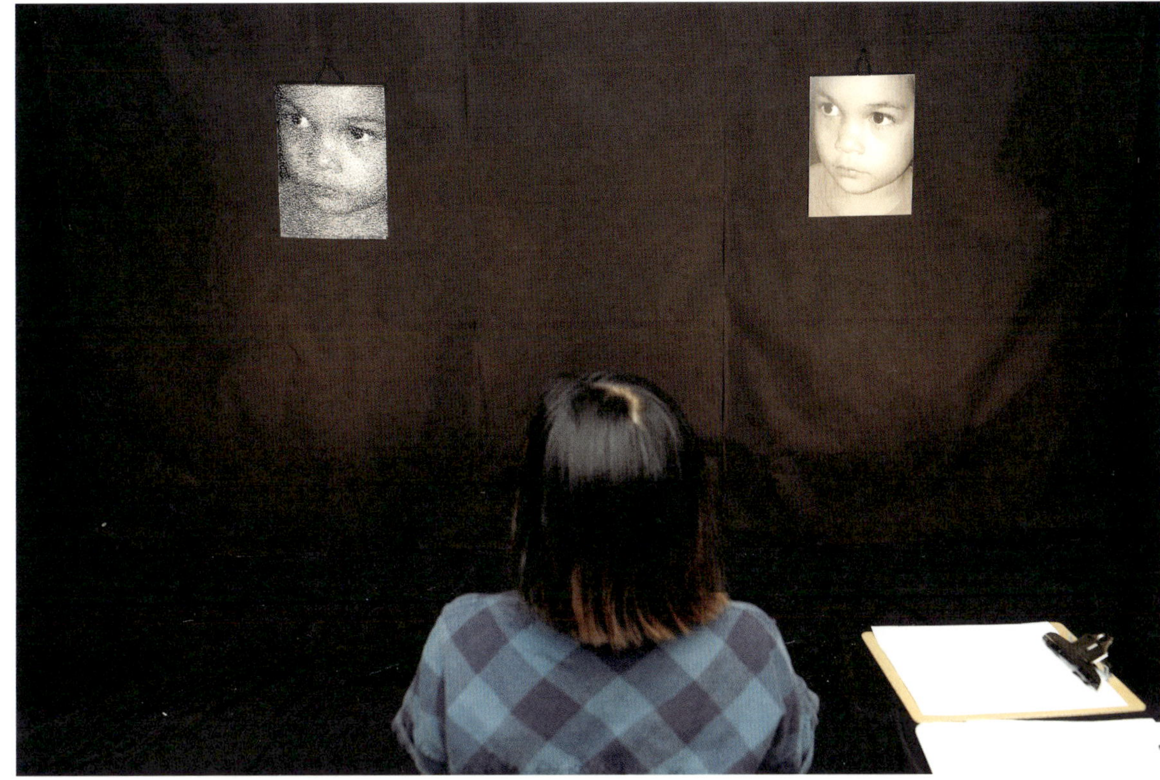

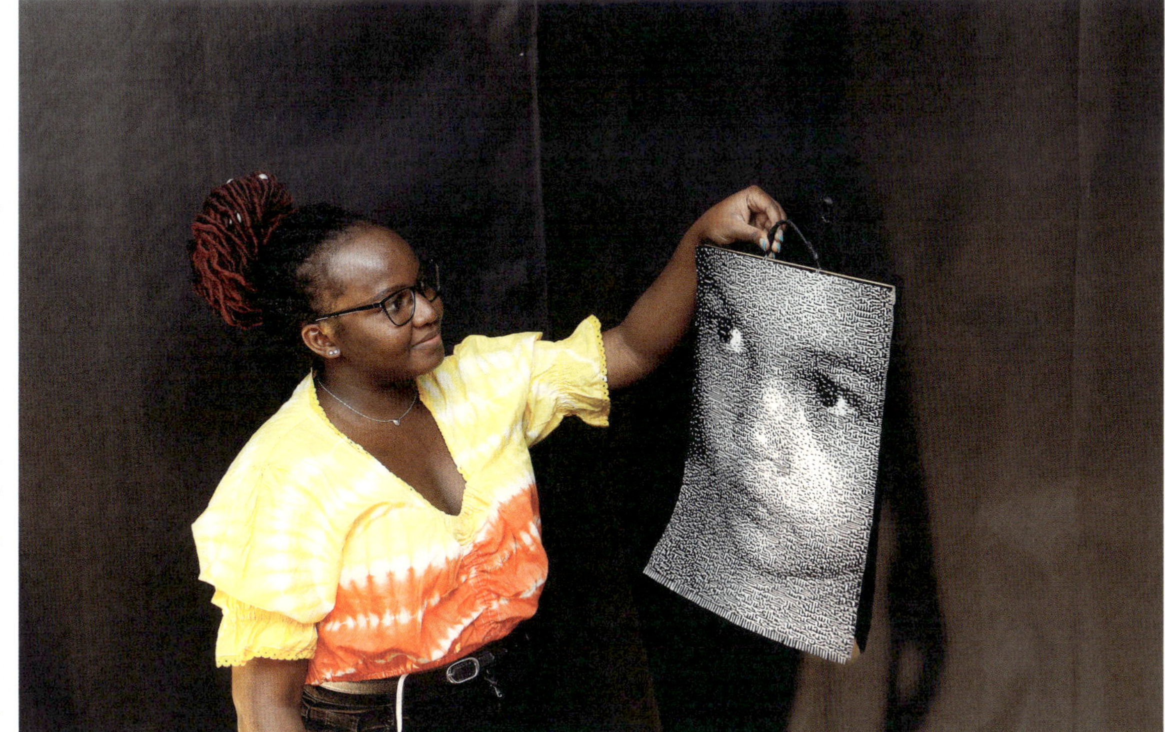

In Denise Evert's Cognitive Neuroscience laboratory, a Skidmore College student researcher (bottom) runs experiments with an anonymous participant (top) to determine how they rate the emotionality of facial expressions in an image displayed in a woven form and in a printed form.

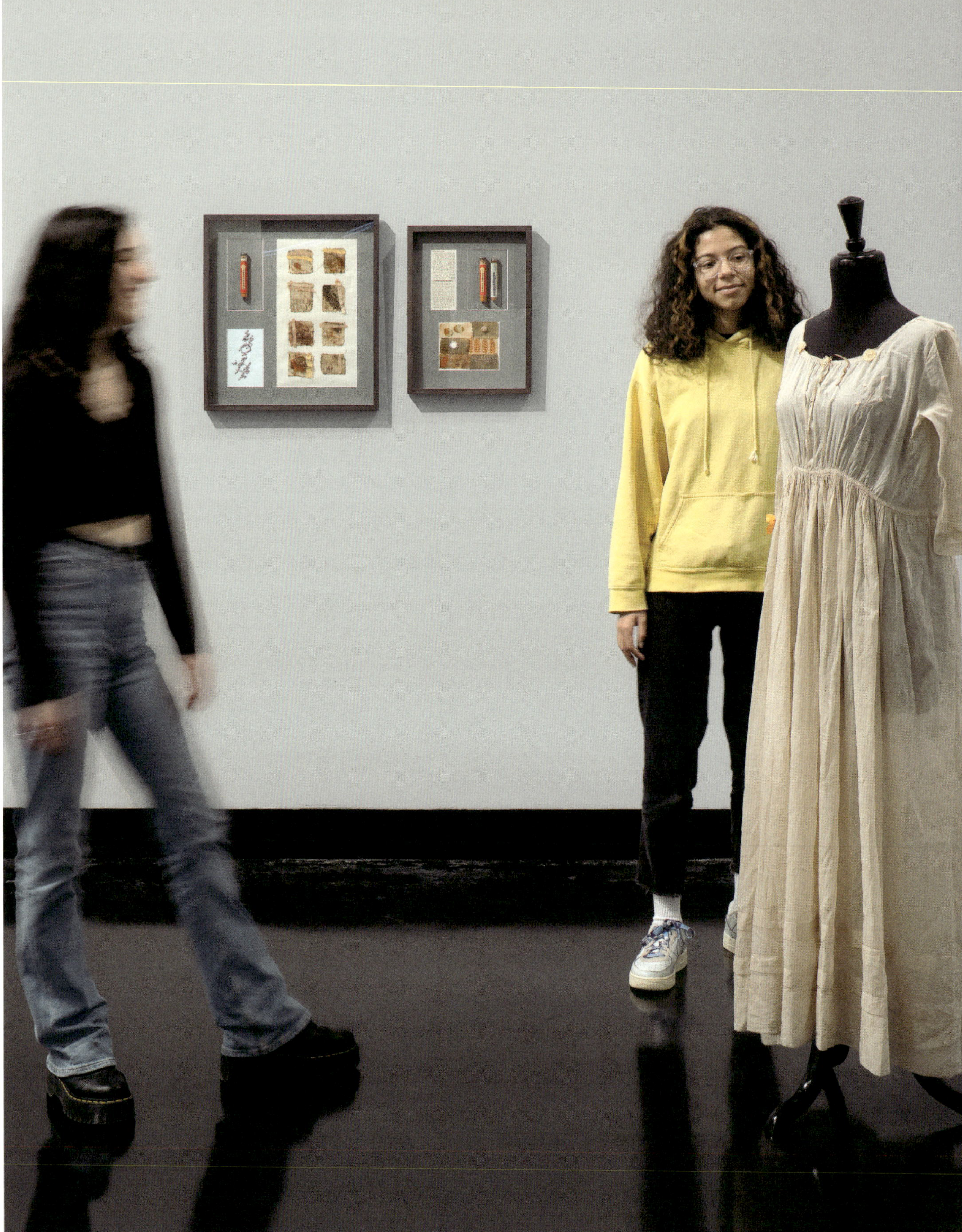

by Elaine Larsen

DYES FOR FASHION AND WELLNESS

Clothing has an intimate association with our bodies. It covers and protects, but it also projects information about how we see ourselves and how we connect to others. These views are in turn influenced by society and by the times we live through. The "romantic disease" of Anna Dumitriu's *The Romantic Disease Dress* refers to tuberculosis. While creating this work, Dumitriu incorporated genetic material from the bacteria *Mycobacterium tuberculosis*, which causes tuberculosis, after it was rendered noninfective. At the time this garment was originally made, probably in the early nineteenth century, tuberculosis was epidemic; in the seventeenth through nineteenth centuries, it caused 25 percent of deaths across Europe.[1] Mortality rates were highest among young adults, and the weight loss and anemic paleness associated with the progression of the disease influenced the fashion of the times.[2] Into this era, when tuberculosis was referred to as "the white plague" and "the good death," came a new industry with the discovery of synthetic dyes.

Anna Dumitriu's *The Romantic Disease Dress* was created using natural dyes—walnut hulls, madder root, and safflower—as well as prontosil, an early synthetic dye. For centuries, plants had been used as both medicine and color. Natural dyes provided a palette for fashionable clothing and textiles. Obtaining a full range of colors, however, required knowledge and skill, and the expense of certain raw materials made some rare colors luxuries. For instance, blue dye was obtained from indigo-containing plants, and the best of these are tropical or semitropical crops. Beginning in the early to mid-eighteenth century, wealthy European nations forced colonized people in various regions, including throughout the Caribbean islands and in South Carolina, Bengal, Bihar, and elsewhere, to produce indigo crops.[3] The few red dyes were also cash crops from southern colonies. It wasn't until 1856 that the first synthetic, laboratory-made dye was created. That year, English scientist William Henry Perkin (1838–1907) was working to synthesize quinine when he discovered aniline dyes, which he patented, developing a new synthetic-dye industry. The vivid purple hue became fashionable among

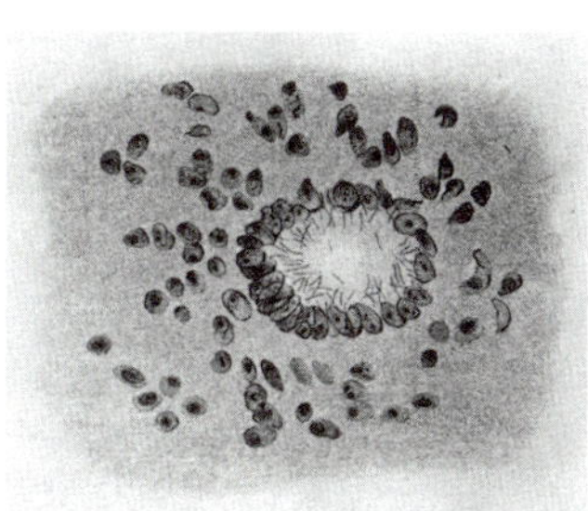

Drawing by Robert Koch depicting *tubercle bacilli* surrounding a giant cell in the lung, published in what became a seminal text on bacteriology. In 1882, Koch first publicly announced his discovery of *tubercle bacillus* (named, in 1886, as *Mycobacterium tuberculosis*) as the cause for tuberculosis at a Berlin Physiological Society meeting, on March 24 of that year, and soon after published the article "Aetiologie der Tuberculose" ("The etiology of tuberculosis").[4]

Elaine
Larsen

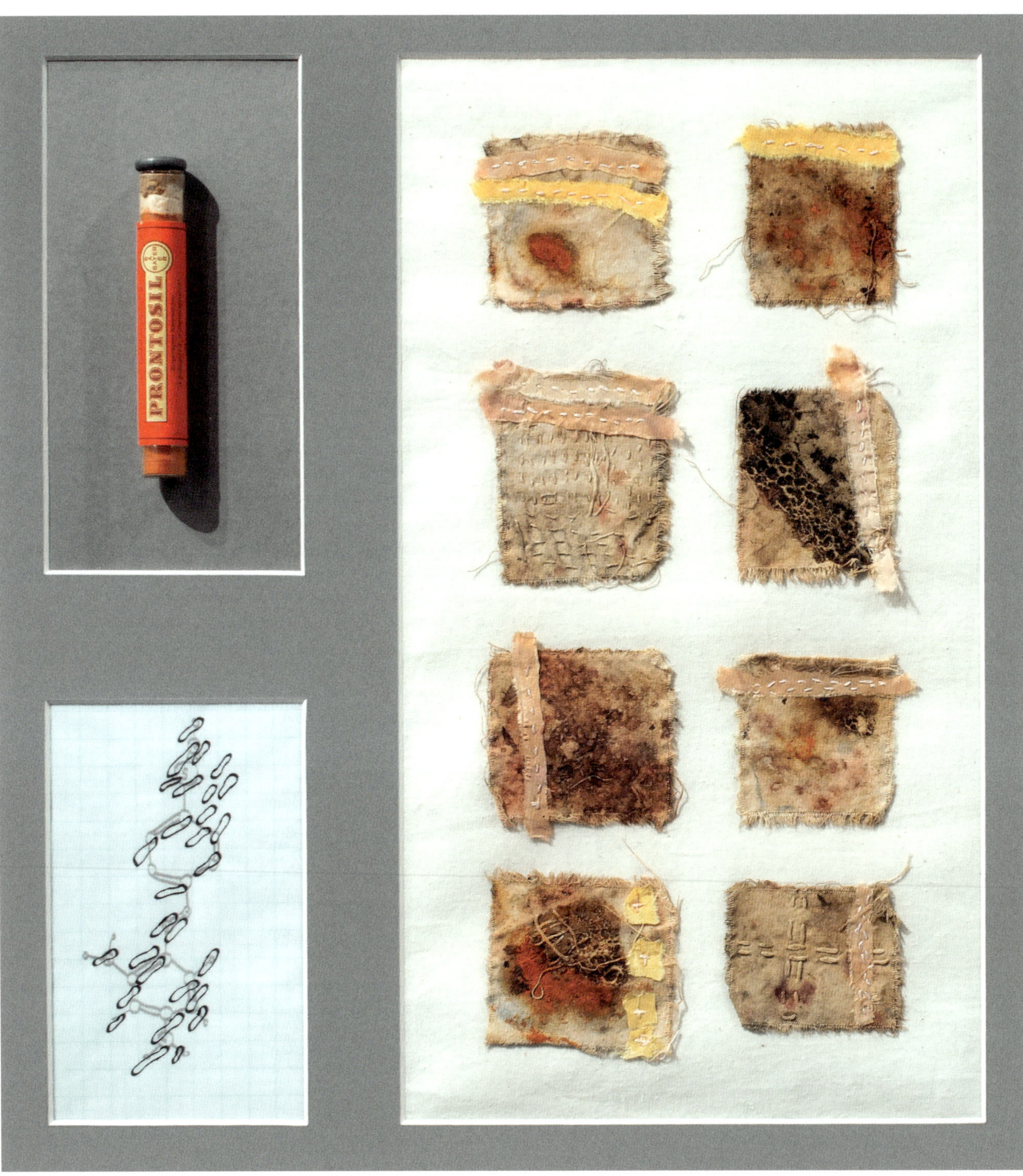

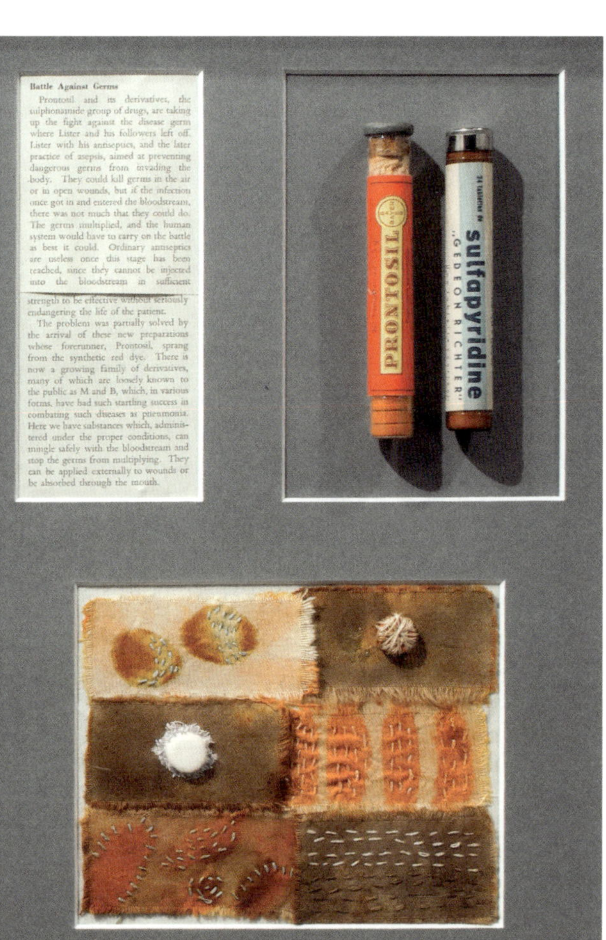

Anna Dumitriu, *Magic Bullet Series* (details), 2015, wood, metal, glass, textile dyed with natural dyes, drawing on paper, vintage printed materials, sulfa drugs in tubes, 20 1/8 × 18 × 1 1/4 inches (left, framed), 17 1/4 × 12 3/4 × 1 1/4 inches (right, framed)

the royal and the wealthy. This success spurred chemists to search for other dye compounds, and their manufacture became the first "chemical" industry—with impacts far beyond fashion.[5]

Bacteriologist Robert Koch learned that some synthetic dyes stained bacterial cells, revealing details under the microscope that were otherwise invisible. He used dyes to photograph bacteria and, in 1882, identified the bacterium that causes tuberculosis.[6] Biologists continued to study chemicals that could bind to bacterial cells, hoping to find compounds that would inhibit bacterial growth—and that weren't toxic, as were mercury and arsenic. This was the concept of the "magic bullet" that could strike disease-causing organisms while not harming the body they infected.[7] In 1932, the red dye prontosil was synthesized, and in 1935, it was used as an antibiotic treatment, curing infections in both mice and humans. Ironically, it was not the dye itself that had medicinal properties, but the compounds that the body formed when it metabolized the dye.[8] Using techniques and knowledge gained from decades of work on chemical dyes, scientists developed compounds similar to but safer and more effective than prontosil, creating the sulfonamide family of antibiotics, a new chemical pharmaceutical industry, and the study of pharmacokinetics (the dosage and metabolism of drugs in the body).[9] Less than a century after an industry formed to make fashionable dye colors, it had diversified into the creation of lifesaving pharmaceuticals. While our recent COVID-19 pandemic is very different from the decades of suffering caused by tuberculosis before the advent of antibiotics, there are echoes from the time of the "romantic disease" in our struggles with how to present ourselves while masked and in our search for more "magic bullets" through vaccines and antiviral drugs.

1   CDC Division of Tuberculosis Elimination, "History of World TB Day," Centers for Disease Control and Prevention, www.cdc.gov/tb/worldtbday/history.htm, last reviewed February 15, 2023.

2   Emily Mullin, "How Tuberculosis Shaped Victorian Fashion," *Smithsonian Magazine*, May 10, 2016, www.smithsonianmag.com/science-nature/how-tuberculosis-shaped-victorian-fashion-180959029. Also

see Carolyn A. Day, *Consumptive Chic: A History of Beauty, Fashion, and Disease* (London: Bloomsbury Academic, 2017).

3   Catherine Legrand, *Indigo: The Colour that Changed the World* (New York: Thames & Hudson, 2013), 21.

4   Robert Koch, "Aetiologie der Tuberculose" ("The etiology of tuberculosis"), trans. Berna Pinner and Max Pinner, *Berliner Klinische Wochenschrift* no. 15 (April 10,

1882): 221–30.

5   Penny Le Couteur and Jay Burreson, *Napoleon's Buttons: How 17 Molecules Changed History* (New York: TarcherPerigee, 2004).

6   Sam Wong, "Robert Koch," *New Scientist Magazine*, www.newscientist.com/people/robert-koch.

7   Le Couteur and Burreson, *Napoleon's Buttons*.

8   Ibid.

9   Ibid.

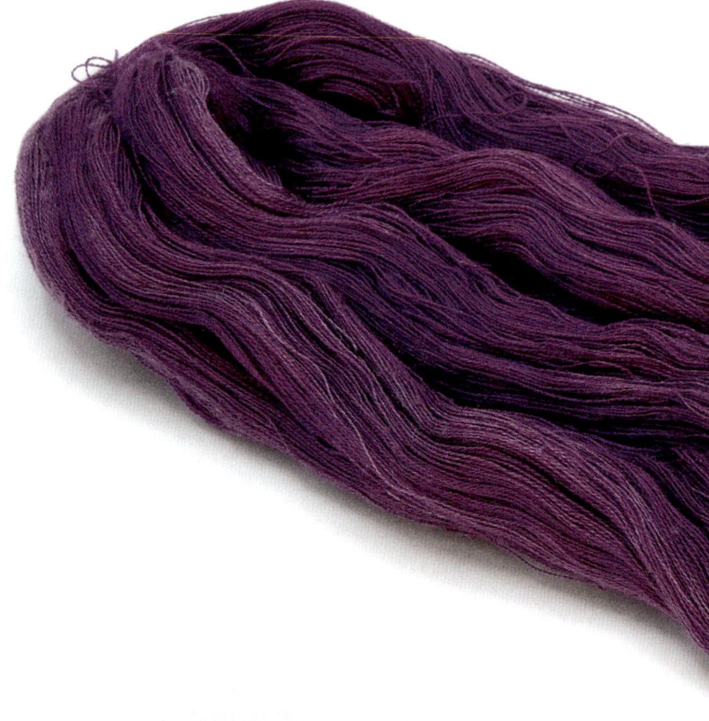

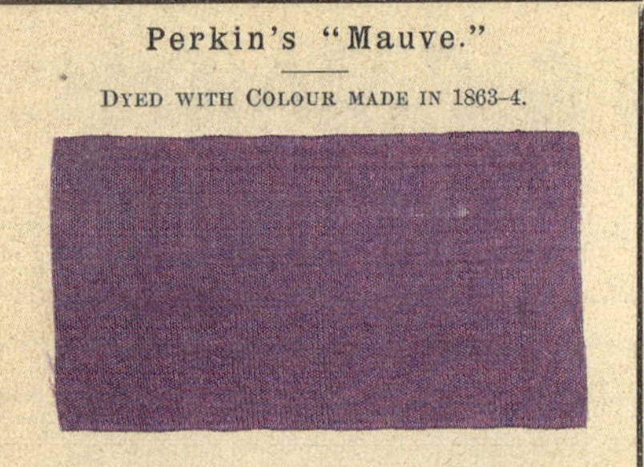

The Perkin Jubilee.

Perkin's "Mauve."

Dyed with Colour made in 1863–4.

Great historical importance attaches to this, the first coal-tar dye, and universal interest has been aroused in it not only amongst dyers and colourists, but amongst the public generally, by reason of the recent Jubilee celebrations. At the same time there appears to be much confusion as to the exact hue of the original mauve. It therefore seemed of interest to insert a dyed pattern in the Journal illustrating this dye, and Dr. Perkin has very kindly supplied a small quantity of Mauve manufactured in 1863 or 1864 for that purpose. The pattern has been dyed in the Bradford Technical College dye-house.

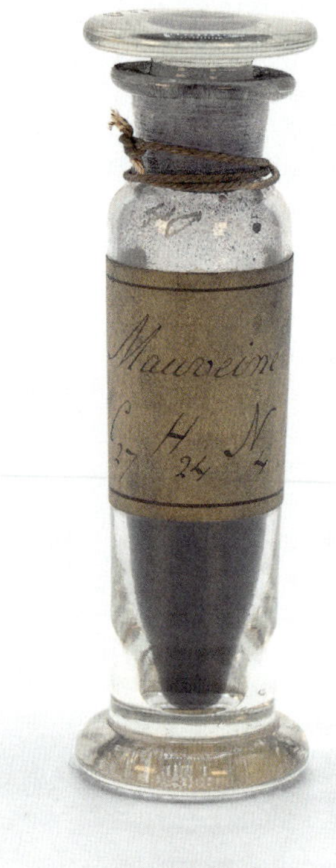

Left to right: Sample of Perkin's mauve, *Journal of the Society of Dyers and Colourists*, November 1906, published by the Society of Dyers and Colourists, England–Bradford (West Yorkshire), 11 × 7⅞ inches (page); Henry Edward Schunck, mauveine dye, a sample chemical, late 19th century, mauveine dye, glass bottle, paper label, 3¾ × 1 × 1 inches; Imperial Chemical Industries, Dyestuffs Division, hank of yarn dyed with mauveine,

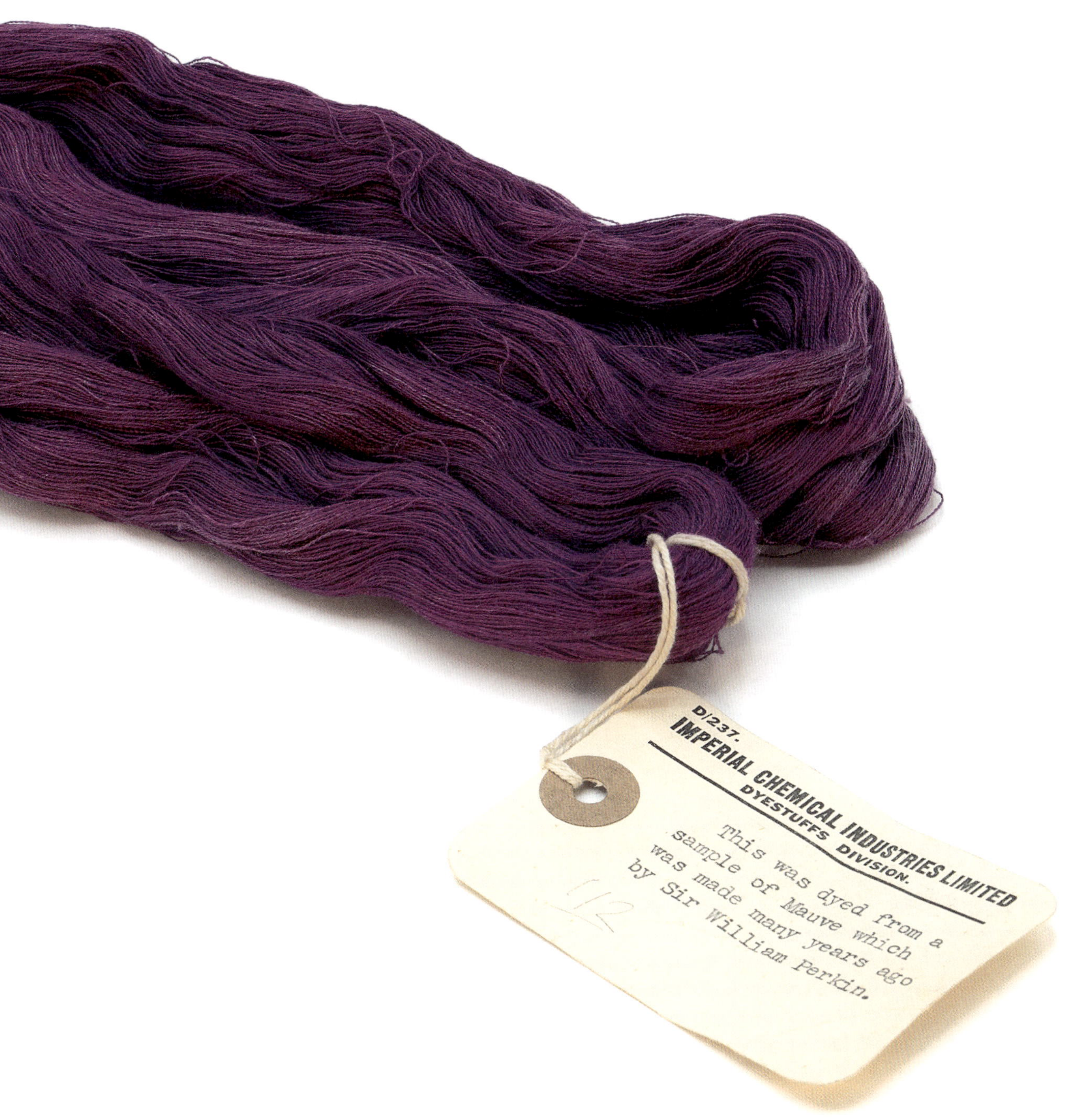
D/237.
IMPERIAL CHEMICAL INDUSTRIES LIMITED
DYESTUFFS DIVISION.
This was dyed from a
sample of Mauve which
was made many years ago
by Sir William Perkin.

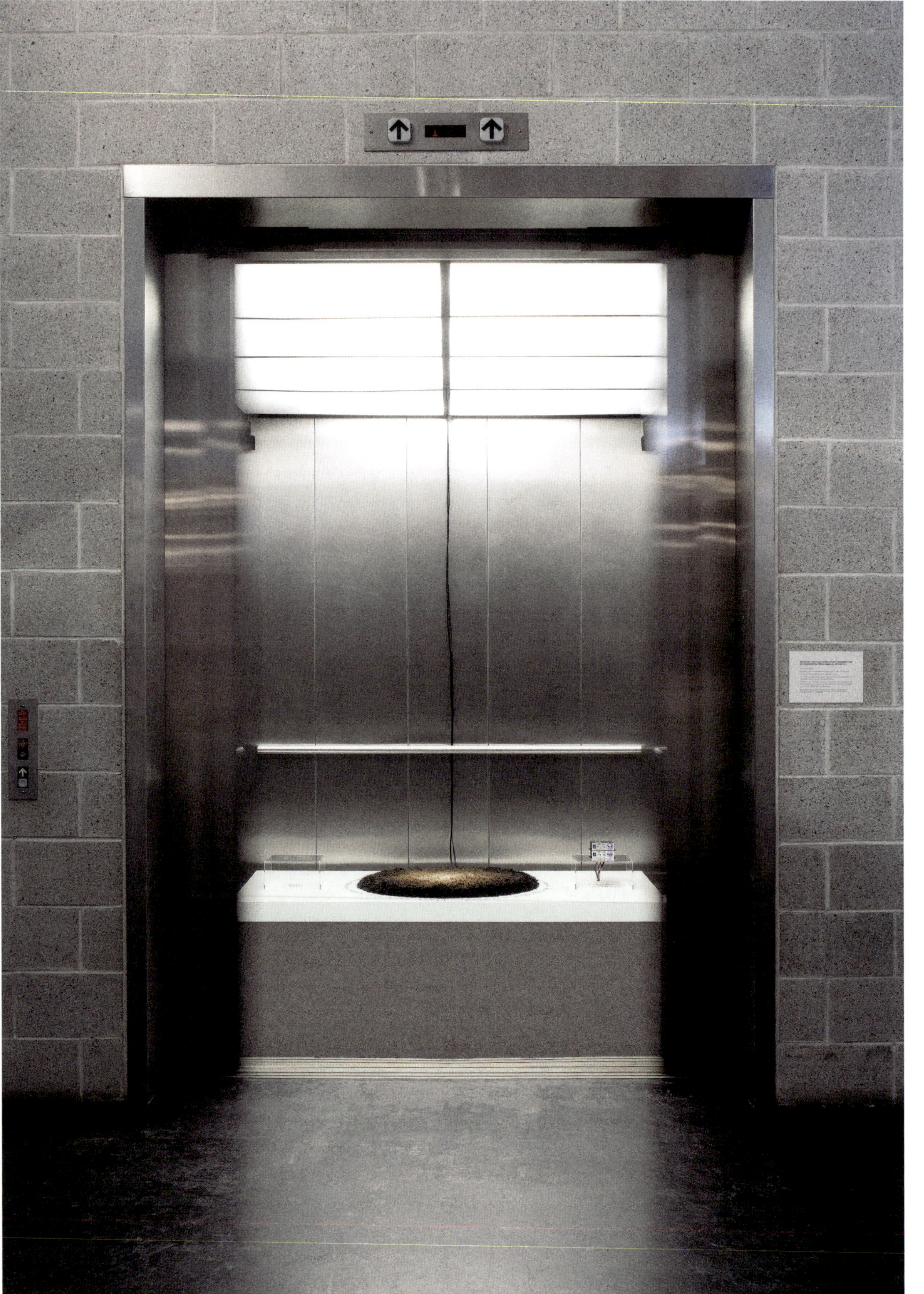

Installation views, *Elevator Music 42: Laura Splan—Rhapsody for an Expanded Biotechnological Apparatus*, Tang Teaching Museum. This exhibition, which ran concurrently with *Radical Fiber*, presented a sonic and tactile journey through a biotech lab where Laura Splan spent several months in spring 2018 as artist-in-residence. In the sound work *Chaperone*, the buzzing of specialized laboratory equipment, chatter of a rug made from the wool of laboratory llamas and alpacas, offered visitors the opportunity to embody the folding of proteins inside a cell's lumen. Through this multisensory experience, visitors were encouraged to contemplate the unseen labor of animals—whose unique antibodies, a type of protein, were commonly used in biological research and medicines—and the day-to-day procedures of scientists

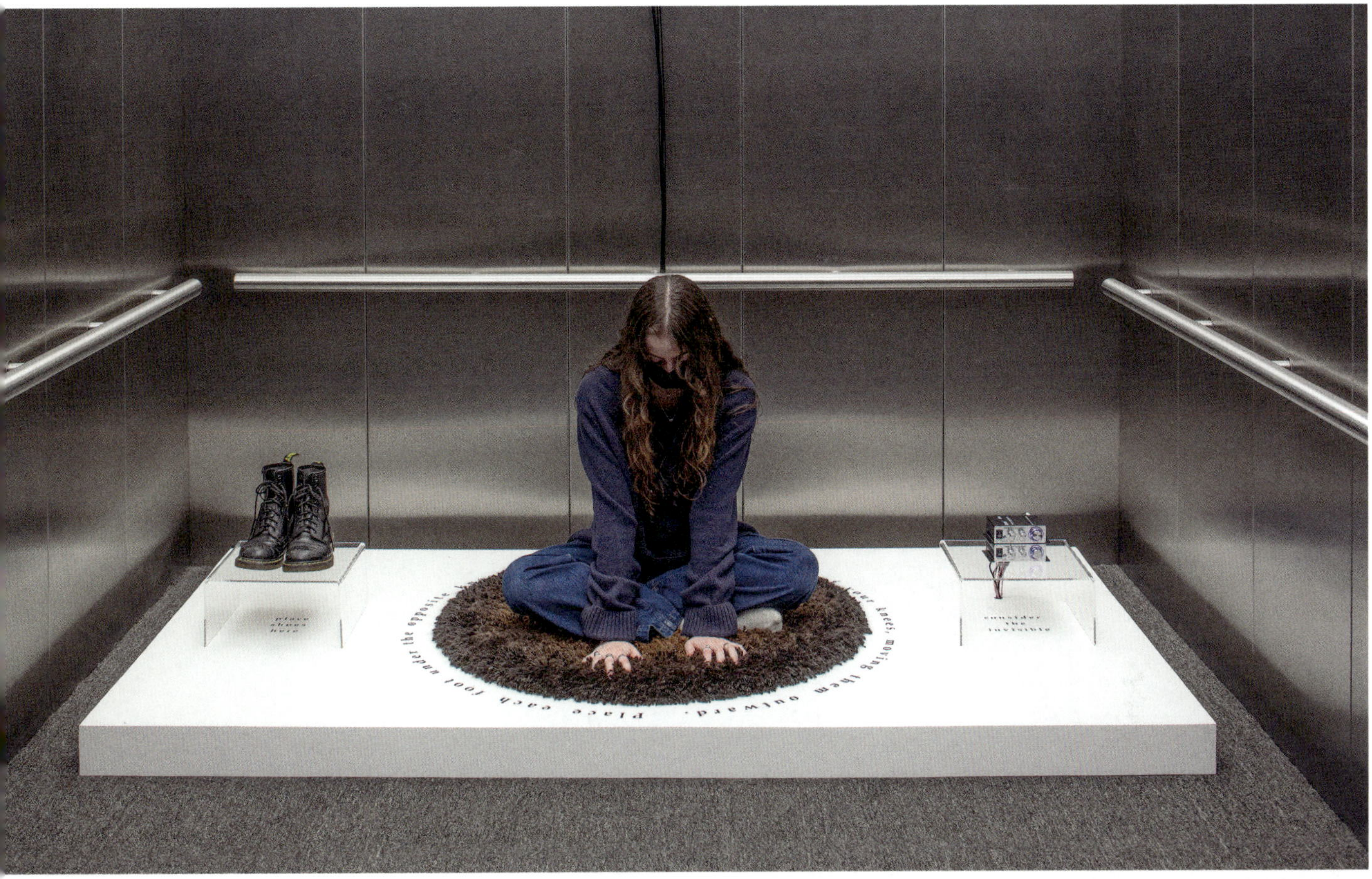

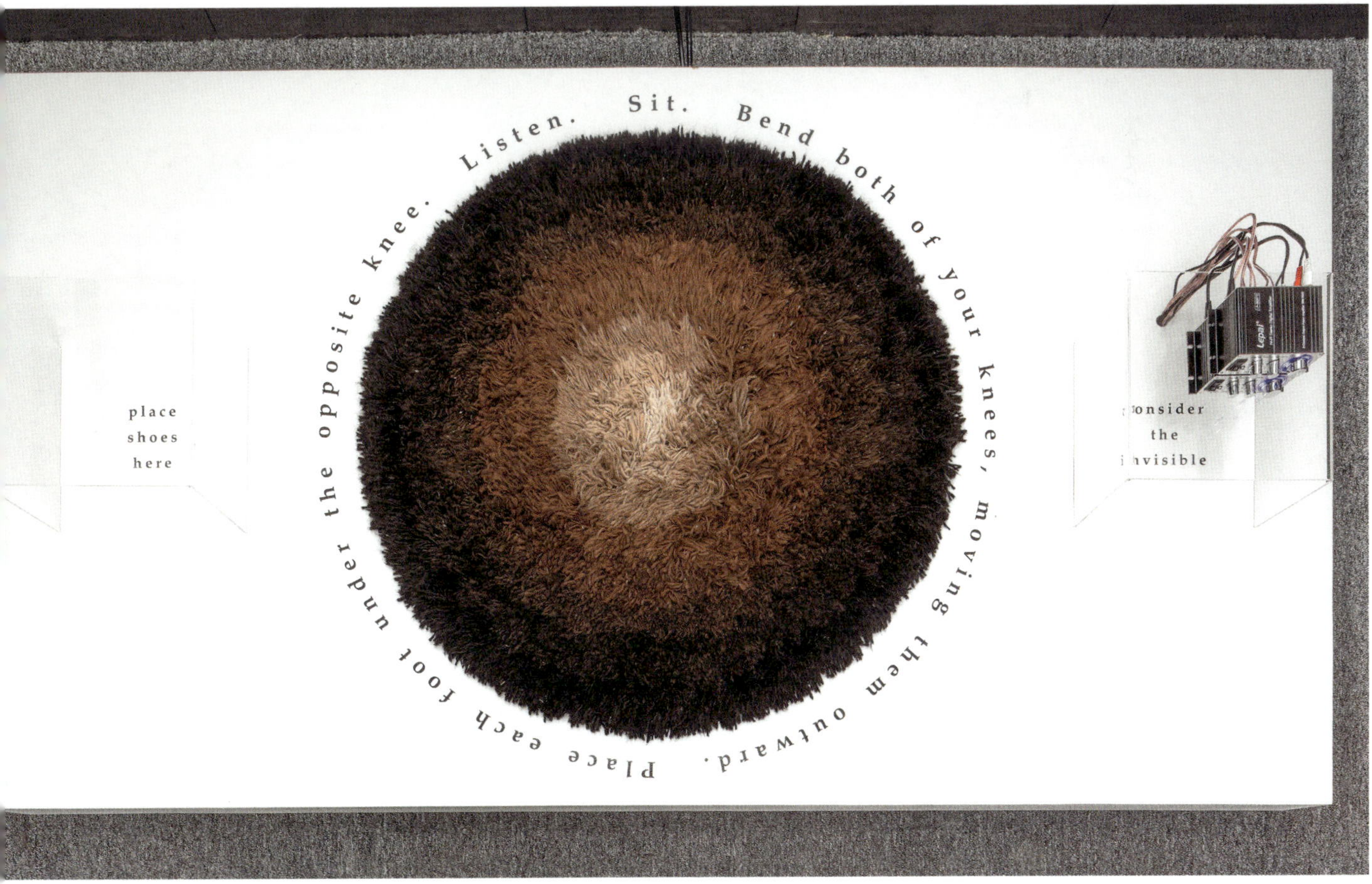

place
shoes
here

Listen. Sit. Bend both of your knees, moving them outward. Place each foot under the opposite knee.

consider
the
invisible

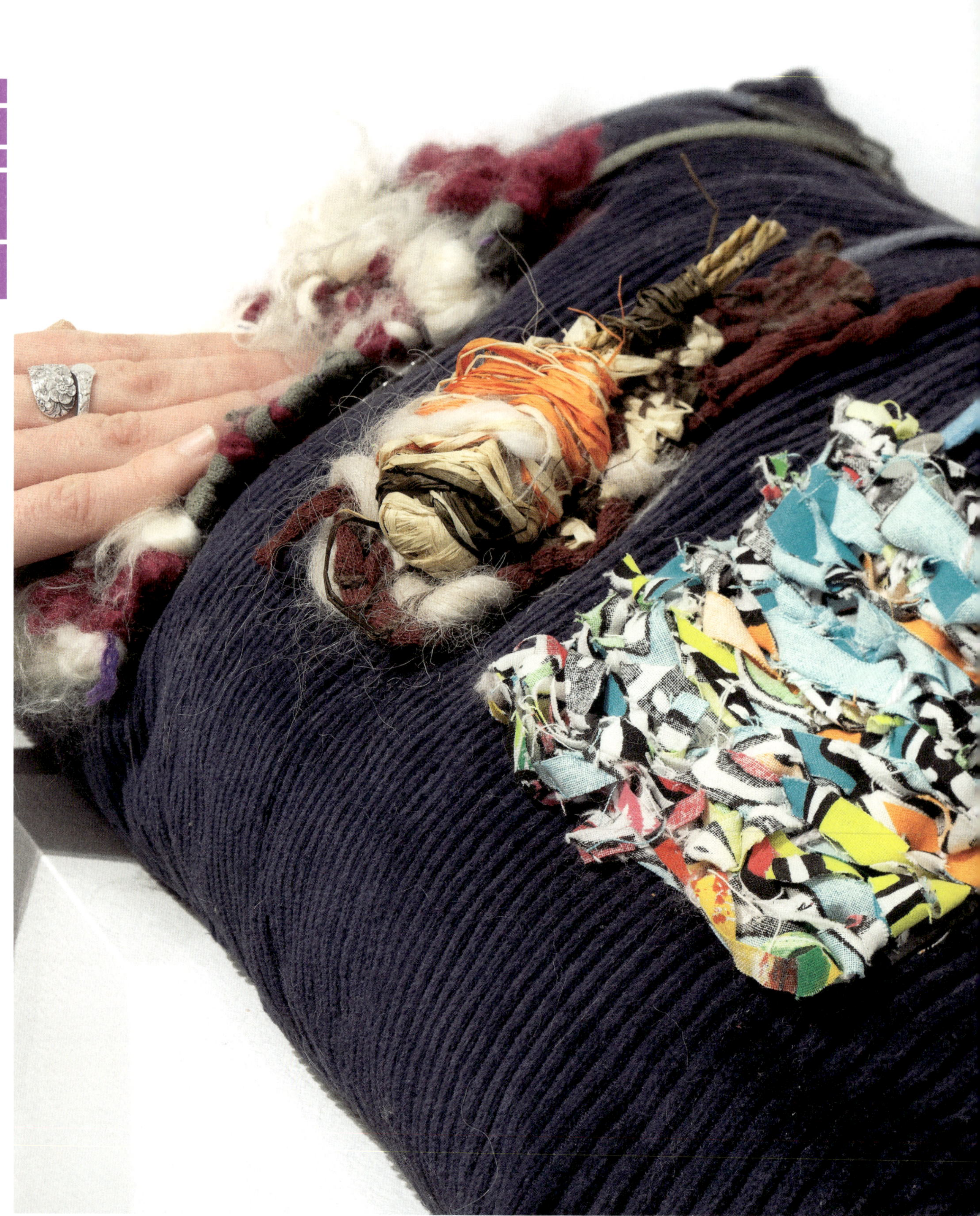

TEXTILES, TECHNOLOGY, AND SOCIAL GOOD

In Conversation: Trisha Andrew, Emilie Giles, *and* Ursula Wolz *with* Aarathi Prasad

...ry, Nature Calls (with visitor interaction), 2011, cushion pad, ...abric, wool, cord, twine raffia, fabric scraps, wire, 12 × 22 × 8 inches

AARATHI PRASAD: My research focuses on how people use and don't use smart devices. Even though I've been coding for more than twenty years, I was unaware of the role that weaving played in the origins of computer science. Through *Radical Fiber*, I've learned so much about textile and science connections, and I'm really inspired by how researchers in computing around the world are using textiles.

TRISHA ANDREW: I have a PhD in chemistry and electrical engineering, and I've been in the part of the field that uses dyes and pigments for explosives detection, microelectronics fabrication, and energy harvesting since about 2005. I stumbled into textiles coating in about 2014, and I've been in that field ever since. My research lab focuses on understanding how to bring together microelectronics fabrication techniques with traditional textile processes, like sewing, embroidery, and weaving, so we can convert the fibers in fabrics into electronic circuits, then use the fibers, or the fabrics themselves, as sensors. This is something we've been focusing on for the past five years.

My research lab developed a kind of suite of conductive coatings that we can put on fabrics and fibers, which can be sewn, patched, or woven together to make very different kinds of fabrics that are actually electronic devices. We've made patches that can harvest solar light and shirt collars

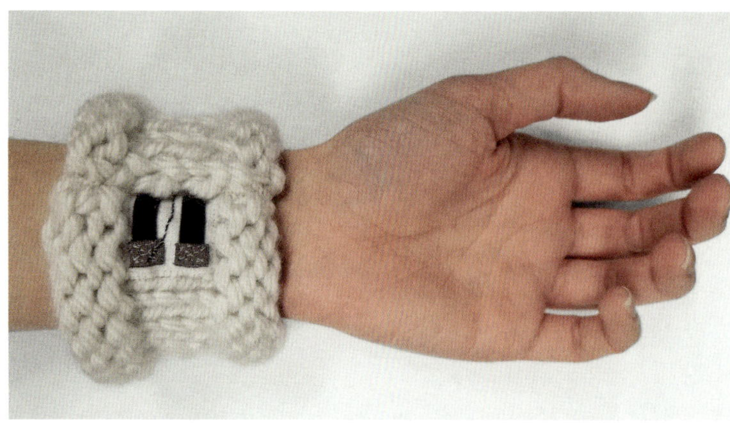

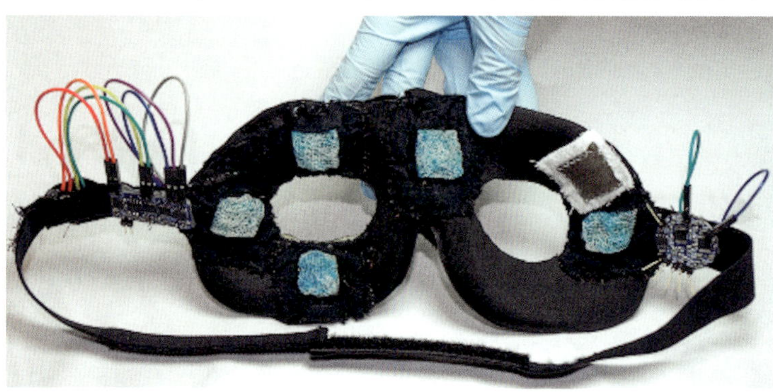

Top: Using wool and cotton, which have low heat-transport properties, Trisha Andrew and PhD student Linden Allison, both working at the University of Massachusetts–Amherst, developed a fabric that can harvest body heat to power wearable microelectronic devices.[1] *Bottom:* Chesma, a bimodel eye mask with embedded fabric hydrogel electrodes, pulse detector, microcontroller, and wiring, wirelessly tracks eye motion and pulse.[2]

and wristbands that can harvest your body heat to provide a small amount of wattage for you to charge a battery or small device. These cotton patches are actually charged storage devices, basically batteries, and you can have these patches hooked up to solar cells, for example, and harvest that energy and then put it back into whatever device you need to store it for the future.

Some of the work that we've been focusing on for the past three years has been on garments that are sensors—things like health monitors in your shirt or in an eye mask. We made a pajama shirt that has a sensor suite of fabric patches that tell you which positions you're sleeping in throughout the night and that reveal your heartbeat and your respiration: for example, it will show you if you stopped breathing in the night, or if you were breathing differently in different sleep postures. We amended some fabric masks that you can buy on Amazon by sewing on our fabric sensors, and now they enable

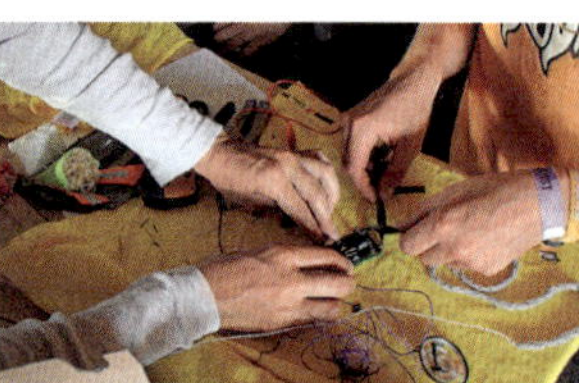

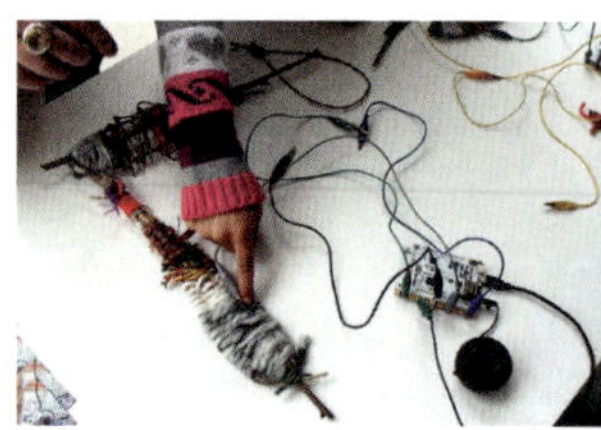

Emilie Giles's work involves organizing and leading workshops to teach programming and craft-making skills to a wide variety of audiences. Here, participants build and interact with conductive thread and sound circuit boards in a workshop entitled "Hacked Human Orchestra," with MzTEK, 2012 (top), and in another workshop entitled "Electric Willow Weaving," with Codasign, 2015 (bottom).

you to track your eye motion. They can also look at your brain waves as you're sleeping in order to understand your sleep health much better than you can by wearing a tight wristband.

We've been lucky that our work in the research lab has found commercial applications and has generated demand. In 2018, I started a company called Soliyarn, based out of Boston, to commercialize our coating technology and bring these fabric patches onto the market. And lastly, something that has been really exciting and rewarding for me is the company's focus on transforming the environmental impact of textile manufacturing. Recently, we have been trying to address the fluorine contamination issue, which is pernicious and affects our groundwater. And we developed some completely fluorine-free coatings for textiles that we're really excited about.

EMILIE GILES: I'm an artist, a researcher, and an educator. I have a PhD in human-computer interaction, but I come from a creative background in media arts. My research focuses on empowerment, working with people, teaching them creative technology skills—coding, using things like Processing or Scratch if working with kids, and circuit making through creative methods, such as electronic textiles or conductive paint. I see myself as a maker. A lot of the research I do is research in action, or research and reflection: prototyping, making things with people, and overseeing participatory making processes.

While I was doing my master's in interactive media at Goldsmiths, University of London, I became involved in a women's tech group called MzTEK. It was all about looking at communities of women artists who might want to start using technology in their own work, getting female artists together who were already working with these things to teach other people, and doing exhibitions. MzTEK started off my passion for looking at how can we get creative communities involved with coding and electronics.

I ended up corunning a company, called Codasign, with a friend of mine, Dr. Becky Stewart, who's based at Imperial College London. We were interested in working with families

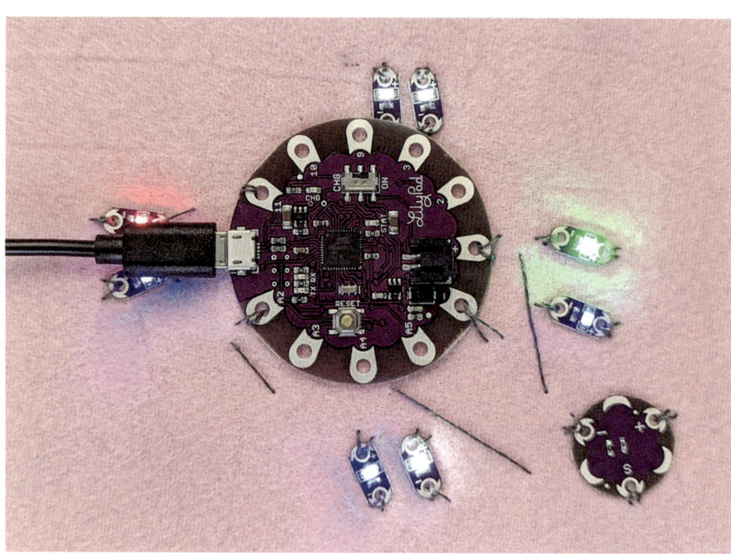

```
/* Blink
Turns an LED on for one second, then off for one second,
repeatedly. This example code is in the public domain. */
// give pin 10 a name:

int led = 11;
 // the setup routine runs once you press reset:
void setup() {
 // initialize the digital pin as an output.
 pinMode(led, OUTPUT);
}
// the loop routine runs over and over again forever:
void loop() {
 digitalWrite(led, HIGH);  // turn the LED on (HIGH is the
                              voltage level)
 delay(1000);              // wait for a second
 digitalWrite(led, LOW);   // turn the LED off by making
                              the voltage LOW
 delay(1000);              // wait for a second
}
```

and looking at how kids could start to use this kind of technology. For example, kids at a forest school made a woven leaf using conductive and nonconductive yarns and materials. It is connected to a Bare Conductive touch board, and as the student touches the leaf, sound files are triggered. I was very interested in the questions: How we can start to work with people? How can people tell their stories?

We started working with arts charities and connecting with people with disabilities and impairments. I started working with Professor Janet van der Linden at the Open University, and we created weaving workshops for people with a visual impairment using e-textiles. My human-computer interaction background has involved working with participants and getting them to envision how they can express themselves through stories. We helped them to make interactive e-textile art pieces to envision how they can be the makers themselves; then we, as researchers, learn from them.

URSULA WOLZ: My background is in artificial intelligence. I have a PhD from Columbia University, from back in the dark ages before we knew what machine learning was. But I also have a master's degree in computing and education, also from Columbia. I have this technical degree and I have this humanistic degree, and they inform each other. I've been teach-

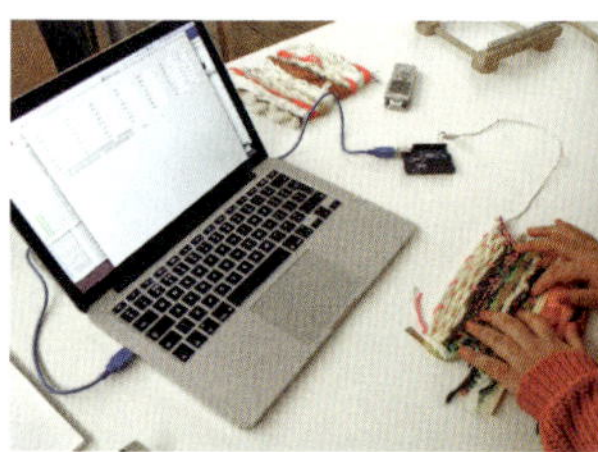

A participant experiments with a weaving at a 2014 workshop organized by Emilie Giles and Janet van der Linden, who demonstrated how to weave e-textiles to create touch-based technologies with and for people who are visually impaired or blind.

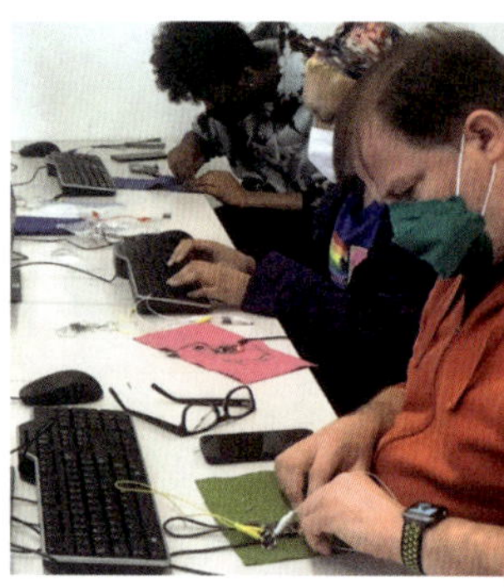

"Wearable E-Textile Programming with Arduino Lilypad," organized by the Tang Teaching Museum and Skidmore's IdeaLab and Computer Science department with Emilie Giles, design a simple circuit using fabric, conductive thread, LEDs, and light sensors to create light-sensitive wristbands for use while walking or biking at nighttime.

ing computing since 1978. I was part of the MIT-initiated Brookline Logo Project, which proved that middle-school children could code. What a concept! I've been doing game development since about 1980, first at the Children's Television Workshop, building some of the first games for kids. My work has always been focused on environments in which users are creators, rather than consumers, of technology, which is especially true for microworlds and educational games. Throughout all of this, I've been a crafter. I learned to do needlework from a cousin when I was seven; I knit; I crochet; I sew; I quilt.

About five or six years ago at Grinnell College, where I was a visiting faculty member, I was asked to create a course that would attract nonmajors in a radically different way. We looked at how crafting, crocheting, machine embroidering through TurtleStitch—which is a variation on Scratch—and quilting could come together and inform the way we think about the process of coding. There is an afghan hanging at Grinnell that was assembled by a small group of students who supervised everyone else's attempts. These are not perfect squares. The crocheted pieces were made both by individuals who were very skilled, who had been doing it since they were five, and those who were not. It was interesting to see the switch between the men and the women in the class, which was three-quarters women. The women felt very competent about what they were doing, and the men were worried that they weren't going to get an A. That situation has occurred in code-crafting classes ever since.

Both in formal classes and informally in groups, I've worked with machine embroidery through TurtleStitch, the brainchild of Andrea Mayr-Stalder in Austria. At a workshop at the Association for Computing Machinery SIGGRAPH conference a couple of years ago, work was developed by individuals who didn't know anything about coding and who were shown how to code in TurtleStitch and Scratch in about fifteen, twenty minutes. Then, within an hour, they produced beautiful drawings. **We use crafting as a vehicle for teaching programming.**

Part of my volunteer work is making hats and raising money for the homeless. My quilting group is looking at how we can use textile arts in order to create social justice

code (right) used during the workshop "Wearable E-Textile Programming with Arduino Lilypad," organized by the Tang Teaching Museum, Skidmore's IdeaLab, and Skidmore's Computer Science department, with Emilie Giles.

at a very local level. And, as a computer scientist, it strikes me that these patterns are marvelously complex, but regular. Quilters modify as they go, and we've also started to explore how classic patterns can inform digital art. For instance, we created a friendship square. By doing some coding, we defined the regions, split up some pictures, started playing with it, and created classic quilt patterns using images.

We need to get technologists and computer scientists to think seriously about what traditional craft communities have known for, literally, millennia. Computer science did not start with a Jacquard loom. Computer science—and the notions of pattern, repetition, and variation, which is all that an algorithm really is—was invented millennia ago by the first weavers, sewers, and embroiderers. This culture is very different from the technology culture that we're all trying to bring everybody into, because it's inclusive, it's collaborative, and it's about sharing. When you look at the history of textiles, you see that ideas and concepts move among local cultures as people say, "Hey, that's a great idea. Let's try that."

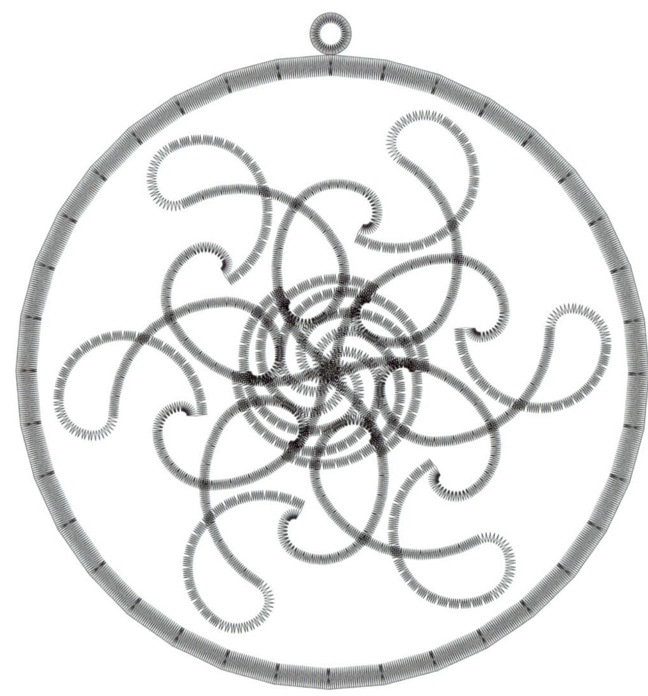

AP: In middle school, Logo was one of the first things that introduced me to computing, and I really enjoyed it. Thank you for your role in making that happen.

It's very common to classify certain fields as belonging to certain genders, and I find that textiles are considered a "women's field" or a "women's science." But what really does not get discussed are the complexities involved in this field, and how they are actually related to science and engineering. Could each of you talk a bit more about whether your experience supports this observation? And if so, what are you doing to change this?

TA: That observation reflects my own process of growing up, to some extent—even I thought that textiles were a women's science. In my youth, it was a point of ego to say that I was in the "hard" sciences, in the physical sciences, because there aren't a lot of women in fields like physics and chemistry. Over the past decade, I've been examining my own biases, my own self-doubt, and the narratives that I tell myself in order to simply function in a misogynistic field that has very few

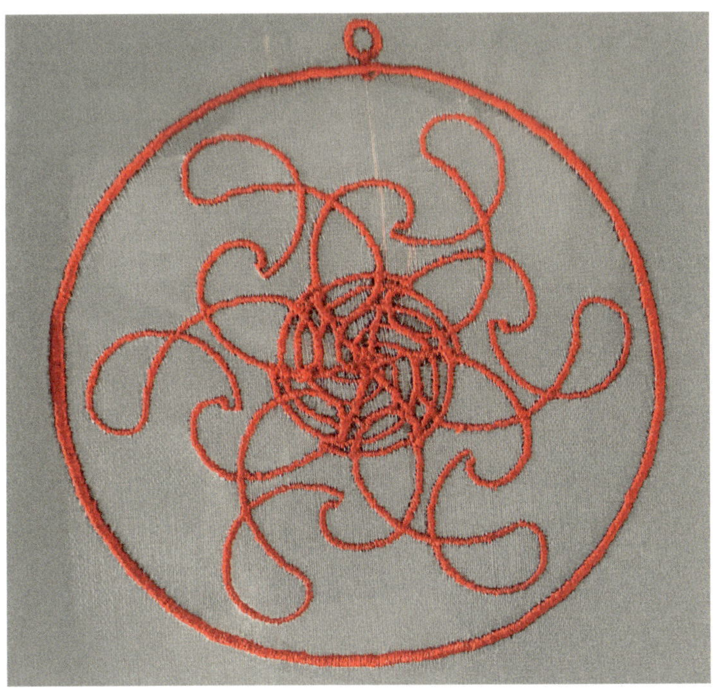

Ursula Wolz, *Snowflake*, 2022, TurtleStitch design (opposite), organza and embroidery thread, 4 ½ × 4 ½ inches (left)

women working in it. And a lot of that self-growth coincided with me stumbling into textile coatings.

When I discuss my journey in textiles, I always talk about it, honestly, as a slap in the face after assuming that I knew everything and textiles were unimportant or uninteresting. We had been making solar cells on paper, and a colleague of mine at the University of Wisconsin–Madison said, "That's cool. But I don't really need a solar cell in my *New York Times*. Have you thought about a curtain?" And my immediate, youthful, hubristic response was, "Yeah, they're just cellulose acetate. It doesn't matter. Of course we could do it."

I went to my lab, and I told my grad student, "Just make the solar cell on fabric." Nearly a decade later, I haven't made a solar cell on a fabric that's as good as the one that I made on paper. The first thing I learned is fibers are not the same as paper. They're not the same as wood. They're not the same as plastic—that's more obvious. But fibers are not the same as something planar that we make out of pulp. Fibers stretch. They have topology. They have a weave structure. And all of those differences create levels of complexity that I learned in material science classes when we talked about crystal structures. And these are all found in everyday garments. In a knitted cotton jersey, you find one of the most complex, crystal-packing structures that you could find in a unit cell. And we don't think about it that way when we teach discrete sciences. I think fiber drawing, for example, is chemistry, but we don't teach it in chemistry. Textile coatings, fiber coatings, fabric coatings...these are all metallurgy and material science. And we don't bring these up when we think about the "noble sciences" and how to conform metals to make our implements.

These realizations have been wonderful and edifying, and also a little bit anger-inducing. Self-knowledge is the start of transformation. Part of that means not shying away from bringing up textiles, fibers, and fabrics, and actually talking about how complex they are, and talking about why textile manufacturing processes are actually incredibly complex chemical reactors that we should think more deeply about. **Textiles are one of the top three polluting industries**

that we have as a society. And we think, just because we buy too many things, that we could address textile pollution by simply recycling our clothing. But that's not true. A lot of the pollution comes from the chemistries that we use throughout the process, from start to finish, not only in the harvesting of the material that we make the fiber with, but how we draw the fiber and how we finish the fiber. Even the locations that we use for the machines that draw the fiber are part and parcel of the pollution that we find in our waterways, as are the dyes that we use to change the color of things. Simply recycling fibers is not going to actually address this chemical pollution issue.

For a lot of people in the chemistry community, green chemistry has been a big thing for a long time. But it has all been directed at making pharmaceuticals or commodity plastics better. No one ever talks about textiles. Everyone in my lab has to at least sew a small patch onto a shirt. We have one shirt hanging in my microfabrication center that has all the devices we've made on it. And the person who makes a device has to sew it on there.

EG: I was always very interested in science. I loved science and math in school. I actually quite enjoyed textiles as well, but I didn't follow either path for my BA. This whole idea of thinking about textiles as a science, then thinking about a more traditional computer science practice…Creative students, design students, they're often quite afraid of both, which is really, really interesting. It's one of those things where, when I work with design students, in particular, and I say, "We're going to do some creative coding, and we're going to do some textiles," they get this slight look of fear on their faces. This is something we're all trying to prevent and make better.

You get this idea that some of the students don't want to do textiles because maybe they don't see it as something that is relevant to them. Maybe these students developed a bias against textiles from school, or maybe it's a cultural thing, which is, obviously, very unfortunate. There is something about textiles being considered a women's science that makes some people look down on it. But I think there is a fear about the complexity as well. How can you make both of these things approachable together? I was probably quite scared of both learning to code and learning to knit, when I first did it, because they're about following patterns.

UW: I was a teacher in the Logo lab back in 1978. Almost all of the hackers, the students who were coders, are now famous computer scientists. One of the things that gets left out of the narrative is that Logo wasn't invented by Seymour Papert; Logo was conceived by Seymour *and* Cynthia Solomon. Cynthia, who is a generation ahead of me, is now playing with TurtleStitch, trying to understand the mathematics behind it. What we commiserate about together is that it's still a man's world, and it's a man's culture. But we both say, "What the hell? I'm just going to do what I want to do." It's hard science. It's understanding how to fashion something. It's an incredibly complicated task with all kinds of intuitions and skills related to how you assemble the pieces.

I've done enough—recently, with crocheting, knitting, embroidery, and quilting, and many years ago, sewing costumes—to recognize that one of the important things is crafting is never perfect. In the Hopi culture and the Navajo culture, when women teach other women how to weave a rug, for example, they leave an error so that the spirit of the cloth can escape. **You never want to strive for perfection.** What I tell my students when they're in tears because they're ripping something out for the sixth time is *you can't be perfect*. And then you juxtapose this with our academic culture, and all we want from our students is perfection. The academic world and the scientific world are phenomenally competitive. And what I love about crafting, especially crafting with my friends, is that we tell one another the truth. We go back and forth and say, "Okay. I like this. I don't like this." Or: "I'm sorry, but you're going to have to rip that out." Or: "Oh, wait a second, you put the pins on the wrong side." We are constantly correcting one another.

What I found most frustrating in my thirty years of teaching was not being able to approach students like that. The guys would get their hackles up, and say, "Why is this old lady criticizing about my coding flow?" Now, the fact that I code significantly better than any of them ever will is a detail. But they would look at me, with my blonde hair, and my comfy-cushy-mommy-kind-of way, and go, "I can't take her seriously as a coder." But I've taught every course in computer science, except operating systems. I can tell you about the complexity of an algorithm. I can tell you about how thinking about the line of a pen on a screen translates into a piece of thread. And my journey has been about feeling confident about what I know, and about relating that rich complexity in the fabric that Trisha and Emilie are talking about to thinking about coding. The big lie we tell right now is that technology, computing, and coding are about predicting that a solution will work, then selling the solution, getting somebody to fund it, and making it happen. That's not the way the world works. Trisha, how many of the things you've predicted would work didn't?

TA: Too many.

UW: I'm having a lot of fun right now playing with embroidery. I'm doing this cool consulting job in the textiles industry, but I'm also looking at things like data structures for sensing. And when I need to take a break, I go sew a seam. Then when I get to the bottom of the seam, I say, "Oh, expletive, and rip it out."

> **AP:** I had never made a mitten before. And I had this really thin, really pretty yarn that I found at a store when I started knitting. It's fingering weight yarn; it's the tiniest yarn. And I read somewhere that you need to use a small needle, I think size 2. And so, I started working on this mitten. Suddenly, I think something is wrong. I don't think it will stretch out, and it turns out it might fit a doll. So, I ripped the whole thing apart yesterday. It's complicated, right? You have different sizes, different materials . . . The yarn comes apart so quickly. Give me a code with the corner cases to write any day—that is so much easier to handle.
>
> There is that subconscious bias against textile science—that it is less than noble. How can the next generation of fiber scientists, artists, technologists, and computer scientists address this challenge?

UW: We are all involved in some sort of interdisciplinary approach to things. We have to start talking to one another and pulling down the silos. We need people who are cross-disciplinary and have open minds. And we need to define a culture of inclusivity. Some of the most famous women computer scientists out there are clandestine quilters and embroiderers.

EG: Inclusivity is so important for all communities. I work with a lot of older people who are visually impaired. And I think it's important to make sure that everyone feels they can be a maker, and actually be part of this movement. If they want to start working with an Arduino board and make something interactive and combine that with, say, finger knitting, why not do that? It's just making sure that all communities— literally all communities—are having these conversations, and that people feel like they are the maker as much as somebody that might be teaching them the methods.

UW: When my son was in sixth grade, his class had a quilt project where each of the kids had to hand-sew a patch for a quilt. It was part of the colonial history curriculum. Some of us who sew went in to help. We all assumed the girls would do fine and the boys wouldn't. And there was this group of boys who were sitting there and working very methodically. They were Lacrosse players. They knew how to stitch from repairing their Lacrosse sticks. We have a lot of cultural biases, especially about gender, that are patently false.

TA: My current mindset is very capitalistic. At some point someone told me the best way to effect change is just to make a bunch of money. You get power. I'm an academic. But I am a weird academic, in the sense that I care a lot about applications. And I set the bar for myself very high, as I think in general most women do, especially women in academia. When I started my academic career, I said, "If I don't make something, or if my research doesn't end up affecting an industrial process or something that we do in our civilization—a device, or whatever—I'm not going to feel satisfied." So, a few patents came out of my lab, but the patenting process is incredibly bureaucratic. And it's very unequal, too, because the patent process can be seamless and easy—such as for institutions who have supported money-making patents previously—or incredibly onerous, long, brutal, and frustrating.

The easiest way to fix a problem is for women to just kick butt and make it better for others who come behind you. This is something that I've also heard from Frances Arnold, who won the Nobel Prize in 2018, and from pretty much every female mentor who I've had at various levels of my career. They've all said something to the effect of: "**Do your best to be the top in your field, and be sure you make it better for the women coming behind you.**" That's all you have to do.

UW: Trisha, let's talk about the environment. How can we support social good through textiles while the climate is exploding?

TA: The textiles industry hits a completely different part of the ecosystem than we think of when we think about climate change and the environment. Textile manufacturing doesn't produce a lot of CO_2 output compared to other industries. What textile manufacturing does is pollute the water. And that is a deeply pernicious and very, very scary proposition—because even if we address our societal issues around putting out a lot of CO_2 into our atmosphere, as long as we keep polluting our groundwater, we won't have anything to drink, and you're not going fix that.

A single pair of jeans requires 1,600 gallons of water—a single pair, from start to finish. A single white T-shirt, off-brand, whatever, from a two-dollar discount rack, require 600 gallons of water from its inception to its completion. And we're not talking about the gas required to transport it to the store, just the production. The reason that these numbers are so important is that you're taking fresh water and contaminating it. The best-case scenario is that it has soap in it, but more likely it has industrial lubricants, it has dyes that you don't want to be drinking, and it has other surfactants that are meant to make micelles. You are robbing the 600 gallons of fresh water from someone who could drink it, and it comes out as industrial waste. That terrifies me. This is why we need more chemists and chemical engineers to be thinking about this issue—because these are industrial waste products. This is not something that can be addressed by recycling fibers, as I said, because well before you recycle that fiber, the fact that it exists in a factory means that you've already consumed polluted water.

Scientists have been working on microelectronics since the 1950s, and they don't pollute that much. We have the normal gas emissions, but, with a silicon microchip, the problem of pollution arises if you just throw it away and it ends up in a landfill. You don't pollute groundwater during the process of making that hard electronic thing. And the reason is because we use sophisticated chemistries in the vapor phase. During vapor deposition, we don't use water. And there's absolutely no reason why we can't apply those same techniques to fibers. This is the work Soliyarn does, and that's why I started this company. I've been very lucky to meet up with some clever business people who keep the lights on and invest and bring other investors in. **Hopefully, we can convince people to use microelectronics to make fabrics.**

Emilie, when I first was introduced to you, through this panel, I focused on your work with haptics sensing. In one of my earlier lives, it really struck me how important haptics are and also how complex they are—not only for coding, but also for material science. For example, if a material scientist wanted to 3-D print something—for instance, artificial skin, or even something that doesn't need a living component, such as a weave—it's very hard to translate the feel of something using current manufacturing processes. I thought it was amazing that you recognized that and brought in the visually impaired community in order to take advantage of haptics. Could you just talk more about your work in this area?

> **EG:** With fabrics, in particular, haptics are a mega challenge. I've been working on a project for the past year or so using haptics and working with visually impaired people to get them to draw in 3-D space. By using a program that is built on the Unreal Engine, which is middleware, and working remotely with visually impaired people who have a haptic device, they can do their drawings in their space, and then they can actually feel them back. One of the challenges is, if you want to start bringing in other textures, you can replicate things like brickwork quite easily with a haptic device, which is interest-

ing because of the vibrations and other things that you can have in a little device, but textiles are so, so difficult. But people are increasingly working toward improving that.

Jumping back to the e-textile research and working with visually impaired communities . . . Something I've found fascinating is the time and focus that people will take when handling objects. When I was doing my user studies during my PhD work, I would go to community centers and observe how visually impaired people go about doing normal, everyday activities. Whether they were feeling the table for their glasses or just grabbing a cup of tea, they would take a lot of time doing these things. It was the same when I was doing making workshops, or when I observed some nanotextile weaving workshops. They had so much actual focus. I think if you're a sighted person, you don't really spend time actually touching things. A big part of doing the workshops was working with different fabric swatches, getting people to feel them and describe what kind of associations they had with them, and observing how people would feel them. I asked, "Is there a language in this? Is there a consistency in how people might use their hands?" Everyone has a different approach, depending on when they lost their sight, what their background is, etcetera. It is really interesting to think about how we can take that forward into tactile perception, then even more forward into haptics. And we can look at very specific communities that are often marginalized in order to work this stuff out, which is exciting.

Ursula, how do we explore this idea of perfection? If we're looking strictly at academia, how do we get our students to work with things like creative technology and crafting? Being okay with making mistakes is a big challenge, and this is something I found, not only in university teaching, but when I'm working with communities as part of research, too. The thing has to work, right? The crafting has to work as well. Is there a way that we can encourage people to not be afraid to make mistakes?

> **UW:** Yes, we can. We have to give up on grading. I designed a major. I've done curricula. I've gotten computer science accreditation for three different institutions. I know how to do instructional design, and I know how to do assessment. We

have to give it up. We need to step back from these teacher-centered goals and the objectives of the activity. You can't learn how to code by going to the Scratch site, though you can get pretty close. As Seymour said: kids are inventors, but not every kid is a scientist, and not every kid is going to invent a *for loop*. How do you provide opportunities for people?

I have a half-finished design of a placemat that I'm making for my daughter with found material that we augmented. The original design had evenly sized squares, but we only had a small piece of the fabric, and we had to, as quilters say, "fussy-cut" it. We had to pull out the pieces. We ended up with some strips that we could then reassemble. About seven quilting techniques are involved in piecing those things together. There was all this give and take. We had to give up on perfectly even squares because the happy cats, the very symmetrical ones, are much skinnier than the yin-yang cats, and then we had panda images. And we had to deal with varying colors and where the colors go. Then we had to get all the seams right. Then we had to make sure that the seams were flat in the right direction, so things would sew evenly. My friend Arlene Marin and I were working on this project while we were listening to the *Radical Fiber* symposium talks yesterday. And we'd get involved in the talks, and we'd screw up something, then we'd have to backtrack. That whole activity, that craft or maker activity, where you actually learn how to do things—I can't make that happen in a modern classroom and give a grade at the end of the semester. Is it time to radically pull down the institutions? No, it's not. But it is time for us all to push back and say, "We're not teaching them the stuff that we want to teach them."

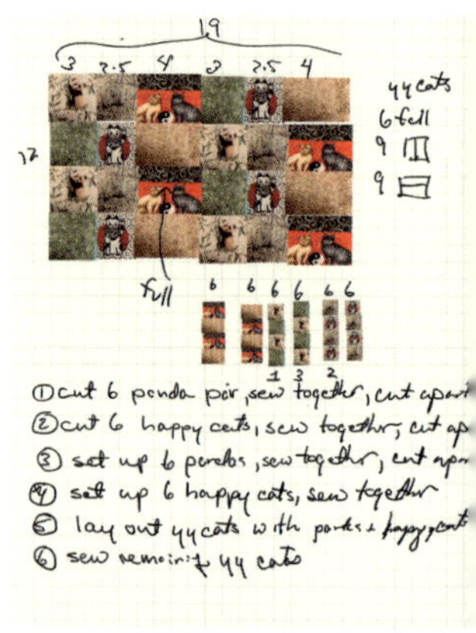

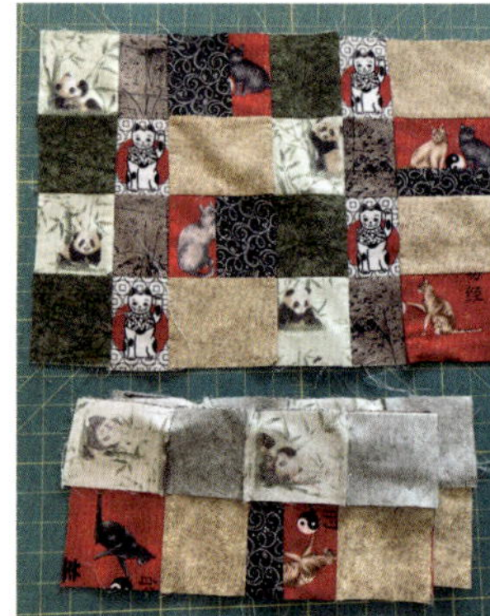

Ursula Wolz's planning document (top) and work in progress (bottom) for creating a placemat.

AUDIENCE: Is there a coding equivalent to the idea of leaving an intentional error, like the weaving example? If not, could that be valuable or interesting to implement into coding culture? Are any of you allowing mistakes to inform your technological work?

UW: We leave intentional things in all the time; they're called bugs. **There is never a perfect piece of code, ever.** The game industry calls it an "Easter egg." When you find your way down into the bowels of the game, sometimes there are surprises.

And sometimes we leave them in there intentionally because the program has to go out.

AP: In my intro to computer science classes, I emphasize that mistakes are our friends. It's totally fine to make a mistake; that's how you learn. The error messages actually tell you how to fix them. If you don't make those mistakes enough, then you are not reinforcing that topic. I've started giving out erroneous code, so that students can see that it's okay to have code that has mistakes and then they actually learn to fix the mistake.

EG: For me, it's about trying to teach students that technology can be a creative tool as much as other tools you might work with. Things might go wrong. But then, once you troubleshoot and fix, you feel empowered.

AUDIENCE: What are some of the early interventions to make crafting gender-neutral? It seems this shift needs to happen early. Perhaps bring more fibers into maker labs, which are largely hard tools and technologies. Incorporating crafting and social justice is another good strategy.

UW: Bring back home economics.

TA: Yes, and woodshop. And don't automatically segregate students based on gender.

UW: Call it something else.

TA: At the University of Wisconsin–Madison, the textiles department is in a college called the School of Human Ecology. That was the erstwhile School of Home Economics, actually. Interestingly enough, this department granted the first bachelor of science to a female in the state of Wisconsin. She was going to college to understand how to make a home and learn interior design and fiber arts to make her own clothing. So, yes, home economics was the way that women got their degrees in the Midwest. But I think maybe we should rebrand "home ec" as "human ecology." Maybe we can bring back human ecology.

1 Linden K. Allison and Trisha Andrew, "A Wearable All-Fabric Thermoelectric Generator," *Advanced Materials Technologies* 4, no. 5 (May 2019), doi.org/10.1002/admt.201800615.

2 S. Zohreh Homayounfar, Soha Rostaminia, Ali Kiaghadi, Xingda Chen, Emerson T. Alexander, Deepak Ganesan, Trisha L. Andrew, "Multimodal Smart Eyewear for Longitudinal Eye Movement Tracking," *Matter* 3, no. 1 (October 7, 2020): 1275–93, doi.org/10.1016/j.matt.2020.07.030.

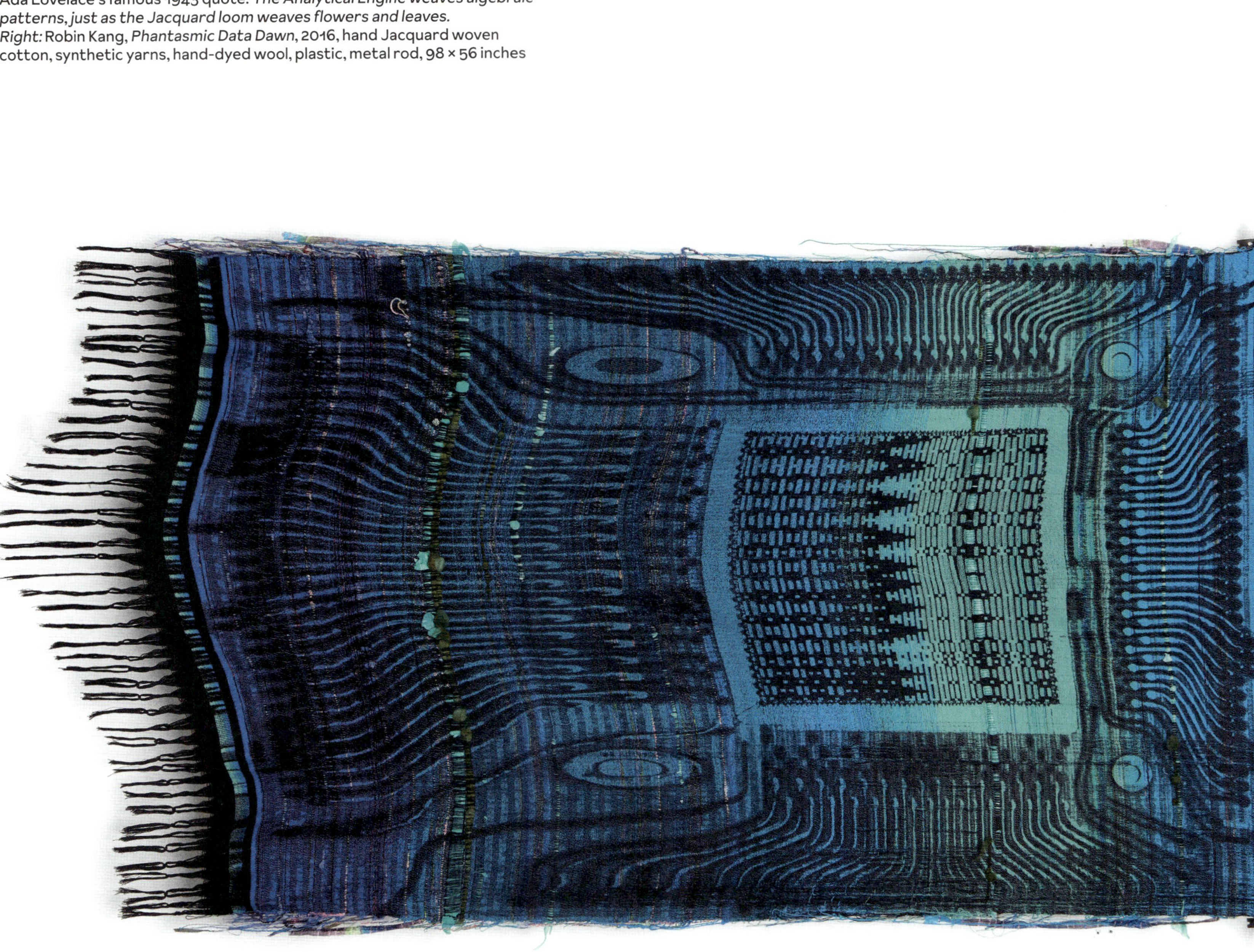

Left: Ruth Scheuing, *The Ada Lovelace Code*, 2017, woven thread, 30 × 32 inches; using ASCII's 0s and 1s as equivalents for weaving's over/under structure, Scheuing generated a pattern for British mathematician Ada Lovelace's famous 1943 quote: *The Analytical Engine weaves algebraic patterns, just as the Jacquard loom weaves flowers and leaves.*
Right: Robin Kang, *Phantasmic Data Dawn*, 2016, hand Jacquard woven cotton, synthetic yarns, hand-dyed wool, plastic, metal rod, 98 × 56 inches

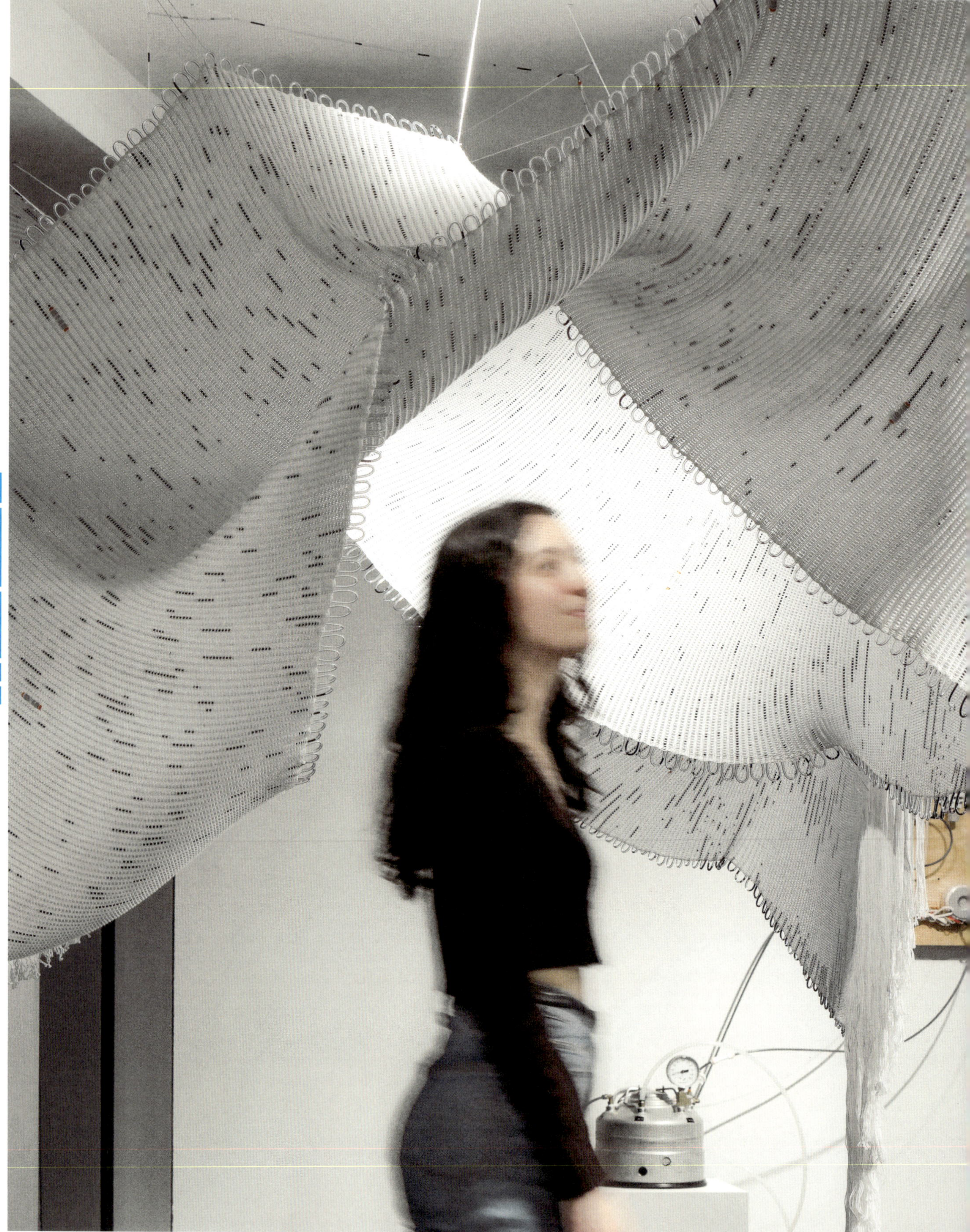

by Sang-Wook Lee

ART AND TECHNOLOGY, INTRINSICALLY WOVEN

A product of the Neolithic Revolution, the earliest looms were simple frames made of branches on which warp yarns were hung and stretched taut or weighed down with stones, while weft, or filling, yarns were interlaced by hand, over and under the warp. This interlacing forms the cloth's structure, is visually articulated on its surface, and is still the primary compositional element of woven art. Following the increase of machine-based industries in the eighteenth and nineteenth centuries, new technologies have facilitated the rendering of complex weave structures in cloth. In particular, the invention of the Jacquard loom by Joseph-Marie Jacquard in 1801 spurred a revolution in textile manufacturing. The Jacquard loom contains a Jacquard "head" attachment that, by utilizing the first punched data cards, mechanically controls the raising and lowering of specific warp threads in a sequence that allows for the creation of any given pattern. Small sensing pins detect the presence or absence of a hole in the card and determine whether a needle should pick up a thread. This mechanism enables a single weaver to have endless ways to "program" their loom and create intricate tapestries and complex patterns by entering a binary pattern that is presented on a series of punch cards.

Essentially, woven structures are where programming was born; before the advent of computers, people programmed Jacquard looms. Nearly two centuries later, in the 1980s and 1990s, the abundant creative possibilities of Jacquard technology were expanded by contemporary artists who use digital Jacquard hand looms to fabricate woven textiles that rise to the level of "great art." Modern technology has significantly increased Jacquard machine capacity to move beyond repeats and symmetrical designs, allowing almost infinite versatility. With digital loom technology, works that were previously produced, stored, and presented exclusively in digital formats can now be expressed in physical woven forms. The new digital domain embraces a focus on visual rather than material effects. Thus, modern textile artists are changing perceptions of fabrics throughout art history yet again by showcasing the vast differences that are possible when dealing with fabric, thread, and yarn.

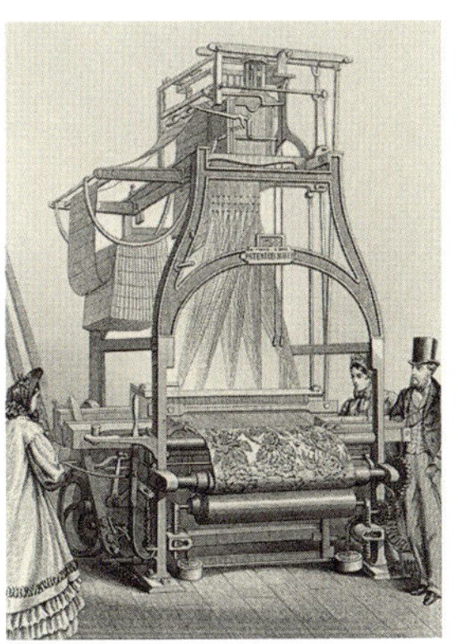

Engraving depicting a Jacquard loom with additional patented technology by William Smith and Brothers in Heywood, England; this loom received a prize medal at the 1862 International Exhibition in London.

93

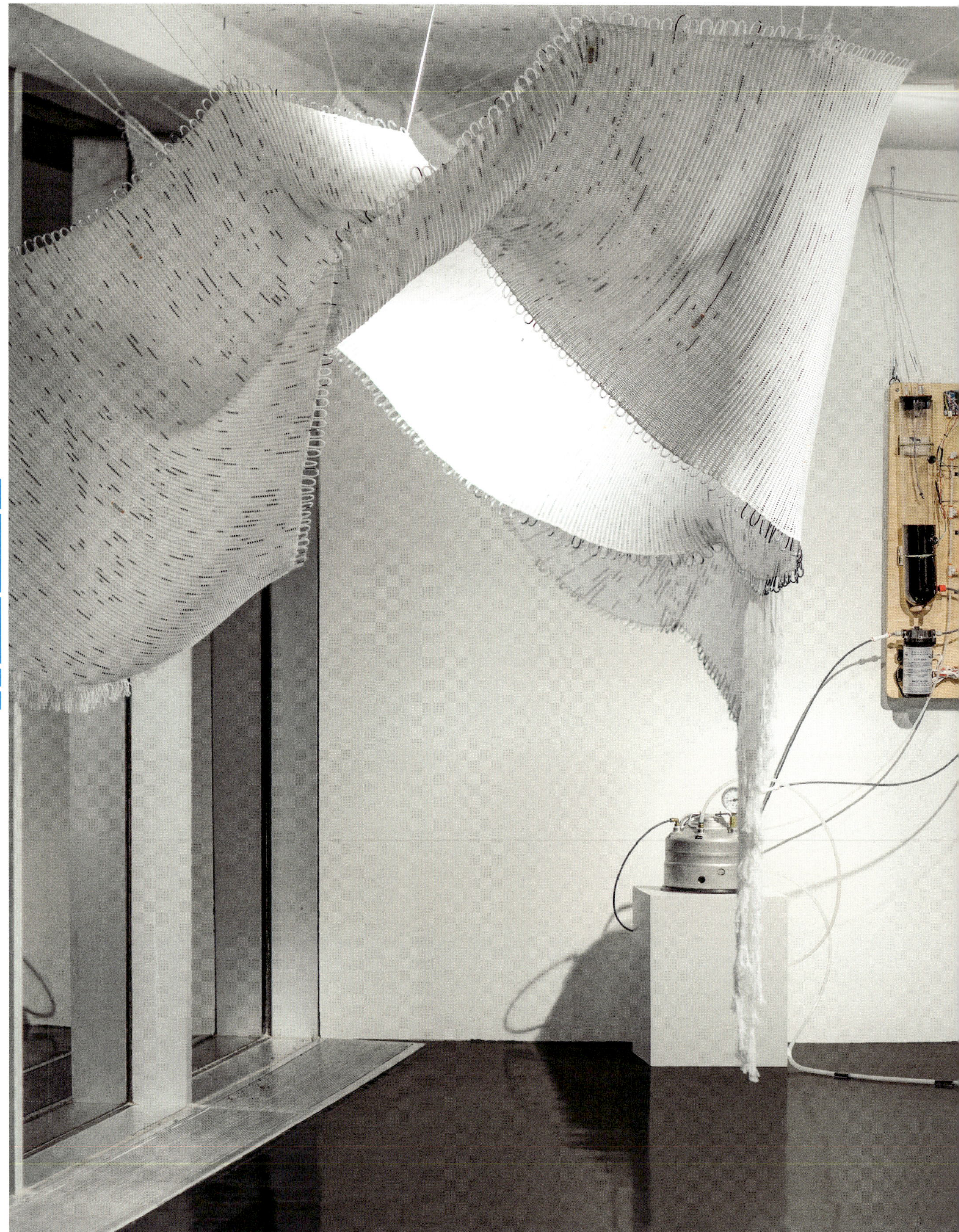

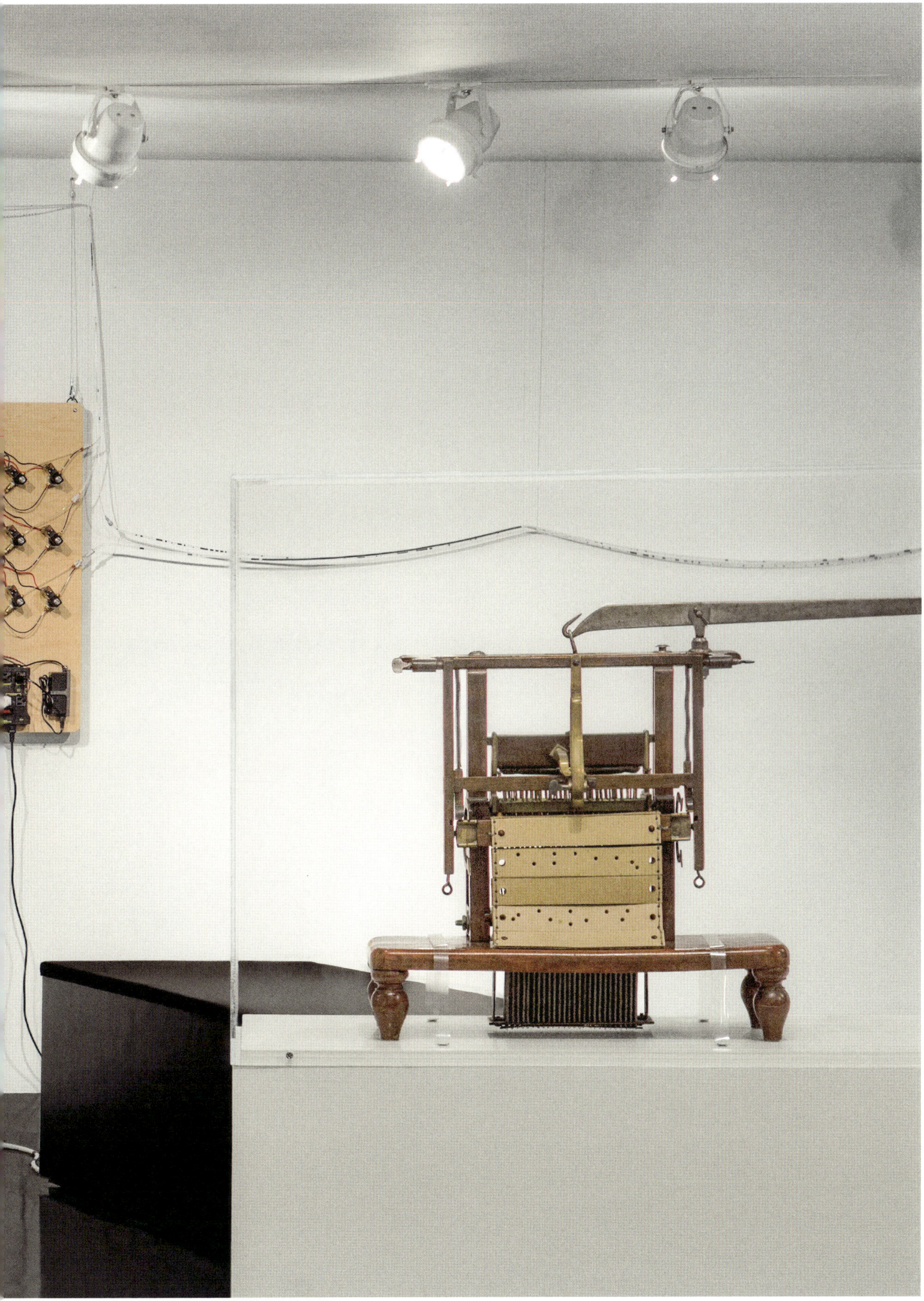

Installation view, *Radical Fiber*, featuring work by Soft Monitor (left and background) and a model of a Jacquard loom (foreground)

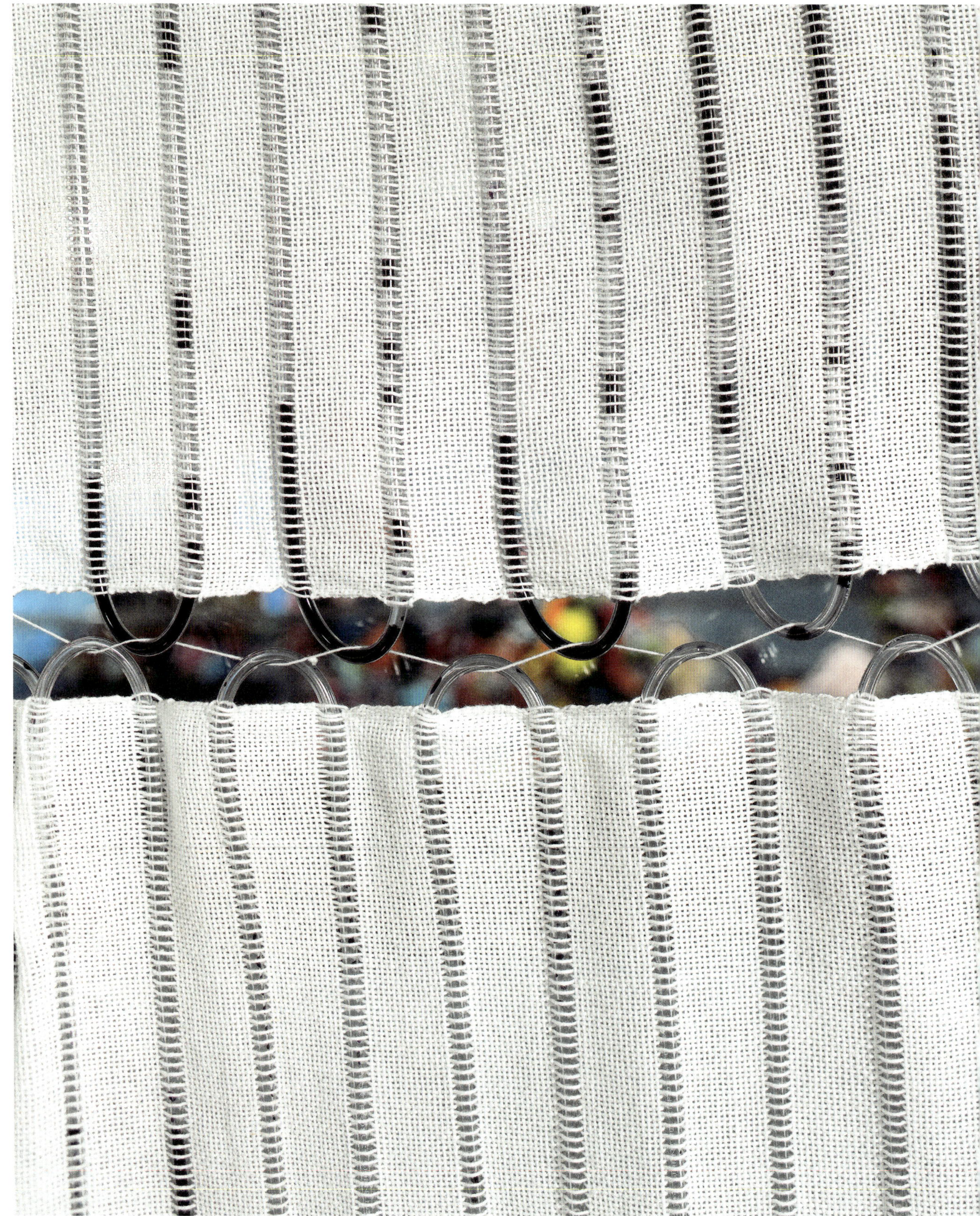

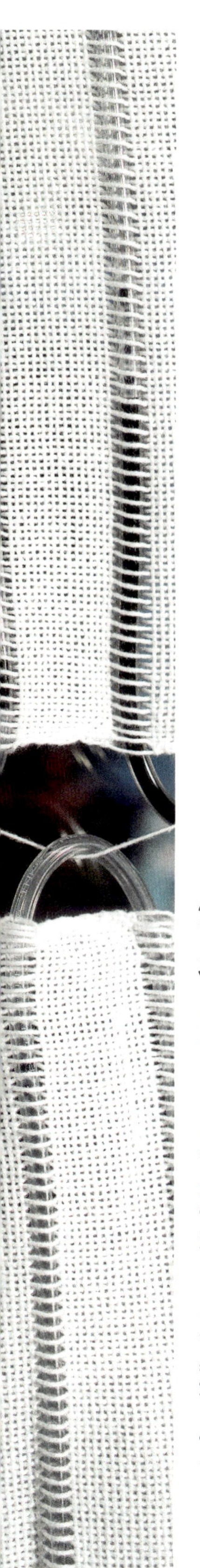

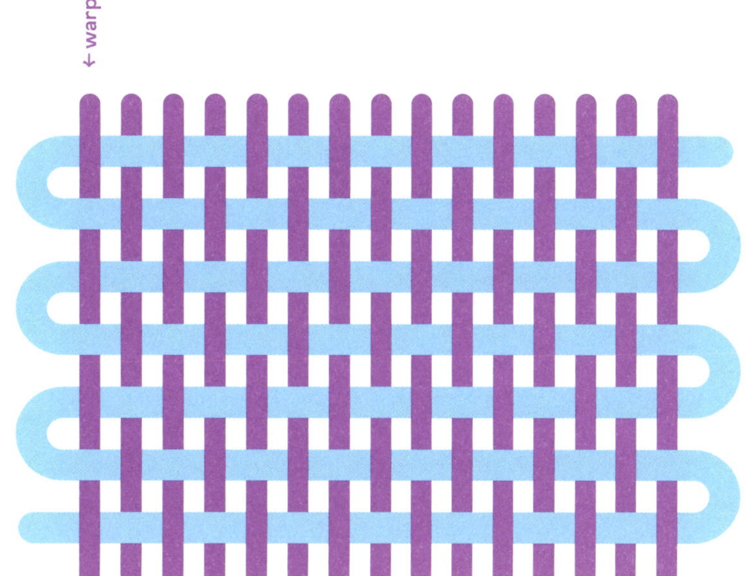

Soft Monitor, the artistic duo of Victoria Manganiello and Julian Goldman, was inspired by the invention of the Jacquard loom and its connection to computer development. Featuring traditional weaving in combination with digital coding, *c o m p u t e r 1.0* examines the relationships between woven art and technological innovation, both of which use binary systems. The title alludes to modern computer history but also recognizes the work itself as a lo-fi computer display. By replacing light pixels with drops of liquid and pockets of air, they have developed an ancient computer, one from natural, analog materials like water and cotton. The duo skillfully connects the foundational patterns of warp and weft thread to the extensive lines of binary code, and as a result, the balanced weave evokes the relationship between the structured and the idiosyncratic.

Natural fiber thread is woven with hollow polymer tubing containing a series of colored liquid "pixels" that traverse and animate the weft thread in an otherwise simple structure. The overall effect conveys a sense of organic unpredictability, as the near-solid color of the textile is interrupted seemingly at random by the high-contrast color of the pixels. However, within the chaos there is order, and the dyed liquid and air pumped through the textile are programmed patterns that are uploaded to the cloth. These anomalies speak to our experiences of the digital age, as complex pattern is created through the binary system flowing within the weaving. The computer language of 1s and 0s can be traced directly to the binary system of weaving itself, harnessed and enhanced by Jacquard technology and its use of punch cards. In essence, weaving has been a binary art form from its inception, applying patterns of code in thread for millennia.

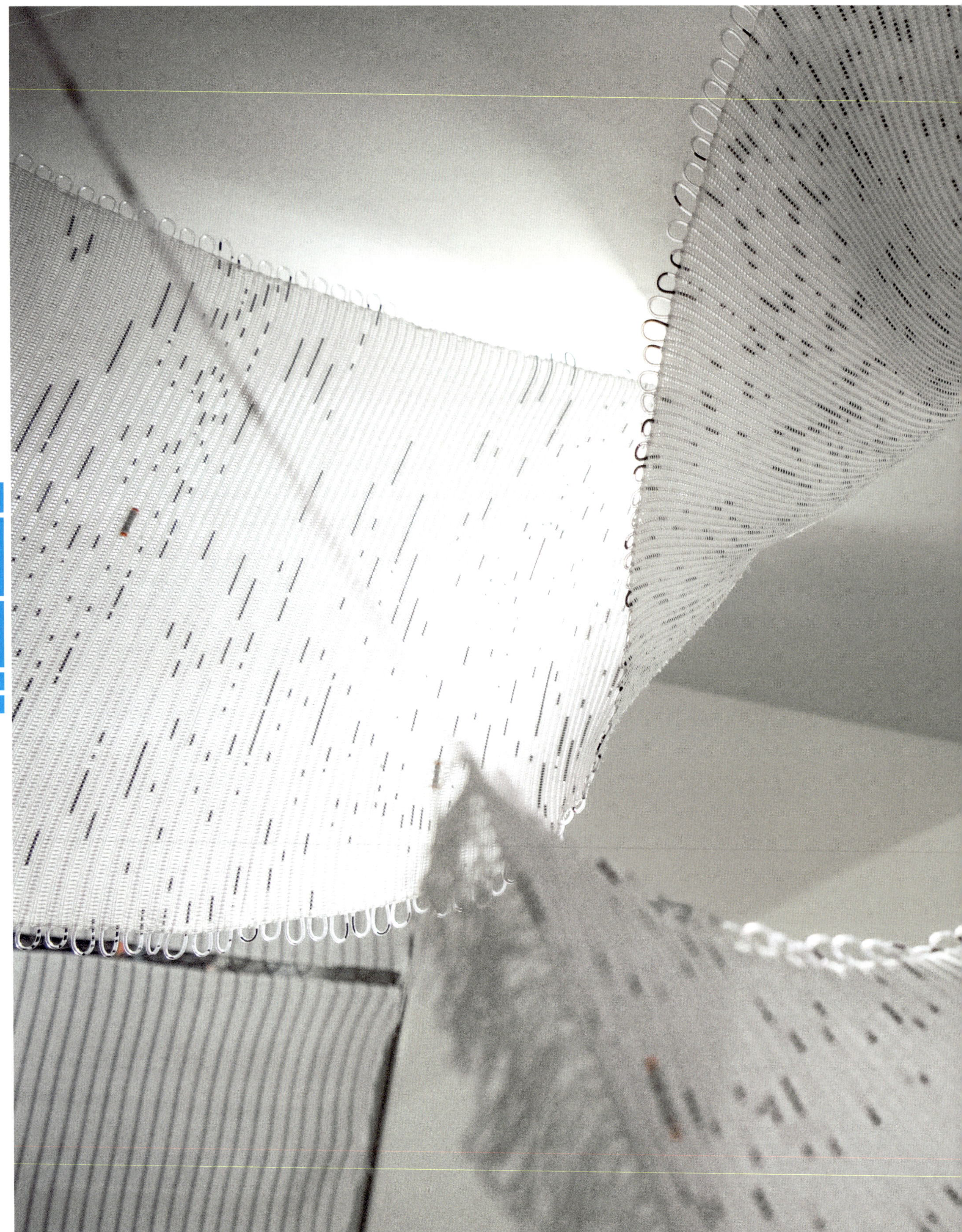

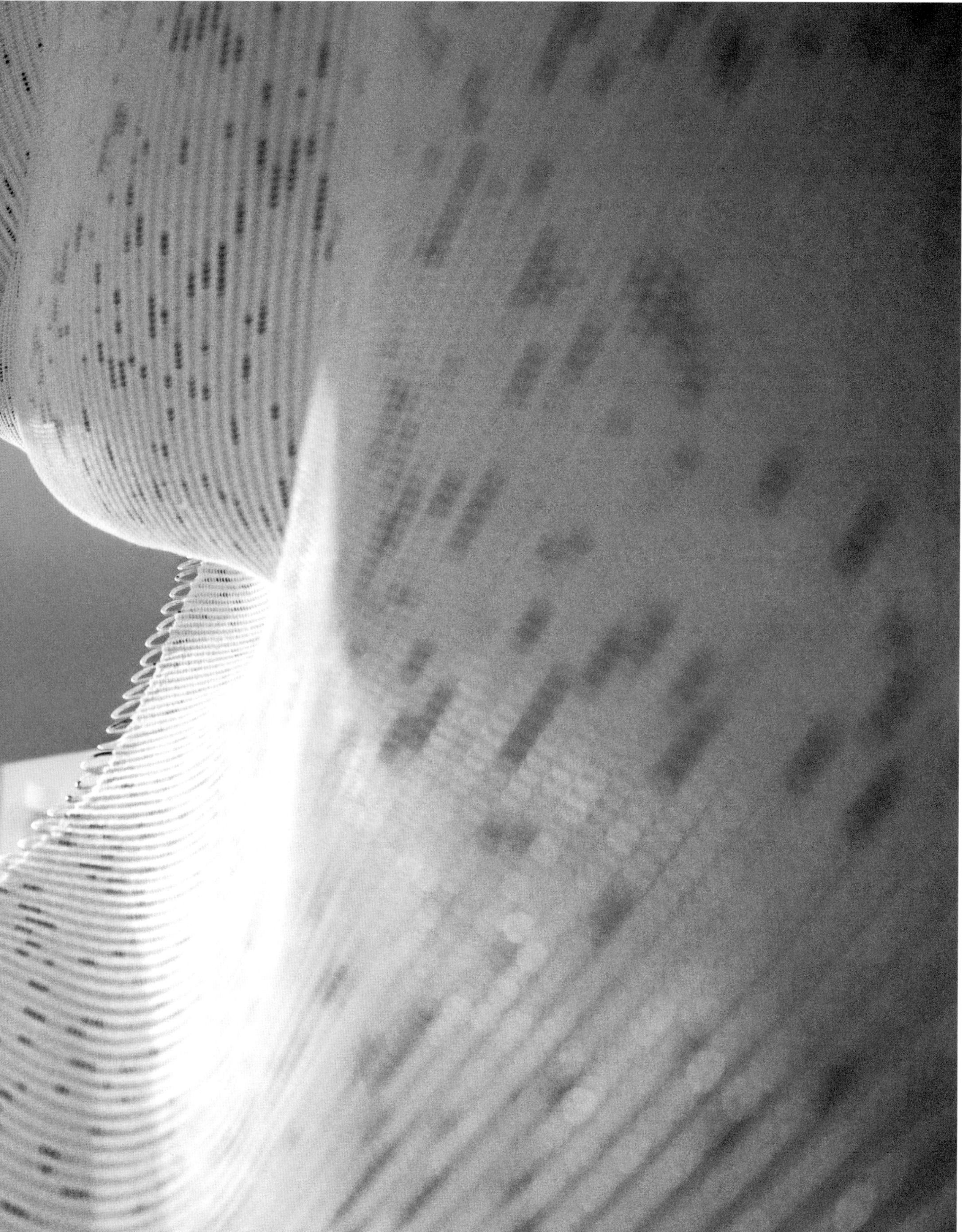

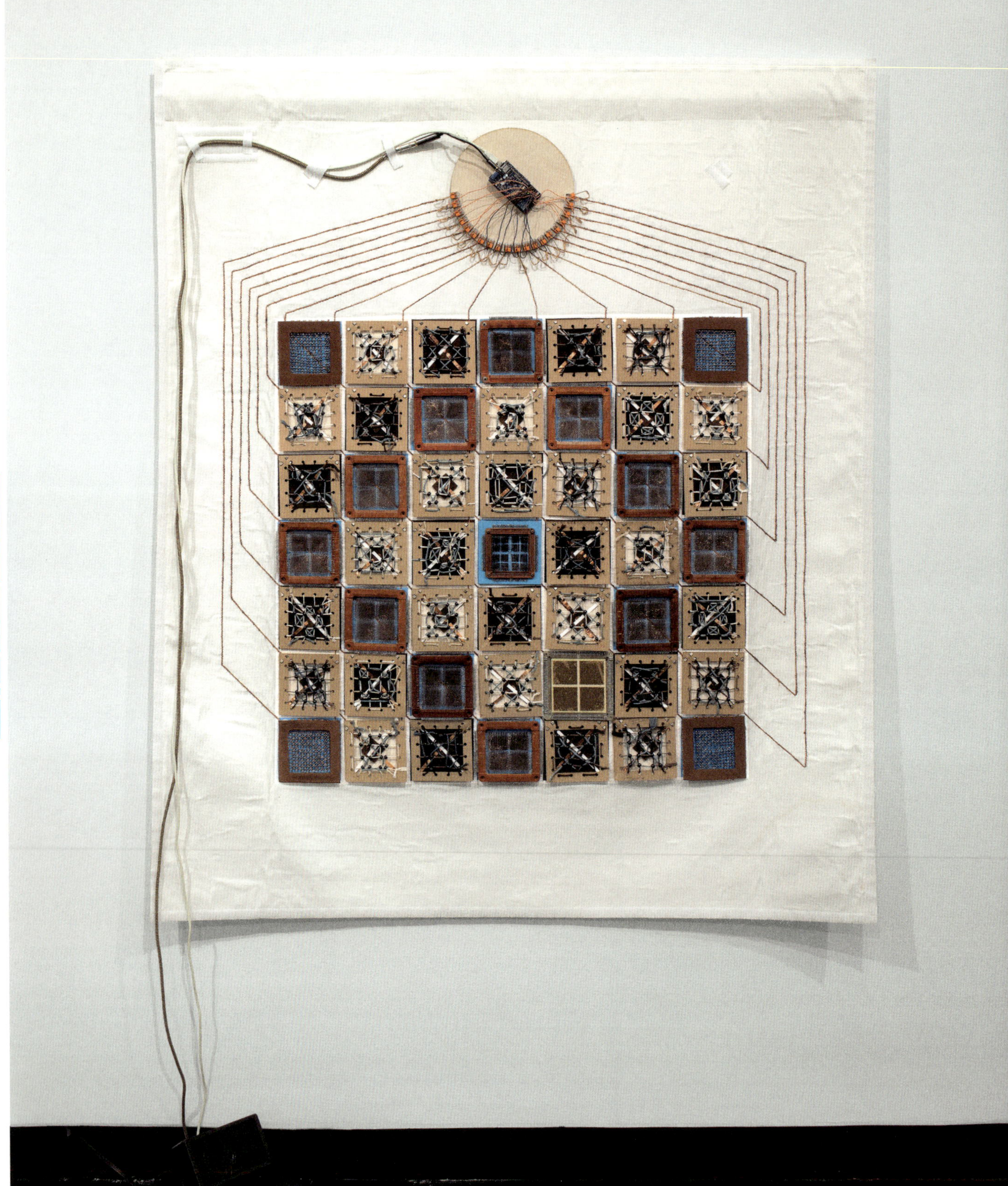

TOWARD A MORE ACCESSIBLE SOFTWARE INDUSTRY

When many people think of who writes software code, the image that comes to mind is often the one portrayed by popular media—a socially inept white male. One way to address this stereotype bias is to highlight the contributions made to computer science and engineering by women over the years.

Core Memory Quilt, created by the interdisciplinary team of Samantha Shorey, Helen Remick, Brock Craft, and Daniela Rosner, achieves this goal by bringing attention to the weavers involved in the Apollo missions, many of whom were women of color and central contributors to the advancement of computing.[1] In its nascent stage, computing was influenced by the Jacquard loom, patented in 1804, which uses punch cards that hold instructions for weaving a pattern. Ada Lovelace, often considered the first programmer, wrote the programs for Charles Babbage's Analytical Engine in the 1840s and famously claimed that "the Analytical Engine weaves algebraic patterns, just as the Jacquard loom weaves flowers and leaves."[2] The software used for the Apollo missions, represented in the *Core Memory Quilt*, was also built on this concept of weaving: when a wire passes *through* a magnetic ring, it is read as a 1; when a wire passes *around* a magnetic ring, it is read as a 0. This binary system is still used today, where data is represented in the computer as 1s and 0s. For example, the letter *A* is stored as the number 65, which in the binary number system is *1000001*. Demonstrating the connections between software programming and weaving can help change the narrative surrounding the image of a software developer as well as rectify falsehoods about textiles.

Perhaps it is also time to rethink the structure of software code. Soft Monitor, the collaborative duo of Victoria Manganiello and Julian Goldman—

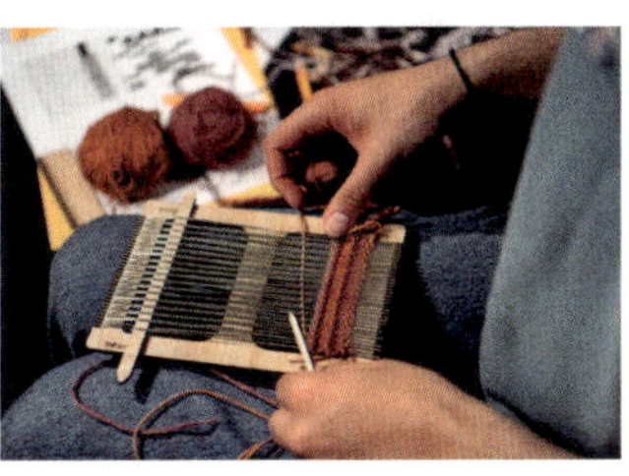

In the "Weaving Codes with Soft Monitor" workshop at the Tang, participants learned how to embed messages into cloth using ASCII character codes and gained material understanding of the intersections of weaving and digital technologies.

who designed a large-scale textile woven using polymer tubing and natural fiber thread, with dyed water acting as pixels—dares us to envision alternate computer displays of the future. A pixel is a picture element, the smallest information in an image and the smallest element of a picture represented on a computer screen. For example, when we talk about 4K televisions, the 4K resolution implies that approximately 4,000 pixels can be displayed horizontally; the majority of televisions come with 3,840 × 2,160 pixels. Often when we look at a television, computer monitor, or smartphone screen, it is not immediately evident how what we see is generated because the code is hidden. But in *c o m p u t e r 1.0*, the viewer can immediately see and hear the computer-controlled valves and pumps that push liquid and air through the tubes in the woven cloth. The lack of transparency in technology encourages people to form their own conceptions—or misconceptions—about how technology works, based on their biases and prior experiences. This hurts adoption of technology: for example, consider the low adoption rates of contact-tracing apps that were deployed to reduce the spread of COVID-19.[3] When surveying two thousand Americans in June 2020 about a hypothetical app for contact tracing, researchers found that participants did not have a good understanding of how contact-tracing apps work.[4] Can we make technology more transparent, and make a person who is not a software developer understand how the code works? Will doing so improve our trust in technology?

Emilie Giles's work considers whether we can build software that is accessible to everyone—especially those for whom coding has been opaque. Giles conducted workshops with people who are blind and visually impaired, where they built interactive electronic textiles that could produce sound when touched or squeezed, allowing them to explore how technologies could become a part of their lives. To make technology accessible to all, we need to increase diversity in the software industry, so voices of all races, genders, abilities, and socioeconomic backgrounds can be involved with software design. And to increase diversity in the software industry, we first need to increase diversity in computer science classrooms—an achievable goal for those willing to expand traditional disciplinary boundaries and make room for diverse interests. For example, in the late 2000s, Leah Buechley built the open-source LilyPad tool kit and later discovered that women were building 65 percent of the projects created using this kit, including interactive clothing, plush toys, sculptures, and biking accessories.[5] Another example comes from Ursula Wolz, who dared the computer science education community to change how we teach computer science by using embroidery to teach programming.[6] Her students use a visual drag-and-drop programming language called Snap to generate patterns for embroidery machines. This creative thinking can open the gates to future computer scientists who may have otherwise felt unwelcomed or uninvited into the industry. By changing the narrative of who a software engineer is, and by encouraging more diverse voices in the software industry, we can strive toward building transparent and accessible technology in the future.

1   Samantha Shorey and Daniela K. Rosner, "A Voice of Process: Re-Presencing the Gendered Labor of Apollo Innovation," *communication +1* 7, no. 2 (2019); doi.org/10.7275/yen8-qn18.

2   Yasmin Kafai and Jane Margolis, "Celebrating Ada Lovelace," MIT Press, October 14, 2015; mitpress.mit.edu/blog/celebrating-ada-lovelace.

3   The Associated Press, "Despite Promise, Few in US Adopting COVID-19 Exposure Apps," December 7, 2020; www.nbcnews.com/tech/tech-news/promise-us-adopting-covid-19-

exposure-apps-rcna189.

4   Baobao Zhang, Sarah Kreps, Nina McMurry, and R. Miles McCain, "Americans' Perceptions of Privacy and Surveillance in the COVID-19 Pandemic," *PLoS ONE* 15, no. 12 (2020); journals.plos.org/plosone/article?id=10.1371/journal.pone.0242652.

5   Leah Buechley and Benjamin Mako Hill, "LilyPad in the Wild: How Hardware's Long Tail Is Supporting New Engineering and Design Communities," *DIS '10: Proceedings*

of the Eighth ACM Conference on Designing Interactive Systems, Association for Computing Machinery (August 2010): 199–207; doi.org/10.1145/1858171.1858206.

6   Ursula Wolz, Michael Auschauer, and Andrea Mayr-Stalder, "Programming Embroidery with TurtleStitch," *SIGGRAPH '19: ACM SIGGRAPH 2019 Studio*, Association for Computing Machinery (July 2019): 1–2; doi.org/10.1145/3306306.3328002.

Helen Remick, Daniela Rosner, Samantha Shorey, and Brock Craft, *Core Memory Quilt* (detail), 2018, woven patches, core memory artifacts,

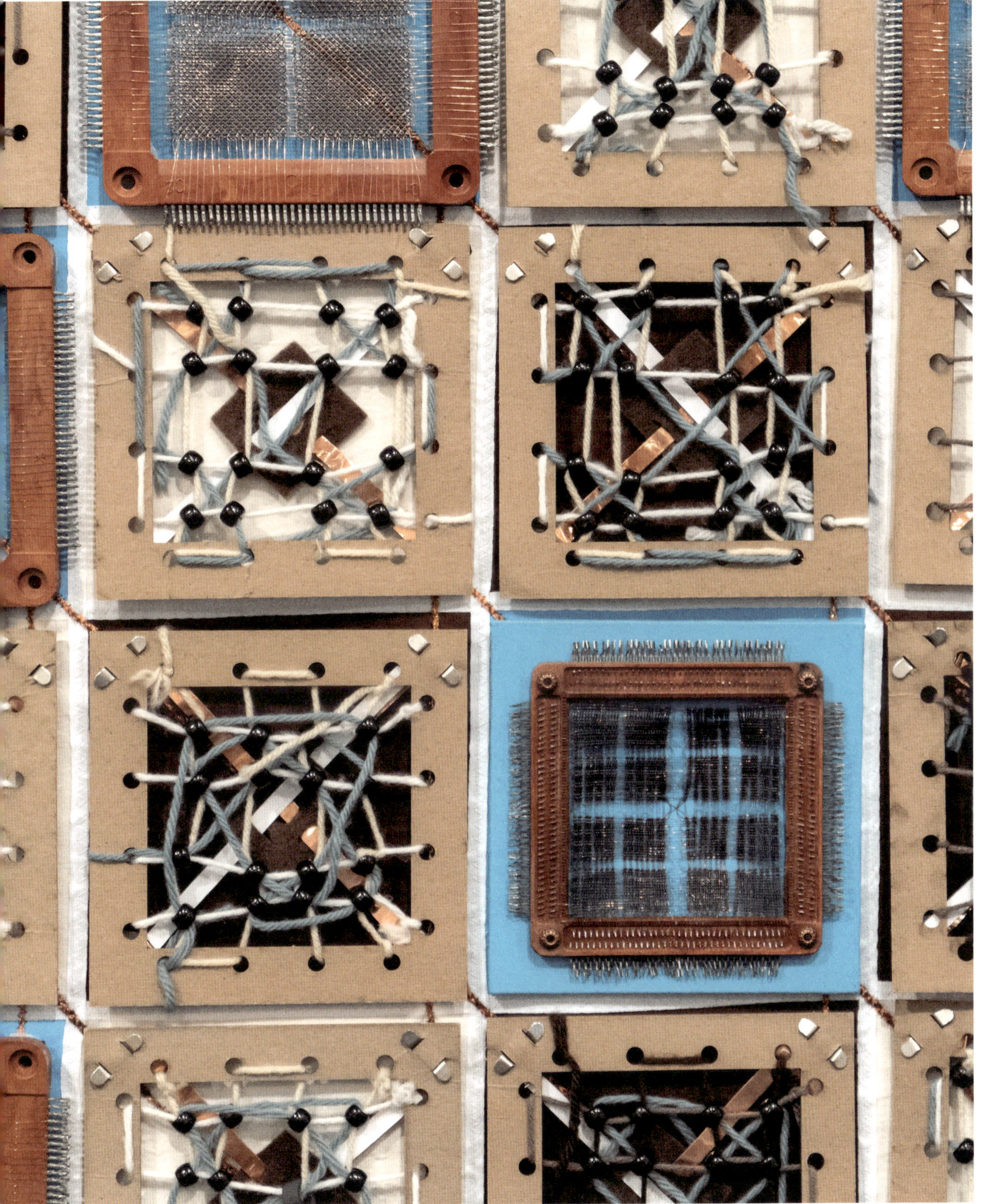

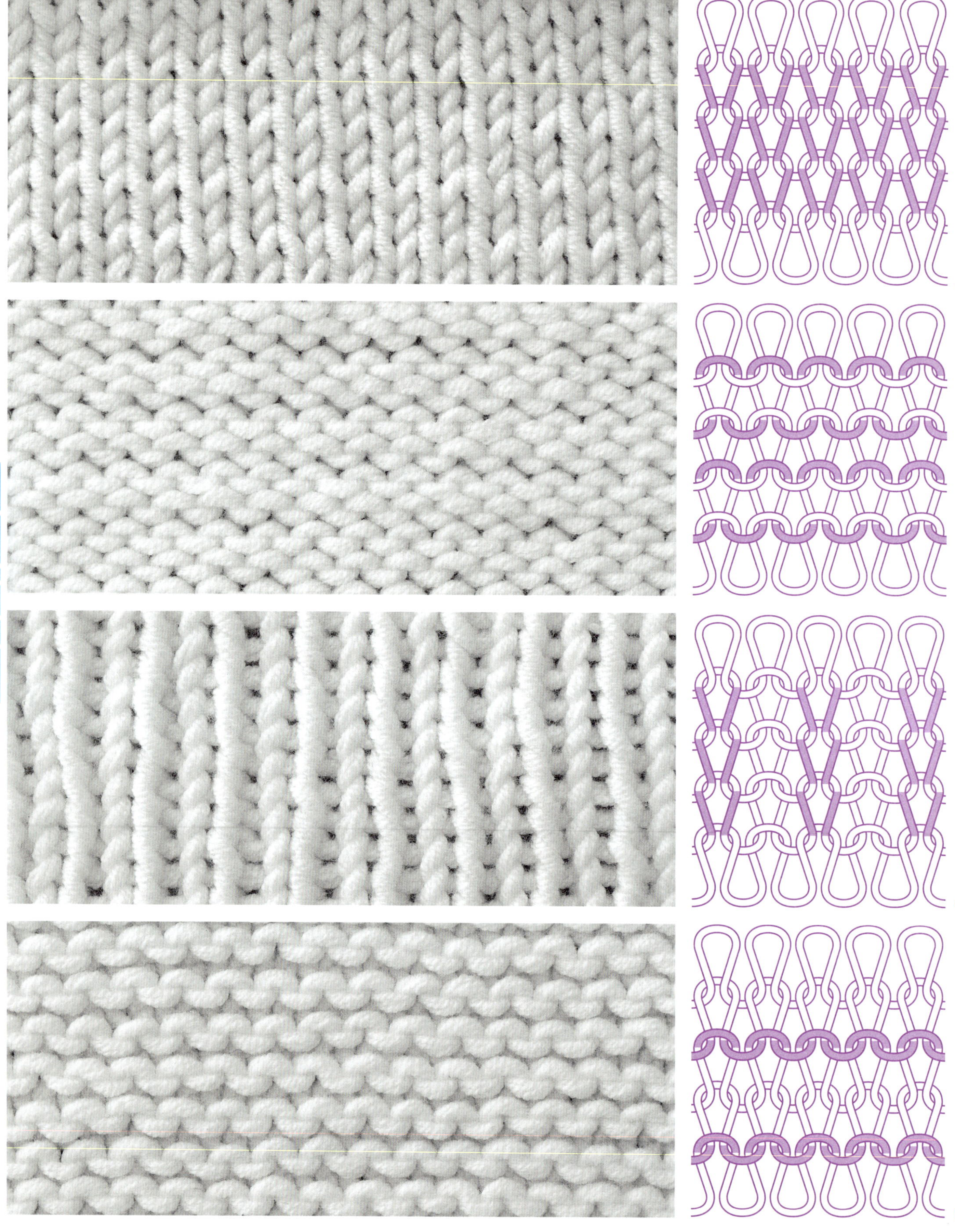

KNITTING, FROM FIBER TO DESIGNER TEXTILES

Knitting is a relatively young fiber art with origins that date to around 1000 CE in North Africa.[1] Initially, hand-knit garments were made from yarns spun from locally available animal fibers (such as wool, cashmere, or silk) and plant fibers (such as cotton, linen, or hemp). Centuries later, in the mid-1800s, researchers developed the first cellulosic fibers in the laboratory—an artificial silk that attempted to bypass the laborious process of harvesting and processing the fiber from the cocoons of silkworms. This class of fibers mimicked the silkworm's method for producing silk: plant matter (cellulose) was processed into a liquid solution and then extruded and hardened into a fiber that could be used in textiles. By the early 1900s, various methods for creating artificial silk were under development, and in 1924, rayon, a cellulosic fiber derived from wood pulp, became the earliest human-made fiber to be marketed for textile use. Today, cellulosic fibers are created from a variety of plants, including bamboo, soy, and corn. The development of rayon convinced chemists that it was possible to create wearable fibers from plant matter, and it raised the question of whether fibers for textiles could be created from nonplant-based materials as well. By 1935, a research team at DuPont Company had created nylon, the first entirely synthetic fiber, extruded from melted plastic-like polymers. Other synthetic fibers, such as polyester and acrylic, followed.[2]

For the rest of the twentieth century, scientific research played a key role in the development of textile fibers. While some of the resulting applications were not designed with the fiber artist in mind, cellulosic and synthetic fibers have crossed over into the artistic domain and are readily found in yarn stores and hand-knit garments. Our understanding of the large-scale properties of knit garments, such as their drape and stretchiness, that makers empirically observe when working with yarn are influenced by both the small-scale (i.e., at the fiber level) and intermediate-scale (i.e., at the stitch level) structure of the yarn.

Long strands of yarn used to make textiles are created by twisting together collections of fibers. The resulting yarns are largely inextensible, although modest variations attributable to the fibers and their interactions do exist. Sketches of common natural fibers from animals and plants as viewed under a microscope are shown on the next page. Natural animal fibers come from the hair of a variety of animals, including sheep, goat, alpaca, musk ox, and yak, and while their length, texture, and color vary, all have fibers that are covered in scales (see G on next page). Many of the properties attributed to wool and other yarns made from animal hair—warmth, breathability, and the ability to wick moisture

Fabric samples made by the author, using the following stitches.
(A) stockinette, (B) reverse stockinette, (C) ribbing, (D) garter

away from the skin—are due to the scales that cover the hair fibers.[3] They create a tremendous amount of surface area, permitting the fibers to trap large amounts of air, which in turn provides good insulation and the ability to absorb and release moisture. Scales on animal fibers also serve as miniature hooks where neighboring fibers can latch onto one another, resulting in strong, durable yarns. Fibers from plants such as linen (E) and cotton (F) as well as silk (H), cellulosic fibers, and synthetic fibers do not have scales. Their smooth surfaces give rise to different properties. For instance, the lack of scales permits neighboring fibers to readily slide past one another with little resistance or tendency to latch on to one another. This results in yarn that has little springiness and feels slippery in hand. The ability to explore the microscopic structure of natural and laboratory-made fibers combined with experiential observations of makers who create garments using yarn made from these fibers permits us to develop a more complete understanding of how the small-scale structure of yarn fibers contributes to large-scale properties of an entire garment.

What is particularly interesting is that a strand of yarn on its own is largely inextensible; however, entire garments made from the yarn can exhibit tremendous stretchiness. How is it that the manner in which the yarn is manipulated on knitting needles transforms an unstretchable collection of fibers into a flexible garment? When creating a garment, a knitter uses two or more needles to create rows of interconnected slipknots with yarn. There are two basic stitches, the knit stitch and the purl stitch, from which a multitude of patterns can be built. A knit stitch pulls a loop of yarn from the back of the fabric through another loop to the front (see A on previous spread), while a purl stitch pulls a loop of yarn from the front of the fabric through another loop to the back (B). Each stitch is interlocked with its nearest neighbors within the row as well as to the stitches in the rows immediately above and below it. Knit materials are prized for their ability to stretch and return to their original shape, a property known as elasticity. A portion of a garment's elastic behavior comes from the small-scale properties of the yarn fibers, but the choice of stitch pattern dictates how the loops of yarn are joined together, which plays a significant role in the overall elasticity of the garment.

When the same yarn is used to create different fabrics, quite different elastic properties are observed in the resulting garments, confirming that pattern choices at the stitch level (intermediate scale) significantly influence the large-scale properties of the garment. A stockinette fabric, often used for sweaters, hats, and mittens, is created by stacking rows of knit stitches (A). Turning a swatch of stockinette fabric over reveals a reverse stockinette, which

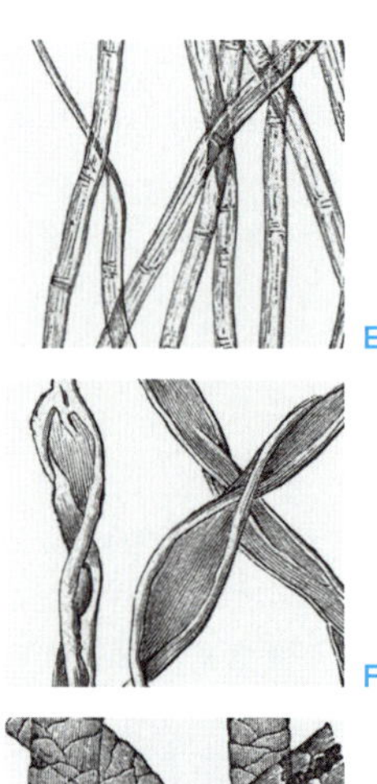
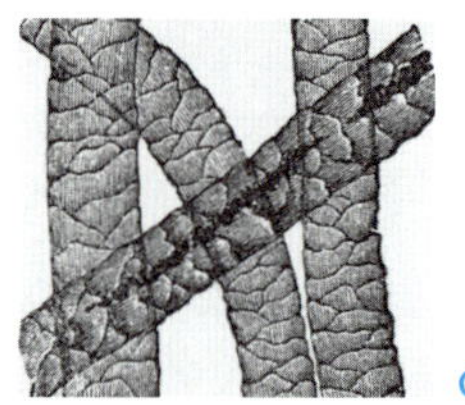
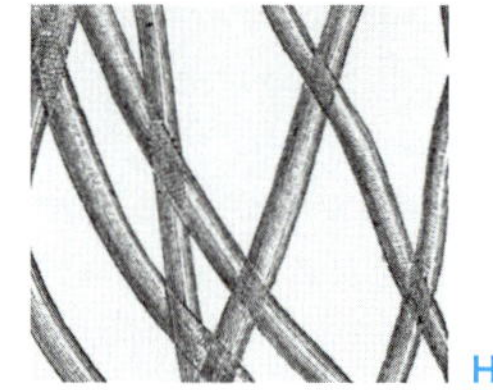

Magnified natural fibers from (E) linen, (F) cotton, (G) wool, and (H) silk

could also be created by stacking up rows of purl stitches (B). These fabrics stretch quite a bit in the horizontal direction (along the rows) but not in the vertical direction. Garter fabric, often used for scarves because the pattern is reversible, is created by alternating rows of knit and purl stitches (D). This fabric has notable stretchiness in both the horizontal and vertical direction. A ribbed fabric is created by alternating columns of knit and purl stitches (C), resulting in a very stretchy material that is often used in sock and sweater cuffs.

Elisabetta Matsumoto, a physicist, applied mathematician, and lifelong knitter, leads a research lab at the Georgia Institute of Technology that explores the mathematics and physics that underlie these interesting elastic properties of knit garments.[4] Matsumoto's experiences in fiber arts provided the spark for her scientific research and an experimental testing ground for her theoretical models.

Her research uses theoretical mathematical models and experimental physics to explore how properties related to stitch formation at the intermediate scale influence the large-scale elasticity of knitted garments.[5] The theoretical work appeals to knot theory, a subbranch of topology, to classify the entanglement of stitches in three-dimensional space with the goal of identifying a knot description for each knittable stitch in textiles. Because different knots possess different properties, the ability to classify all knittable knots would permit designers to choose stitch patterns based on the desired properties of the final garments.

Matsumoto's research group has also developed constitutive relations for knit fabrics that predict how fabrics deform under applied stress. The validity of these models is assessed by comparing predictions with experimental stress-strain measurements on a variety of fabrics to measure and differentiate their large-scale elastic properties. In Matsumoto's laboratory, art and science coexist side by side, and one could not flourish without the other. Her early exposure to knitting and the complex fabrics that she learned to create fuel her research interests and provide insights into her mathematical models; in turn, her research is now leading to advances in the design of sophisticated textiles and related materials. The long-term goal of this work is to be able to design programmable soft materials—from high-performance athletic wear to artificial tissues that stretch and twist with the body as movement demands— where specific design choices at the stitch level enable the desired elastic properties in the finished garments.

1 "The History of Hand-Knitting," Victoria and Albert Museum website, accessed November 23, 2022, www.vam.ac.uk/articles/the-history-of-hand-knitting.

2 Clara Parkes, *The Knitter's Book of Yarn: The Ultimate Guide to Choosing, Using, and Enjoying Yarn* (New York: Potter Craft, 2007).

3 Ibid.

4 "Geometry of Materials," Matsumoto Group website, accessed November 23, 2022, matsumoto.gatech.edu.

5 Shashank Markande and Elisabetta Matsumoto, "A Study of Two-Periodic Knitted Fabrics Using Tangles," *APS: Bulletin of the American Physical Society* 65, no. 1 (2020); Michael Dimitriyev, Krishma Singal, and Elisabetta Matsumoto, "Constructing a Constitutive Model of Knitted Fabric," *APS: Bulletin of the American Physical Society* 65, no. 1 (2020); Krishma Singal, Michael Dimitriyev, Elisabetta Matsumoto, "Stress-Strain Studies of Knitted Swatches," *APS: Bulletin of the American Physical Society* 65, no. 1 (2020).

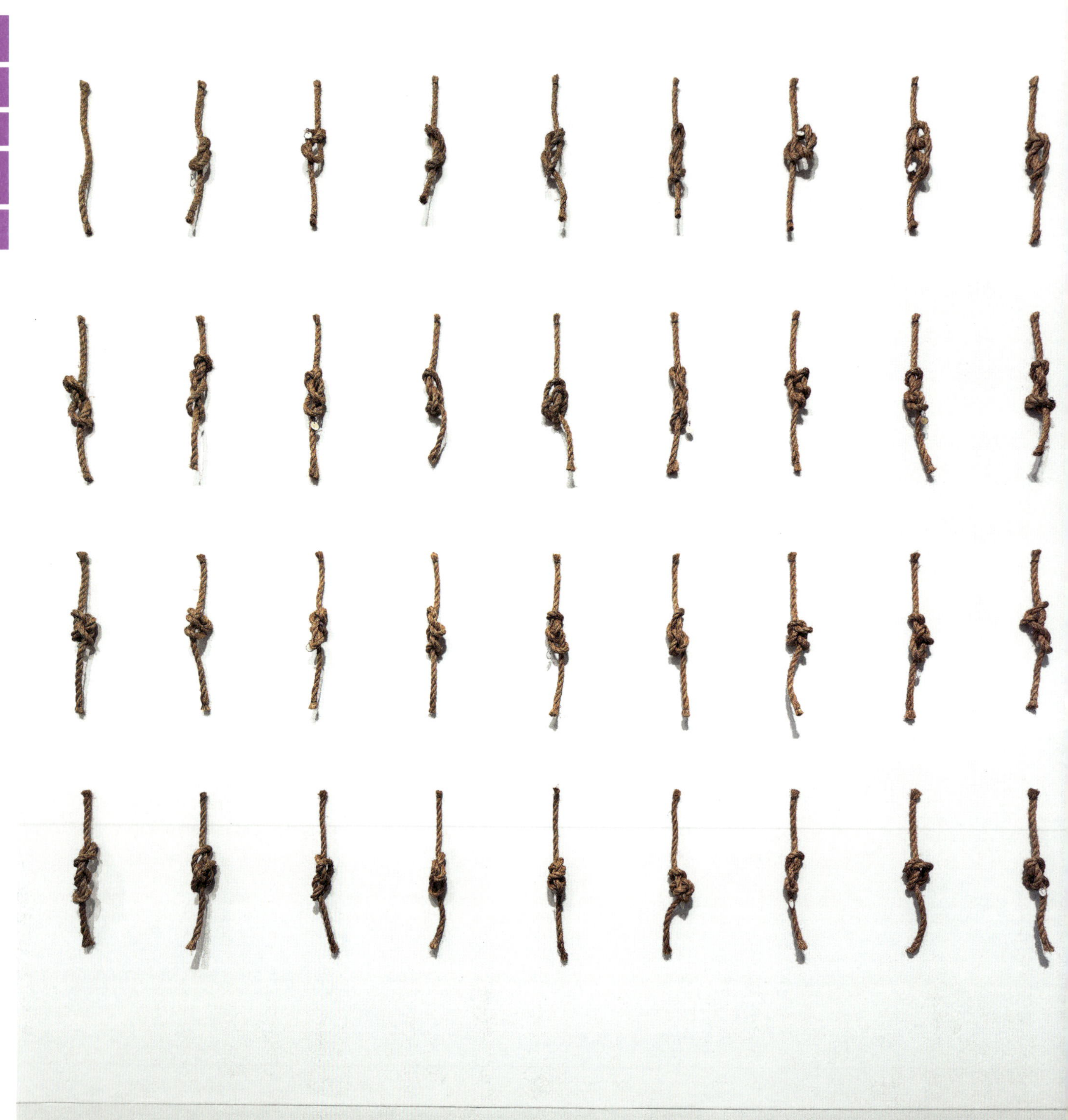

MAKING VISIBLE :
MATH, CRAFT, CULTURE

In Conversation: John Sims,
Jeffrey C. Splitstoser, *and* Daina Taimina
with Stephen Ornes

STEPHEN ORNES: I'm a science writer. I'm particularly interested in writing about the intersection of math and culture and society. This panel is such an exciting way to explore those themes. John, could you start by telling us about your work?

JOHN SIMS: I'm a math-artist, writer, curator, and producer. **Math and art share the language and the process by which to see the world in incredible, impactful ways—in different ways, but also in intersected ways.** In *Square Root of a Tree*, the tree structure on the left is called the square-root tree.[1] In some ways, this is a metaphor for how mathematics can support the root structure, the tree, the drawn tree, the art of the tree. Then, to the right, we do a 180 and we see how art can support the math process, the search-for-structure process; that's called the tree root of a fractal. And in the middle, we have an averaging, and we change the metaphor from a tree root to a right brain/left brain, and now we have the math, our brain. This triad speaks to the ways I look at mathematics and art— in terms of how math can encode the art process and how art can help liberate and express the wonderful world, the language, the reality of mathematics. And right in the middle is this space that gives you a certain kind of consciousness and hypervision to see the world in an incredible way.

For a long time, I was interested in pi—being able to see pi and think about pi. In *Seeing Pi*, I drew the number pi in a rectilinear way, starting in the middle and going toward the outer edges. I wanted to see what the numbers of pi look like, so I color coded each number, and I created a digital image. When I was first working on this, I made some mistakes, and I had to decide whether I wanted to just wave my hand and call it art. But it became important that I be as correct as possible. **The mathematical tendency is to speak with a certain accuracy and precision and to journey toward truth and perfection. In some ways, perfection is a very big part of the math.**

At some point, I wanted to move beyond the digital, so I reached out to Amish quilters in Sarasota, Florida, to help me visualize and manifest these ideas in a craft context. And I learned how to make a quilt. I ended up creating *Seeing Pi* as a quilt. Then I started thinking about pi in different bases. *Seeing Pi* is a base 10, but then I moved to base 2, so we get black and white, white and black. And then I was thinking of pi in base 3, so I color coded those: red, white, and blue became *American Pi.* Black, red, and green became *African American Pi*; I have another piece called *Civil Pi Movement.* Because I'm able to change the bases, I'm able to connect with a certain sociological narrative around communities and identity—Black and White, American and African American— and create this new metaphor about connections.

These one-sided quilts became part of a system of large, square quilts. A thirteen-quilt piece is called *Hyperquilt.* On the outer ring, you have all the pi quilts; the inner ring is a self-portrait. On the diagonals, you have different visualizations of a Pythagorean triple. And then, in the middle, is a quilt that is made out of African patterns and structures—different textiles that I got from Ghana. So, *Hyperquilt* became a superstructure that spoke to these ideas of pi, Pythagorean triples, self-portraiture, and going back to the original space of Africa right in the center.

For *The Hanging of Knots up to Eight Crossings,* I was motivated by the work of Sol LeWitt, who was a mentor of mine. I've always been interested in his conceptual artwork, particularly *Variations of Incomplete Open Cubes.* He was thinking like a math person in terms of classifying, writing out all the possible ways of looking at this particular ques-

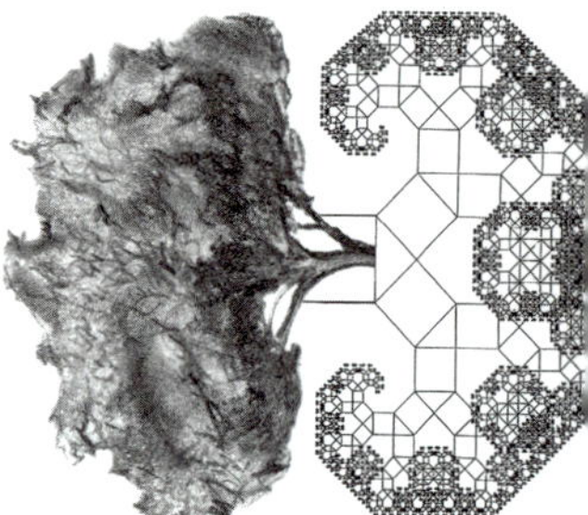

John Sims, *The Square Root of a Tree,* from the *Square Root Tree* series, c. 2005, drawing

Sol LeWitt, *Variations of Incomplete Open Cubes,* 1974, drawing

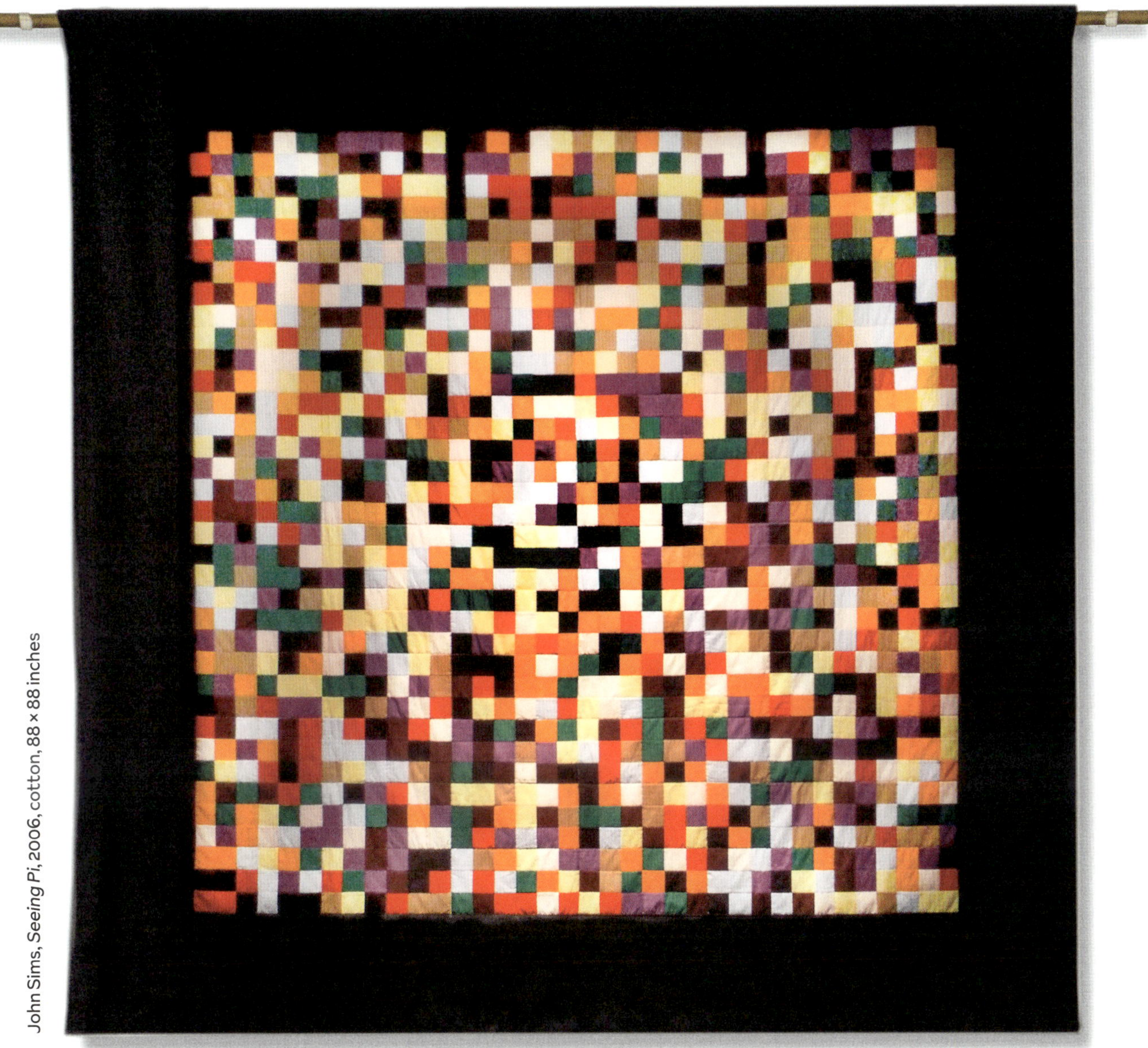

John Sims, *Seeing Pi*, 2006, cotton, 88 × 88 inches

tion, and graphing it in some way. In my work, I look at prime distinct knots, up to eight crossings. Of course, when I cut them, they became more like tangles. It became almost a kind of Sol LeWitt tribute piece as well as connected to my interest in looking at the structure of the hanging noose, and looking at knots in general.

JEFFREY C. SPLITSTOSER: I'm an archaeologist, and my experience with math, craft, and culture involves fiber-based objects from ancient Peru, with a focus on quipus [also spelled *khipus*], which are colored and knotted cords that served as an information system in the pre-Columbian Andes. Quipus are still used today in some parts of the Andes, and we know from firsthand written accounts from the sixteenth and seventeenth centuries that quipus were capable of recording just

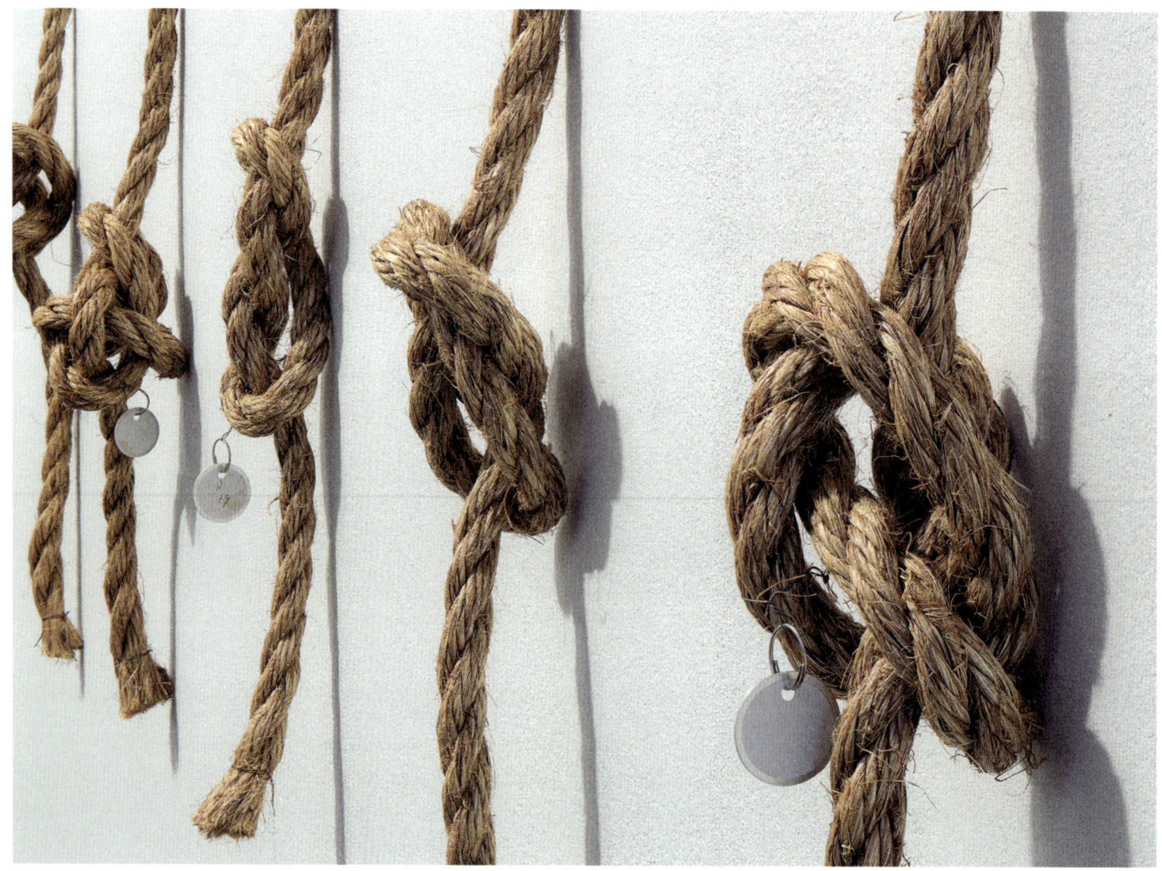

John Sims, *The Hanging of Knots up to 8 Crossings, A Tribute to Sol LeWitt* (detail), 2007, knotted rope, dimensions variable, 120 × 144 inches (*Radical Fiber installation*)

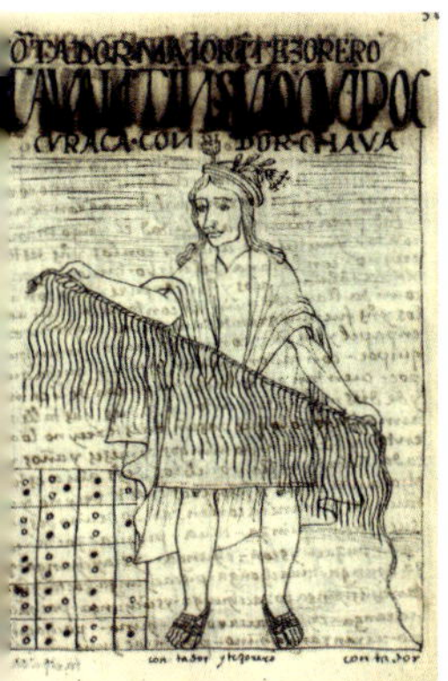

...lipe Guamán Poma da Ayala, ...awing of chief accountant and trea-...rer, *Tawantin Suyu khipuq kañaka*, ...thority in charge of the knotted ...rings, or quipu, of the kingdom.

about anything that could be written. The Inca recorded their laws, histories, and religious practices, as well as statistical information, such as censuses, tribute information, and the recording of time, on quipus.

An early seventeenth-century drawing by an Indigenous governor named Felipe Guamán Poma de Ayala of the chief Inca accountant and treasurer in Cusco, the Inca capital, depicts a *yupana,* which is an abacus-like device that was used to perform addition, subtraction, multiplication, and division with perfect accuracy—at least that's what the Spanish tell us. The results of yupanic calculations were then knotted into quipus.

I began making quipus in order to formulate research questions that aim to infer the cultural values and agency of the individuals who made them, called *quipukamayuqs,* as well as of the communities within which they operated and lived. In other words, I wanted to know what motivated the quipu makers and how their communities influenced them. One quipu I made is a pseudo-replica of an actual quipu that anthropologist Sabine Hyland, a colleague of mine, studied from a town called Collata in Peru. This quipu was probably a missive sent by a man named Topa Inca Yupanqui during the 1780–83 Inca rebellion against the Spanish. It served as a letter; people could read these. Hyland discovered that the last three cords in this quipu had syllabic meaning, in that the colors literally represented syllables that spelled out the name of the community. **By making quipus like this one, I've learned many things that I could never have learned any other way.** My work includes growing, processing, and using my own cotton and dye plants to make and dye quipu cords.

DAINA TAIMINA: I made my very first hyperbolic plane piece twenty-five years ago. Since then, I have crocheted a lot of these—at first, for teaching purposes, but then, just to go further and further with the form. In 2005, I was first invited to participate in an art exhibition, so I started to think about what else I could do with these forms. My husband was a topologist, and he liked to play with them. We took these hyperbolic planes, and we made them into various forms. In topology, these forms are called manifolds. It is fascinating how open hyperbolic planes can be configured to make different shapes.

I read the philosopher Gilles Deleuze's book *The Fold*,[2] and there were particular quotes that I liked and I thought they described those manifolds we created. For instance, Deleuze wrote: "The problem is not how to finish a fold but how to continue it, to have it go through the ceiling, how to bring it to infinity." He has also written: "I am forever unfolding between two folds, and if to perceive means to unfold, then I am forever perceiving within the folds." I am forever with these hyperbolic planes.

I also started to give talks and workshops for general audiences of various ages. As participants of the *Saratoga Springs Satellite Reef* have said, craft projects involve the community and offer a way of connecting with other people. In 2014, I was invited to participate in a show during the European Capital of Culture festival in Riga, Latvia, organized by the RIXC, The Center for New Media Culture. It was my largest exhibition to date in which I was showing all of my work. We also did a project with the community, *The Cloud of White Thoughts*. We asked people to make white hyperbolic planes. I am Latvian, and in Latvia, *white thoughts* mean good thoughts, encouraging and positive thoughts. It's a very poetic idea. The idea in the call for participants was that to find encouragement, you need to look inside yourself; likewise, don't create the hyperbolic plane from something store-bought, you have to find this white thread in what you already have. As a result, the work features various repurposed materials like recycled yarn, strings, even shredded T-shirts. The installation process connected me with a lot of people, and I visited some of the participating communities. The most emotional was a visit to a village for blind people. They were crocheting these crenelated forms, too. And they said something like, "Thank you for including us, because nobody ever thought we could be involved in an art project."

In 2019, I was invited to participate in the Riga International Biennial of Contemporary Art. It was a very hard time for me because I had just lost my husband in a tragic accident. I didn't want to do anything, but my family

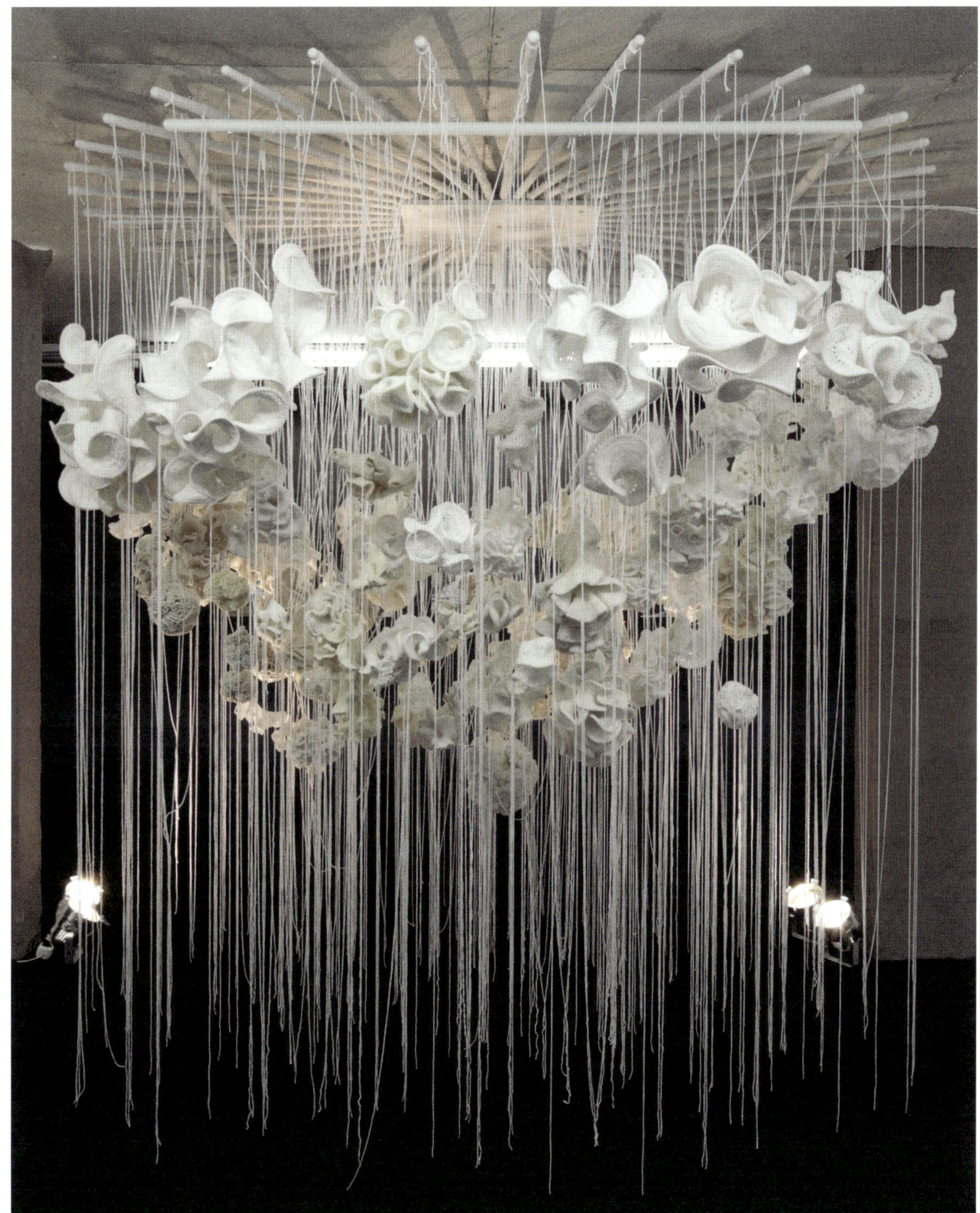

Daina Taimina, *Cloud of White Thoughts*, 2014, crocheted yarn, dimensions variable, installation view, Kim? Contemporary Art Centre, Riga, Latvia

encouraged me. So, I decided I was going to make a work in his memory. *Dreams and Memories* is my largest installation to date, and, of course, it included other people. I put out a call for entries that read: *If we have a dream, it can't get lost. It could be that a dream just turns into a memory. But, nonetheless, that dream made us more secure and stronger and helped us move forward. What was a dream yesterday may turn into a reality tomorrow. As Albert Einstein once said, 'The person with big dreams is stronger than the one with all the facts.' And it's okay if some dreams are not fulfilled, but instead are turned into memories. Memories are the collection of our life experiences. They hold our successes and losses, joy and sorrow. They, like glue, hold our lifeline together. With years passing by, our memories might fade, but they never age. We cannot always express our dreams to others with words. It may be easier to give our dreams or memories some tactile form through crochet—to repurpose the scarf that still keeps a loving memory of a beautiful moment or some hidden tears . . . We will join this all together, each piece in a single color.*

People joined in from all over the world because I posted this call on social media. In 2019, in the fall, I went to Latvia and traveled around, and I deliberately decided to visit marginalized communities, including several schools for children with learning disabilities. The students were just so surprised that they were invited, and they wanted to participate. Those were very memorable experiences; it really meant a lot. I wasn't allowed to go into the women's prison myself, but I reached out to the women there through a

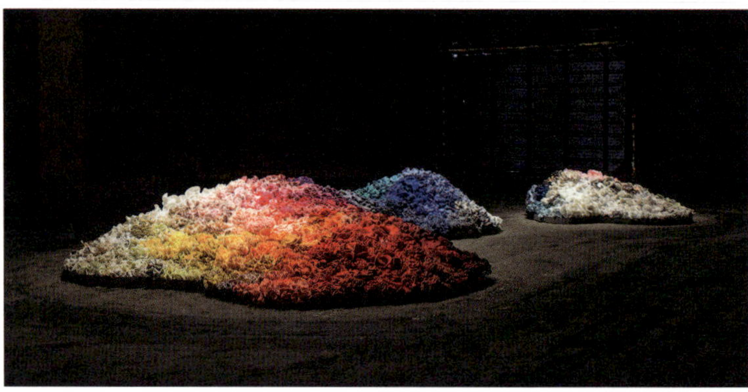

Daina Taimina, *Dreams and Memories*, 2020, crocheted yarn and reclaimed plastic and fabric, dimensions variable, installation view, 2nd Riga International Biennial of Contemporary Art, RIBOCA2, Latvia

friend who worked with the minister. With submitted pieces, some people were also sharing their stories.

In 2020, I was ready to go to Riga and do the installation. I had an idea about how I was going to make it and how it should be displayed. But then, of course, COVID happened and all those plans fell apart. So, instantly, my dream became a memory. Instead of being five months, the biennial was happening for just three weeks. The curator, Rebecca Lamarche-Vadel, decided to create a movie, called *and suddenly it all blossoms*, to show all the works. *Dreams and Memories* is my largest work, and I have never seen it myself. It was not installed the way I imagined. My first reaction when I saw it was, yes, well, this is a pile of my dreams, which has turned into memories.

SO: I hadn't seen the piece with the dreams before. That is incredible. It strikes me that you found that non-Euclidean geometry somehow creates a language for talking about dreams and things that are maybe harder to communicate. One common theme that we've heard is that math, in some way, allows for a kind of communication that might be surprising to a non-math audience. John, you used math to create quilts; quilting has a rich cultural history, and you've been able to reach certain communities. Daina, you've been able to reach a variety of communities and bring them together. Jeffrey, you've been able to connect with ancient cultures and get an inside look into how they communicated.

I wonder if we could go a little deeper into the idea about encoding or decoding. Both John and Daina, as math-artists, can encode mathematical ideas into their art. Whether it's number theory or geometry, it becomes the language that you use for your art. And Jeffrey, in a sense, you are doing both. You're decoding by examining the knots and, when possible, translating them into a language that we understand. But then, by creating quipus, you're also taking what you've learned and encoding. What are your thoughts about this idea of math as communication, that we're encoding or decoding—truths may be too big of a word, but at least information—to reach broader communities?

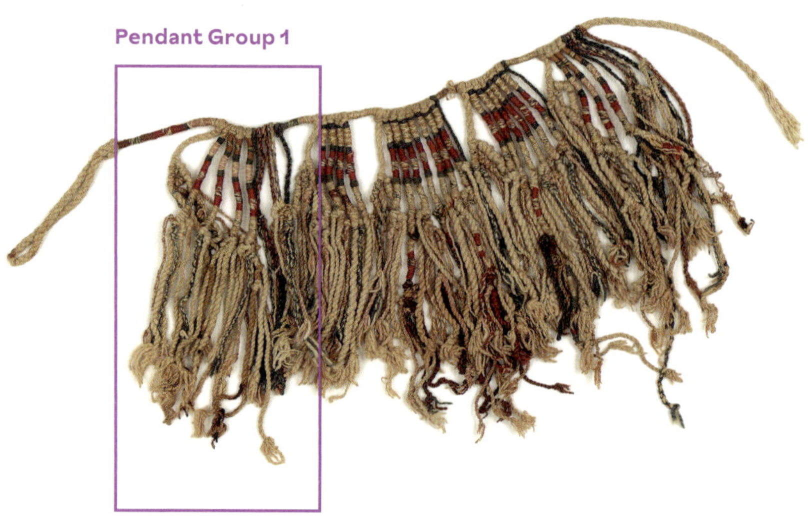

Jeffrey, why don't you start by talking about how math figures into the quipus. We heard you talk about the counting and the strings, but there is also this richness to the spinning itself and to the creation of threads.

JCS: Math can be found in a few different places in quipus. One of the most interesting to me is the number of yarns that makers plied together to make these cords, because there are patterns in them. When you're spinning or plying, you can do it to the left or to the right, making S or Z yarns. The terms don't matter—it's just that they're opposite of each other. Most of the time, when you're spinning, you spin in one direction, you ply in the opposite direction, and then you re-ply again in the opposite direction. So, if you spin S, you ply Z, and then you re-ply S. But in quipus, they often combined both spins in a single yarn or they plied in the same direction twice in order to create these complex numbers in the cords that are not even visible.

The person who made the quipu knew that these numbers were there, but they weren't something that would have been "read." This is a different kind of communication, maybe a spiritual type of communication, as opposed to the most visible part of quipu, which was used to record mathematical, statistical, and other quantitative information in knots. We understand how the knots worked in most Inca quipus. The Inca

Left: Wari, wrapped-main-cord quipu with color seriation, 600–1000 CE, cotton and camelid fiber
Right: Diagrams by Jeffrey C. Splitstoser showing

118

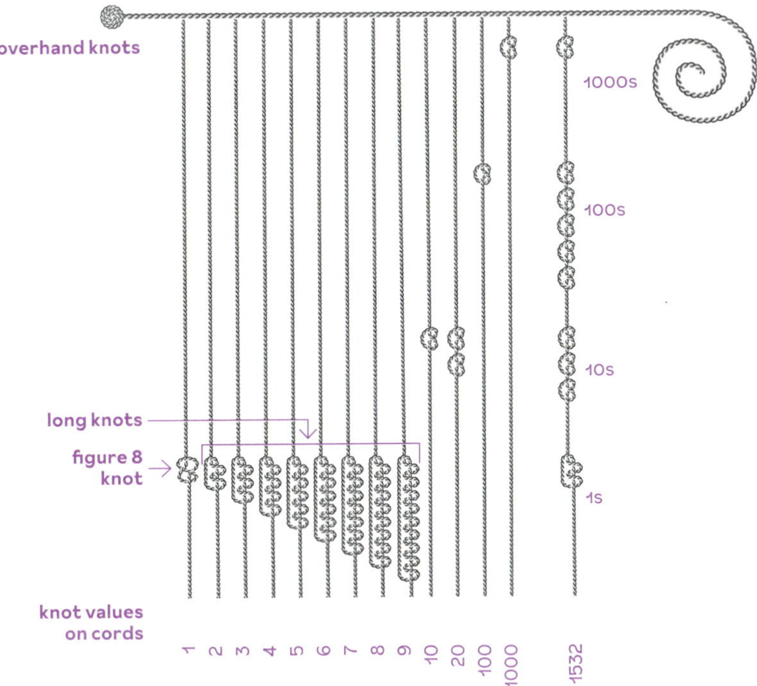

Diagram of Pendant Group 1

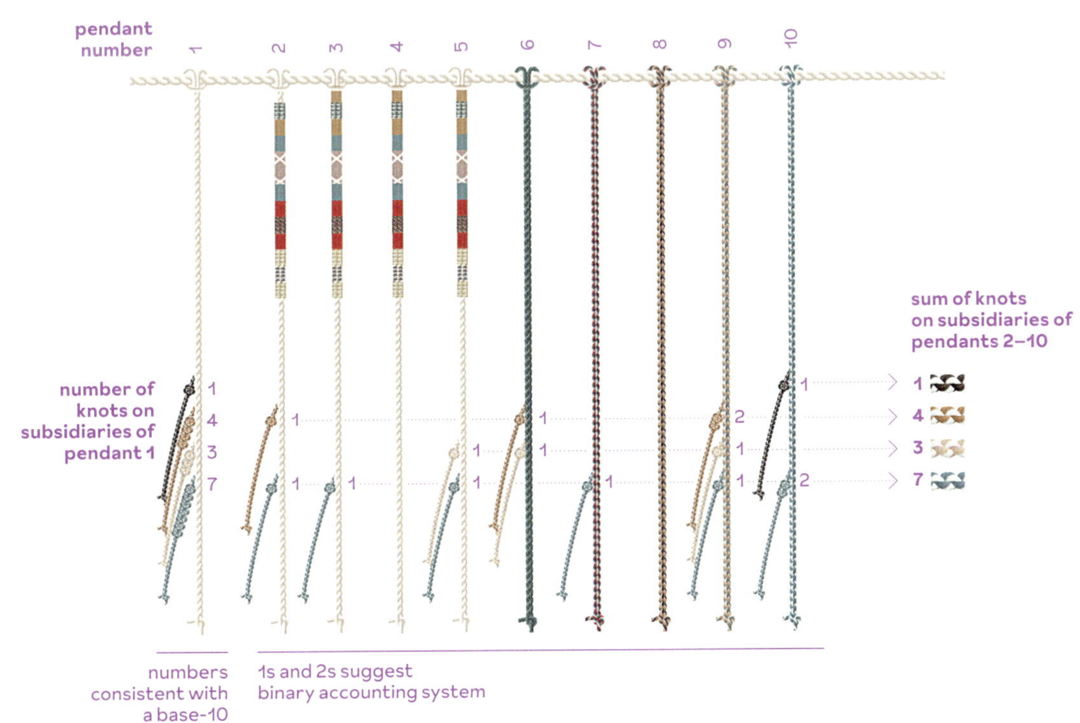

used a base-10 positional system, which means that the position of the knot in the cord determines its numerical value. The lowest register has the ones, the next has the tens, then the hundreds, and the thousands. They would sometimes have cords that stuck up from the top that would summarize the numbers in the lower cords.

My work is primarily with Wari quipus, which were the very first quipus made. The Wari forged South America's very first empire, which was around 600 to 1000 CE, making them more than one thousand years old. Wari quipus are notable for their bright colors and wrapping on their pendant cords. But they used a different knotting system than the Incas. Inca cords use a positional system of ones, tens, hundreds, and thousands, but Wari cords don't; the knots are always tied right after their attachment to a cord, and they don't function the way Inca quipus do. We don't quite understand how Wari quipus work, but that is something I am working on, and hopefully, the process of making them will help me better understand how they functioned.

SO: Daina, what do you think about encoding? What are some of the ways that you encode math into your art?

DT: It all started with making math visible. I started crocheting hyperbolic planes because I wanted to get to a picture of lines on them. The idea of hyperbolic planes is the opposite of the Euclidean plane. For example, Euclid's fifth postulate states that if we have a point outside the line, we can draw one, and only one, line parallel to the given line. In a hyperbolic plane, if we have a point outside the line, we can draw through this point an infinite number of parallel lines to the given line. As a student, I used to ponder how to see that, so I made these tactile hyperbolic planes, which I can fold like a piece of paper and create a line, then I can fold these other lines and see why they do not intersect. So, it raises a question: what does it mean to be parallel? One of the meanings is not to intersect.

The precursor of these manifolds was a so-called hyperbolic *pair of pants*. I recently went through some old videotapes and in one from 1988, my husband, David Henderson, was teaching geometry, and he was expressing that he would

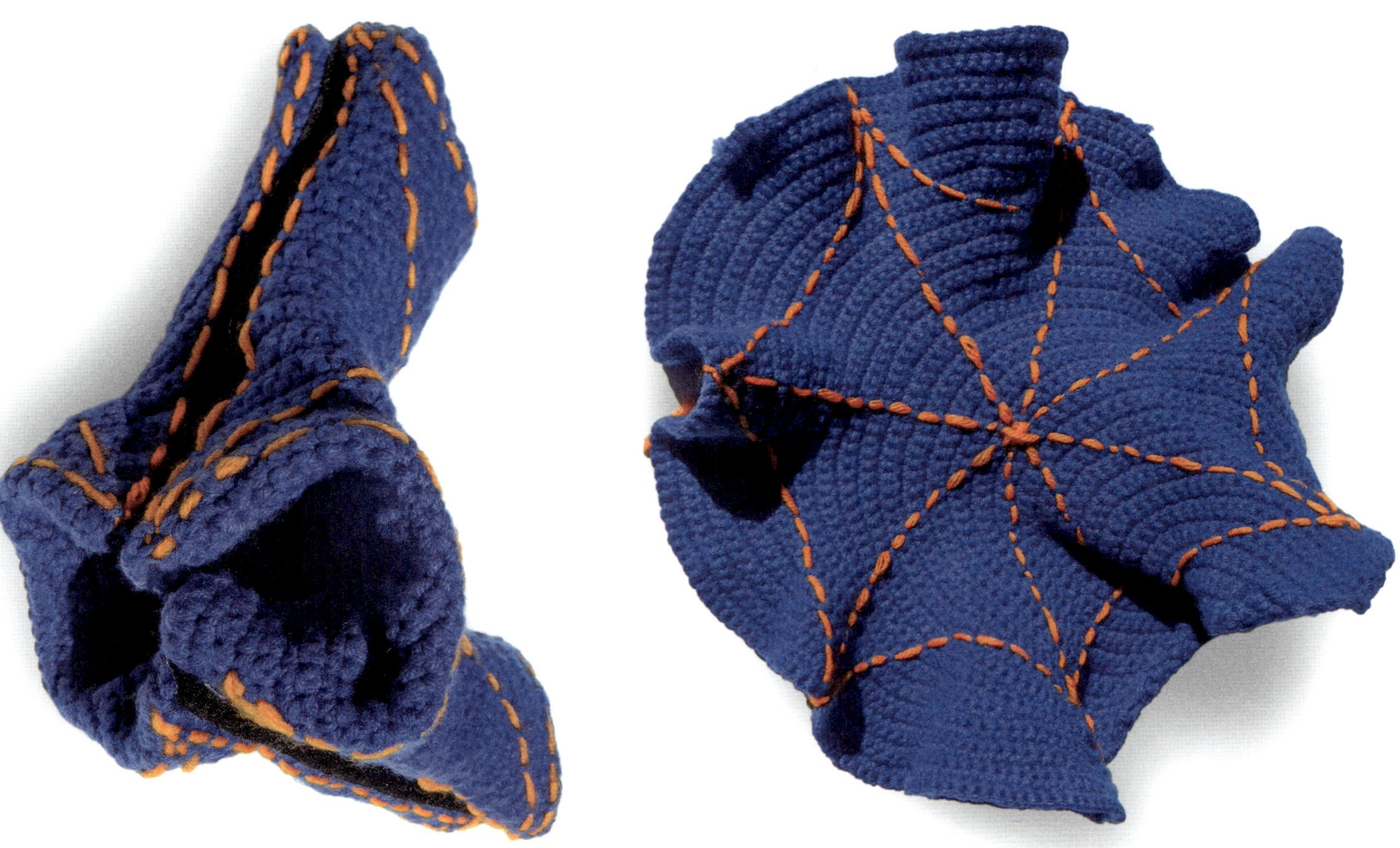

14 × 14 × 1 inches (open), 14 × 7 × 4 inches (closed)

love to see a hyperbolic plane that could be wrapped around a two-holed torus. So, after I made my very first hyperbolic plane, he said, "I want to have hyperbolic pants." How they are done: we know that an ordinary stop sign is a regular octagon. The hyperbolic pants could be folded from a hyperbolic stop sign, which is a regular octagon with 45-degree interior angles. The sides of this octagon can be glued together—mathematicians say "identified"—so I can identify these two sides; they are the same. Then I can identify two other sides, and so on. So, I have two holes here, two holes here, and now, these four created holes are the same. If I had a fourth dimension—because I can't do it in three dimensions—I could identify that this hole goes here and the other hole goes here, and then I would have a pretzel, or two-holed torus. Through the process of making these objects I am continuously discovering what other properties of hyperbolic geometry I can represent in a tactile way.

SO: As someone who studied math and went through that struggle of breaking the rules of Euclidean geometry and talking about Euclid's fifth postulate and wondering what happens if a triangle doesn't follow the Pythagorean theorem and ideas like that—that to be able to hold it, or even just to see what it means for there to be infinitely many parallel lines, I feel like it's just such an expansion of the way we can learn that concept. It's such a useful application.

All three of you create objects that, if we walked up to them in a museum, we would see them as art. And it's only through this deeper discussion that we know there are these rules—non-Euclidean geometric rules, that the knots are communication, or that pi is encoded into these beautiful quilts. John, can you talk more about math in your work?

JS: I want to read a quick passage from a catalog essay I wrote: *Mathematics is a parameter of human consciousness as an indispensable conceptual technology essential in seeing beyond and knowing beyond the intuitive. The language and process of mathematics, as elements of and foundation for art, informs in an analytic expressive condition that inspires a visual reckoning for a convergence: from the illustrative,* *to the metaphysical, to the poetic. And in the dialectic of visual art call and text performative response, there is the interdimensional conversation with a twisting structure of language; vision in human ways gives birth to the spiritual lattice of a social geometry, a community constructivism, a place of connection, where emotional calculations meet structured abstraction.*

Math provides this incredible infrastructure that is able to almost bore itself into various art forms—music, quilt making, dance, conceptual artwork. The math provides a resource, like a mineral, like some kind of ingredient, that gives you insights and methods and processes to be able to experience and see beyond the sensory elements that are in front of you. And I think that's very, very important.

I want to be able to see pi; I want to be able to see and feel the movement of these knots, the growth of their complexity as I go from two crossings to eight crossings. Knowing that things may look the same but be structurally different, or that things may be structurally the same but are manifested in different ways visually—it's very important to be able to have that kind of space to explore.

The mixture of math and art is interesting. If the mathematics is strong, then it becomes visual mathematics. I can draw a circle, and in one context, that's considered art; but in another context, it might be understood that I'm graphing something or illustrating a mathematical property, so it's visual mathematics. On the other end of the spectrum, the circle or the geometrical structure can be a hidden aspect that may inform a portrait or may inform something that is complex on another level.

Going back to Daina's work with crochet, the metaphors are important because that's another mapping system, which helps elevate our sense of consciousness about being able to see in different ways. And math provides an incredible opportunity to do that.

SO: Let's follow up on how the tactile experience can go beyond the visual. There are sensory mathematics, a way to learn things through touch. You each deal with something you can touch and feel as an integral part of your work. How does touching something help amplify the complex ideas that you want to either pull out of something or encode into your work?

DT: It is one thing when you're talking formally about hyperbolic geometry and it is another when you're describing the differences between three possible geometries on two-dimensional surfaces—in other words, planar, spherical, and hyperbolic. For example, on a plane, when you add up the interior angles of a triangle, you always end up with pi or 180 degrees. On a sphere, the sum of the interior angles of a triangle will always be larger than pi. And on a hyperbolic plane, it will always be less than pi; it can even approach zero. And you might ask, how can a triangle have interior angles that add up to zero? Well, one of my models shows a triangle: it is tactile, and we can fold it along the lines I've stitched and see the angle approaching zero. This concept is called an ideal triangle on a hyperbolic plane. And if the hyperbolic plane has a radius of 1, then the area of that triangle is pi (π). So, if the radius is R, then the area of this triangle is πR^2. The tactile model helped my students when I was teaching that concept—they could really touch it.

If you take a hyperbolic plane, and you fold it to make a regular polygon with all right angles, then the smallest one will be a rectangular polygon. On a sphere, you can make a regular triangle with all right angles; in a plane, it will be a familiar square shape. When I show this sequence to people, many times, I get the response, "This is magic!" But this is a tactile thing. You really construct it. You really show it. And they say, "Yes, this is something I can fold. And I can't believe my eyes, but this is what is happening."

SO: That's key: "I can't believe my eyes," but because you feel it, you know that it is a physical reality. John, have you always been working with your hands?

JS: The hands are very important, but you also have to fight against the hands. On a math level, there is this sense of perfection. We can have a perfect circle, we can have a perfect square, we can have this kind of perfect reality. The price you pay to come into the physical reality is that you have to give up perfection.

This relates to issues around memory and copying. For instance, even writing out the numbers of pi—that's a physical, tactile thing, but if I don't watch it, I'll make a mistake. If a computer did it, that would be different. But if I'm going to do it and I'm going to really get into the paper, and the writing, and the paying attention . . . Maybe the humanity lives in the imperfection.

So, what happens when I do make a mistake and I have to pull back? It is not perfect like it would be in the mathematical reality. How do I negotiate that anxiety? I think that's a real thing in terms of moving from the math space to the art space. The hands-on element may bring us certain knowledge and experience, but it can also bring a certain level of contamination, too. This is the human element—we bring our own level of error to the table.

> **DT:** I would object to the idea that mathematics is all about perfection. This reminds me of my graduate studies, which were in theoretical computer science. I was working with problems on various automata and Turing machines that create infinite words. The question was how to know this infinite word—how to have an answer before I reached the end. It turned out that one possibility was to toss a coin. It wasn't perfection; it was using probability to make a decision.

JS: I totally get that. But I'm talking in terms of, for instance, the idea of a square or a circle.

> **DT:** But it's only an image because this is what we are accepting . . .

JS: No, no, the *idea*. The idea of a circle can exist in our mind as a perfect object. But if you try to draw it and if you look hard enough, you can really see how imperfect it is. Also, and maybe more importantly, we witness in mathematics how the *equal* sign or the idea of equality is connected to perfection. For example, the area of any perfect circle is perfectly pi times that circle's radius squared. I am also speaking of perfection in this sense.

> **DT:** My hyperbolic planes are also not ideal. They are not precise models; they are approximations. But still, we can

illustrate the mathematical idea. **We don't need perfection to understand the general idea.**

JS: I agree: we don't need perfection to get the general sense—that's absolutely true. But I think that the beauty of a mathematical reality is that there is a certain level of distance from the physical reality, which allows certain things—such as a circle, or a triangle, or whatever processes—to be perfect in the context by which they are defined. But when you start to create, you're going to have a natural error factor. And we can live with that. But some cultures deal with that level of imperfection in very different ways. Some people round off in various ways and are okay with it emotionally. Other folks, some communities, may be completely anxiety-driven by something that is off a bit. You'll see that in cultures that generate certain kinds of engineering. For instance, when I studied math in Germany and worked in a German factory, I observed that the culture is very driven toward a certain precision and journey to perfection. How close you get to that perfection is part of your culture on some level.

> **DT:** Jeffrey, can you talk about that example from the Andes? How is it that 1+1 can be 3?

> **JCS:** In the Andes, culture is driven by duality—upper and lower communities—and they see everything as paired. And so, if you have 1+1, representing a husband and a wife, they see that marriage as a third component. But yet, the man and the woman have separate identities. In this respect, they see 1+1 is equal to 3. They also see it as equal to 2, because they count to 2 like we do, but they conceptualize it differently.

JS: No one wants to romanticize the fact that mathematics is completely absent of messiness. But I think there is this impulsive desire to incubate things that can grow in near-perfect ways. And then, as we begin to realize them in our own physical reality, we see there's a major gap. Even with computers that copy, you have to make sure that they're not making mistakes.

> **SO:** Let's come back to the tactile, and what that brings to your scholarship and your craft.

Attribute	Decision/Operation	
Material	cotton	camelid
Color Class	bright/shiny	dark/dull
Cord Attachment	recto	verso
Knot Twist	Z	S
Cord Final Twist	S	Z
End-Knot Twist	Z (recto)	S (verso)

JCS: Daina and John are coming at math and tactility from the aspect of making things, embedding it into their creations—and in a very interesting and innovative way. I'm looking at it the other way; **I'm trying to see how tactile information can help me understand a culture.** Color in these quipu cords represents categories of information. For instance, a colored cord might represent a person, or a group of people, or llamas, or alpacas, and then the numbers on those cords represent the quantity. Typically, cords document an obligation. In other words, if a color represents a person, then the number represents the obligation that this person had to the community.

Now I want to explain how the tactile information can work. I work with Wari quipus, which are more than one thousand years old. They are all in museum collections and are strictly off-limits in terms of touching. I can't touch them; I have to show up with gloves on. But Sabine Hyland, who I mentioned earlier, came to this town, called Collata, where she studied a seventeenth-century quipu and actually held

127

it, not wearing gloves. The quipu's caretakers—who are the descendants of the people who made them, and who have kept this quipu for hundreds of years in their community—told Sabine that she had to study it with her gloves off. And that's because the quipu itself is made of multiple animal fibers, including llama, alpaca, guanaco, vicuña, viscacha (which is a rodent), and white-tailed deer. All of those fibers are in this quipu, but the cords all look the same macroscopically. In other words, if you look at one of these cords, you can't tell what it is made of. Yet, each fiber has a different feel. And each fiber carries haptic information, based on its feel, and Sabine was taught how to distinguish each of these fibers by its feel. The Inca were using feel in addition to color to represent categories of information.

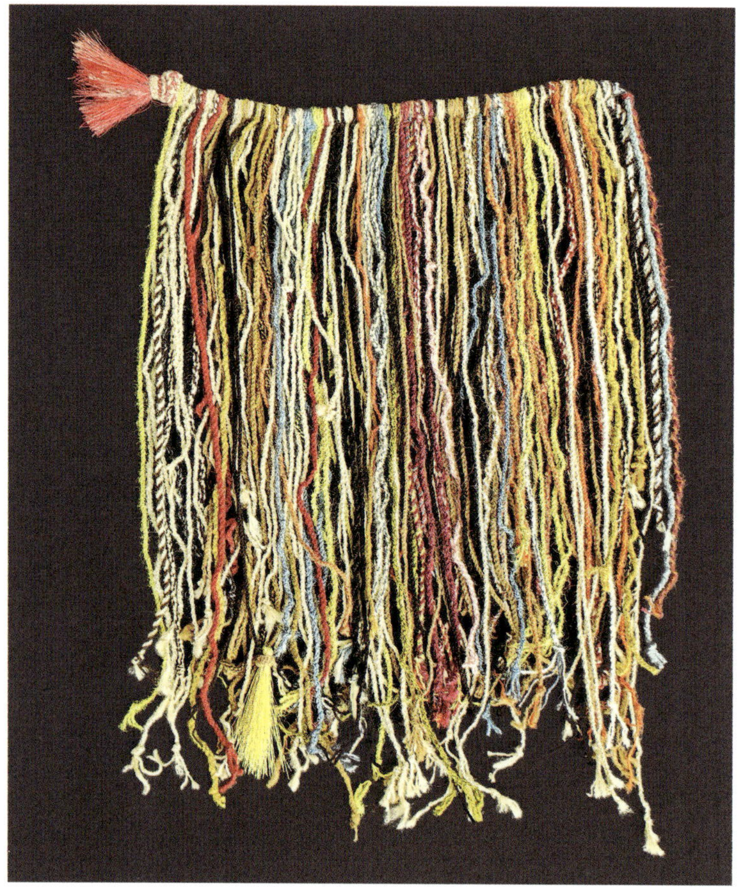

Jeffrey C. Splitstoser, replica inspired by a Collata quipu, alpaca, llama, guanaco, vicuña, and white-tailed deer fibers dyed with black walnut, goldenrod, indigo, cochineal, undyed alpaca, llama, guanaco, and vicuña fibers, approximately 14 inches high

When I was making a replica for her, I could also feel this tactile difference—especially between the deer, llama, and vicuña hairs. Vicuñas are wild animals, and many people consider vicuña to be the softest, finest fiber in the world. I could feel these differences, and they demonstrate yet another way that information was encoded in quipus. So, I'm looking at these ideas from a backward perspective compared to John and Daina, and I'm hoping it's complementary, like the dualism that is so important in Andean traditions.

JS: I think there is something to be said for thinking about this idea of feeling—and not just with your fingers, but also feeling in a more general way. You see that happening with music. For instance, a good jazz musician can feel their way across the score in a manner that makes sense, right? There's this idea of feeling that's very, very important to bring a certain kind of magical insight into sensing or knowing what's going on.

SO: Jeffrey, as you describe the process, you are decoding the colors and decoding the fabrics, you are looking for patterns, you are looking for repetition. I think that's one thing we can agree on: the value of expanding the definition of *math* to embrace pattern recognition and identification. A broader mathematical umbrella includes understanding patterns— what we learn from them and how we use them to communicate in different ways.

JS: I would think that would be part of the current definition of *mathematics*—being able to think about patterns and structures and to know when things are unique or not unique. That's a big part of the work.

SO: Do any of you have questions or suggestions for the audience about how they might explore some of these intersections about math and creating and craftwork and larger cultural messages?

DT: Many people in the audience submitted works for the *Saratoga Springs Satellite Reef*, so in some ways they have explored hyperbolic geometry. It was also a very good exercise to understand exponential growth, which in turn helped

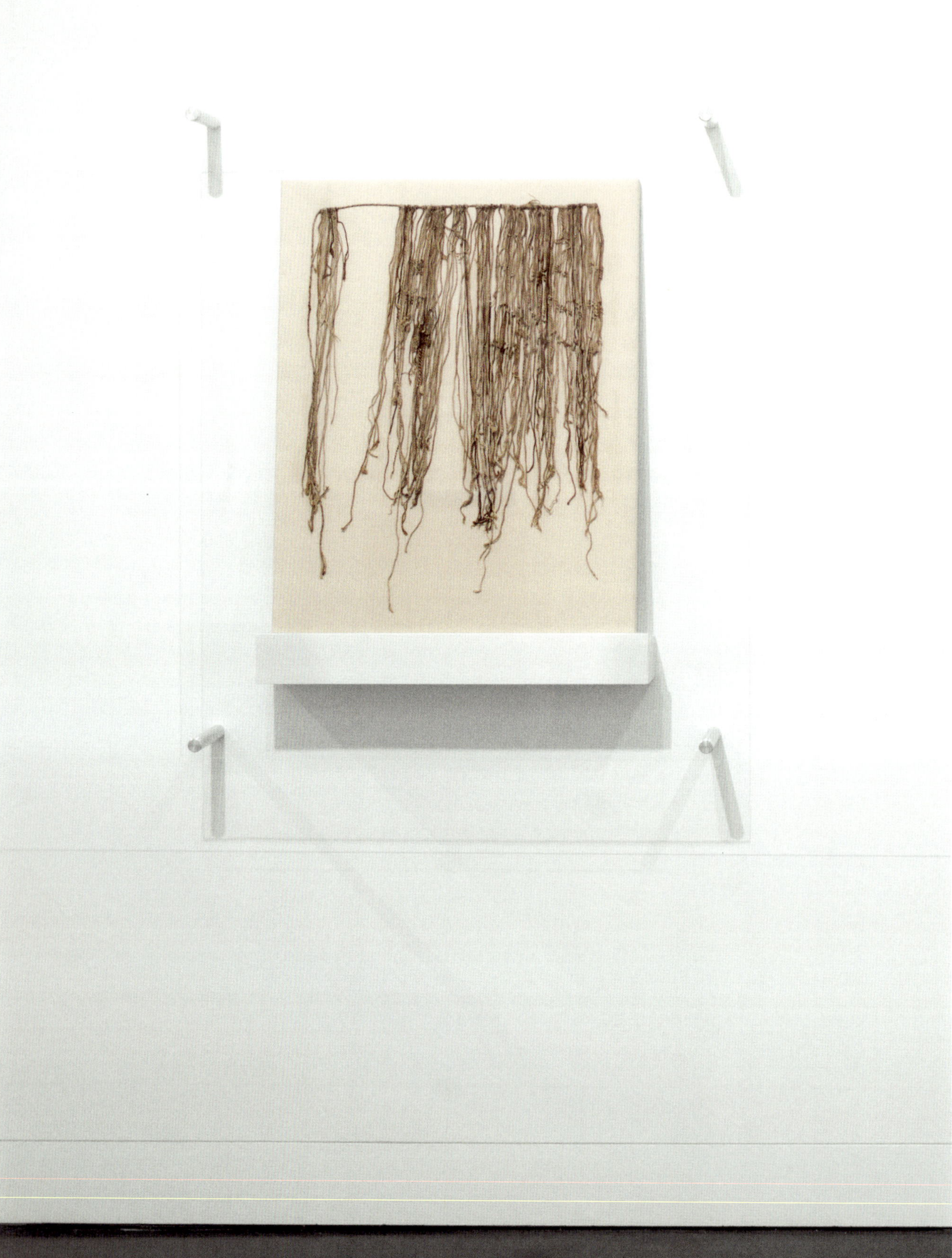

Installation views, Radical Fiber, featuring work by an unidentified Inca quipu maker (left) and Cecilia Vicuña (right)

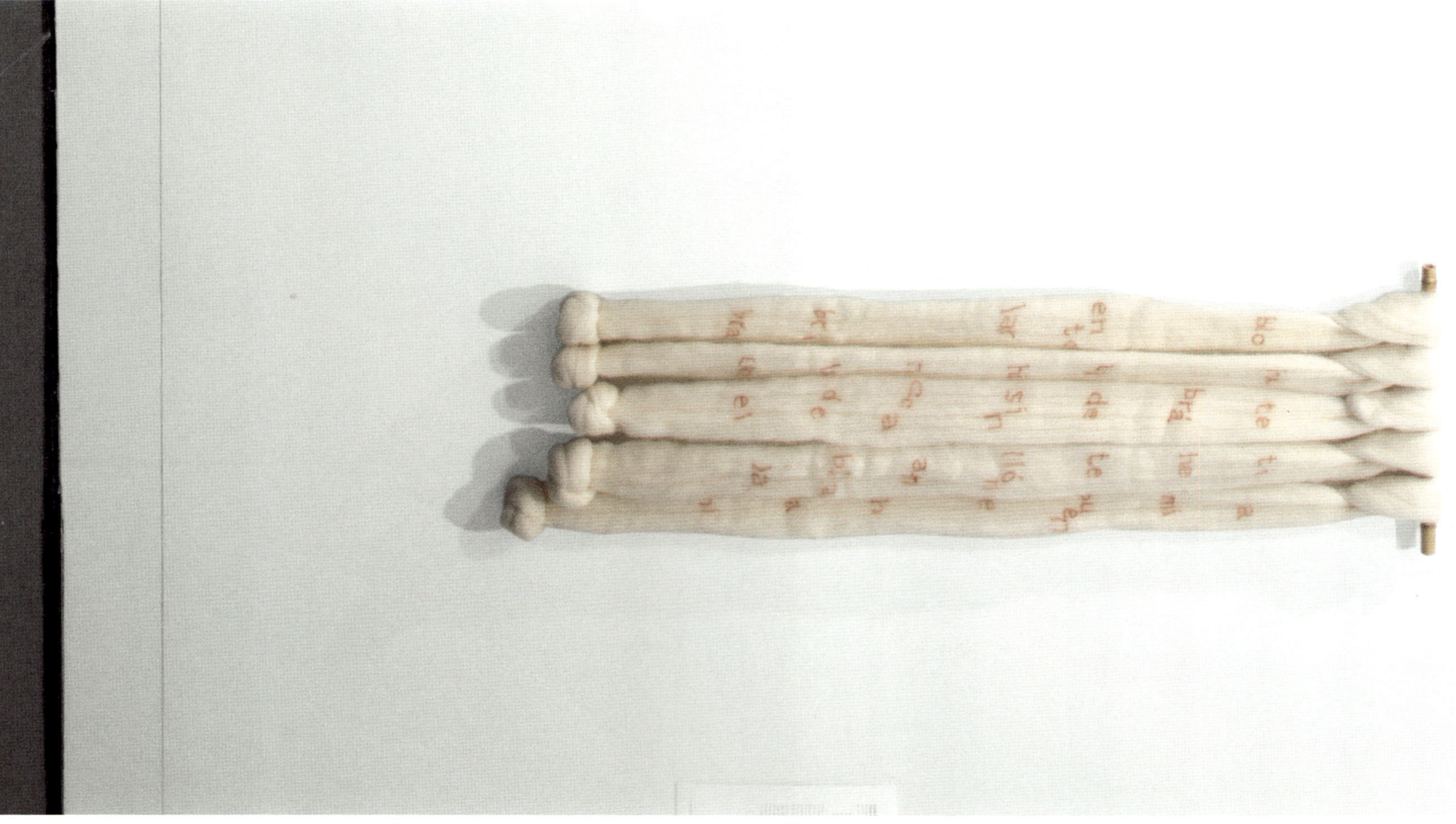

us understand what is happening with the COVID-19 pandemic and how the virus is spreading. For those who are crocheting, I would just say continue exploring—try to make different shapes, fold them and see what else you can discover.

JCS: I really believe, with quipus, we shouldn't think of people from South America as having been preliterate or unliterate. They had a writing system, and it was based on color and numbers. They were surrounded by information. And everything around us is colorful, right? They had an information system that permeated their entire lives. I would like to ask everybody to imagine what it would be like if our world was like that. **How different would our world be if we conveyed our information exclusively through colors and numbers?** Would we be reading in the form of colored barcodes? Would we choose the colors of our clothes differently because everything we wore actually said something? How would our aesthetics be different? Try to imagine a world like that and how sophisticated it would be.

SO: I like that! I'm already spinning out with that thought.

JS: I would like to challenge everyone to go into the visual landscape and see the ways in which mathematical ideas, languages, and processes show up and present themselves. And how math shows up in the culture, or in the architecture, or in the stores—how it's living right in front of you in so many different ways. The challenge is to decode it, and in the process of decoding it, be open to inspiration to encode it somewhere else. Become part of a dynamic that allows the math ideas and structures to flow through your mind into other places, other things, and other objects.

I'm interested in math artwork—being able to encode/decode from this math space/art space, creating a metaphor that moves from that thinking space to that feeling space, from the work of the mind, to the heart, to the hand. That's the challenge I would like to present to the audience.

AUDIENCE: This is the first time I have actually liked hearing about math. Using art to speak the language of somebody who isn't a scientist or a mathematician is really powerful. Could you expand on the power of art as it relates to learning and opening up pathways?

JS: When I was hired to create a visual mathematics curriculum for art students at the Ringling College of Art and Design, part of my duty was to use the languages of art and visual mathematics to get students excited about the beauty of mathematics liberated from the submissive element that is usually presented for the sciences. You learn calculus or other kinds of mathematics to serve the purpose of physics, or chemistry, or economics. To learn math for math's sake—not necessarily to solve a problem for a particular issue in a particular field—but to learn it and love it in itself as its own art form, as its own game, as its own puzzle, and then to be able to use it in creative ways...This is what art offers—this idea of being creative, expressing in different ways. It gives students a certain confidence and empowerment and makes them want to go further and be able to play and exercise in those areas. Art is very, very important for energizing this process. **Mathematicians who work at the highest level are conceptual artists.** And that's an exciting space to be in.

AUDIENCE: How can math-informed artwork make greater inroads as a way to teach math to all learners? This is a perfect example of STEAM—Science, Technology, Engineering, Art, and Math—yet leaders and funders for STEM education have yet to fully embrace or are hesitant to incorporate the art aspect.

DT: I can't speak to the issue about funding, but I'd like to respond to the issue about understanding—and to the previous comment from the person who said this was the first time they liked hearing about math. That reminded me of the beginning of my career. I got into trouble because my colleagues were opposed to the way I was teaching. I always understood math easily. And that's why I became a teacher—so I could make math understandable for everybody else. But my colleagues claimed that I was not teaching math formally enough. I was teaching it informally—and using artwork. But now, these hyperbolic planes have had a life of their own, which has expanded exponentially. Everybody who is following the

Saratoga Springs Satellite Reef has at least learned the term *hyperbolic geometry* and knows that such a geometry exists.

SO: Daina and John are part of a large and growing group of math-artists who bring a voice and a personal expression to the way they create. It really expands the way many of us are taught mathematics, particularly if we weren't taught mathematics in a way that we necessarily responded to. What's great about math art is that it can still welcome us into this world and give us access to some of these big ideas, without having to go through this narrow channel.

JS: When I was teaching math to art students, I had this impulse to illustrate everything for them. And what I real-

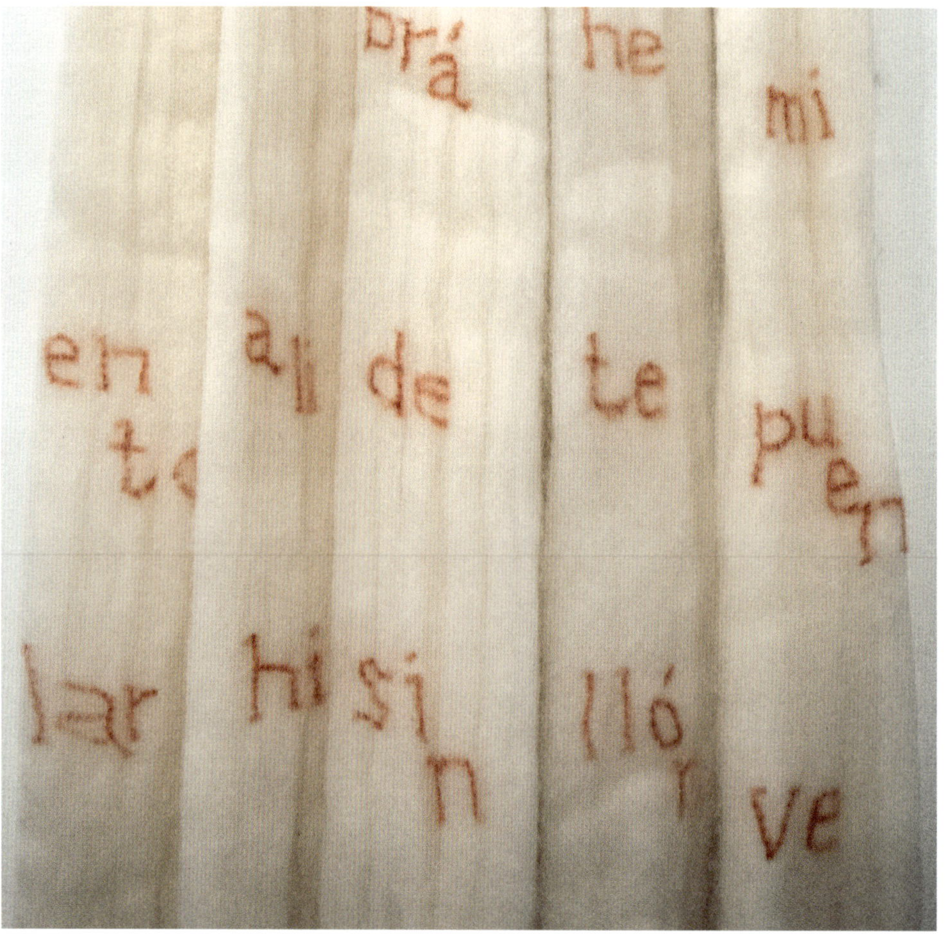

Cecilia Vicuña, *Chanccani Quipu* (detail), 2012, ink on knotted cords of unspun wool and bamboo, 78½ × 17 × 4¾ inches (overall)

134

ized was that I didn't need to do that because the art students could illustrate things themselves. I just needed to give them the opportunity and platform and encouragement to do their own illustrations—to do their own conceptualizations. When you're working with art students, you have to give them space so they can own the pedagogy of the project and be part of the creation. This imprints a sense of learning in their memory in a very different way.

Very often programs are set up in a rigid way: this is what it is, this is what you remember, this is how you get tested. If you want to foster real learning, allow students to integrate the concepts into their sense of knowing and understanding and translate them in different ways. Students need this space, particularly artists. Part of my work is giving students that space and fostering their confidence. When you give students the opportunity to create their own visualizations of a math concept, they own it very differently. When you present mathematics as a creative enterprise, people end up discovering things in a way that makes them feel part of it—as opposed to accepting the knowledge, remembering it, and then applying it.

> **JCS:** As an anthropologist, I see culture at the root of everything. And yet, when I learned mathematics, I was told it was universal—not only universal on Earth but in the universe itself. **Through our work, we're seeing that math is not universal. There's a universality to it, but the way people perceive math and the way they learn math is cultural.** And we need to look at that when we're teaching math.

JS: It's important to put mathematics within its proper cultural context so we don't over-romanticize its disconnection from the physical human reality that gave birth to it. We do that in the way we construct our deities as well: we create these deities that exist beyond us and are more powerful than we are, but we have to be cognizant of our hand in the creation.

1 John Sims, "Trees, Roots and a Brain: A Metaphorical Foundation for Mathematical Art," in *Mathematics and Culture II*, ed. Michele Emmer (Berlin: Springer, 2005), 163–70.

2 Gilles Deleuze, *The Fold: Leibniz and the Baroque* (Minneapolis: University of Minnesota Press, 1992).

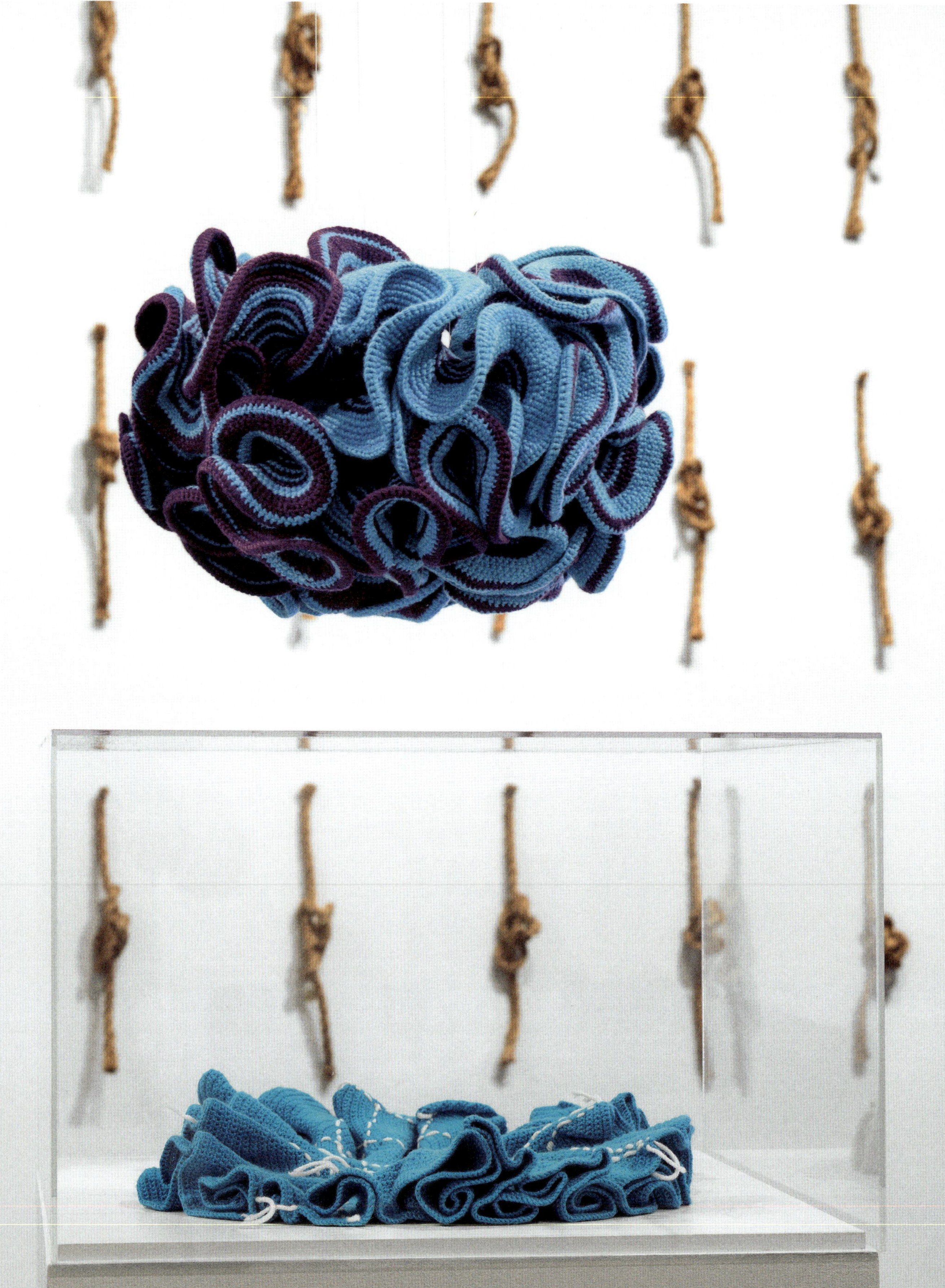

by Mark Huibregtse

THE STRANGE UNIVERSE OF HYPERBOLIC GEOMETRY

What is true about the space around us? For more than two millennia, the geometry presented in Euclid's *Elements* (written around 300 BCE) was considered the last word on the subject. We begin learning Euclidean (plane) geometry in grade school, where shapes such as triangles and circles and the concept of parallel lines (think straight railroad tracks) are introduced. In high school, we study geometry as a deductive science, in which basic truths (axioms) are assumed and other results are derived by logical reasoning. Euclid's *Elements* served as the chief exemplar of certain truths and the deductive organization of knowledge from its writing until the nineteenth century.

Despite the *Elements*' preeminence, a small number of geometers felt that Euclid's fifth postulate—concerning parallel lines—was problematic. The postulate says, in essence, that if two lines *appear* to converge, then they *do* converge and eventually intersect, as the figure suggests. (The precise meaning of "appear to converge" is that the sum of the angles A and B is less than 180 degrees.)

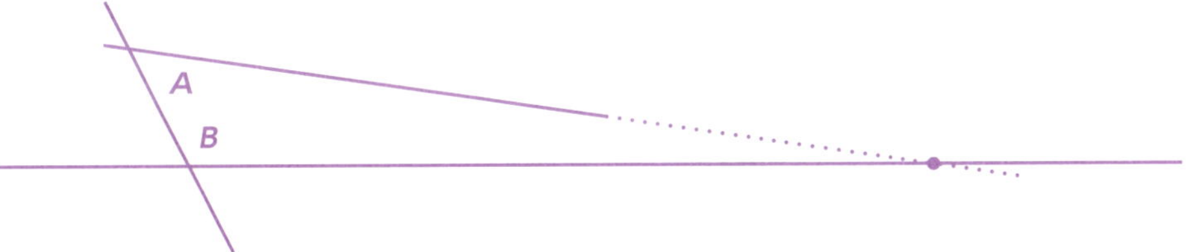

Unfortunately, the wording of the postulate is difficult to follow; it is less immediately understandable and "obvious" as the other postulates (such as "a line segment can be extended as far as desired in either direction"). For these reasons, those attempting to remove this imperfection from the *Elements* strove either to replace Euclid's axiom with a simpler statement, or even better, to prove that Euclid's postulate could be deduced as a theorem from the other (unproblematic) axioms in Euclid's system. Neither of these strategies was successful.

Finally, at the start of the nineteenth century, a few mathematicians (the key figures being Carl Friedrich Gauss, János Bolyai, and Nikolay Lobachevsky) saw the light, as it were. They dared to consider the possibility that two lines might appear to converge in the above sense but would remain parallel (i.e., nonintersecting), while continuing to accept the rest of Euclid's axioms as valid. Many seemingly bizarre logical consequences resulted from these foundations—indeed, Bolyai remarked in a letter to his father that "out of nothing I have created a strange new universe."[1] One consequence of this new system, which came to be known as *hyperbolic geometry*, is the existence of two types of parallels, *asymptotic* and *divergent*. Asymptotic parallels approach each other arbitrarily closely without ever meeting, whereas divergent parallels diverge arbitrarily far from one another in either direction.

The creators of hyperbolic geometry believed in its logical consistency (that is, that no contradictions would ever arise), but general acceptance was delayed for half a century, in part because there were no geometric models that would visually show its reality. In one particularly beautiful model (called the Poincaré disk model, although originally published, along with several other models, by Eugenio Beltrami in 1868), the (hyperbolic) points are the points in the interior of a Euclidean circle. Lines are either diameters of the circle (such as AB) or circular arcs meeting the boundary circle at right angles (such as ZW and RS). In the diagram below, ZW is asymptotically parallel to AB "to the right," and RS and AB are divergently parallel. Angle sizes are "just as they appear" in this model but distance between points is more complicated. The boundary circle is infinitely far away from any of the points in the interior and a hyperbolic measuring rod appears (from "outside") to shrink—there is a complicated formula for this—as it moves toward the boundary. (M. C. Escher's famous *Circle Limit* images are based on this model.)

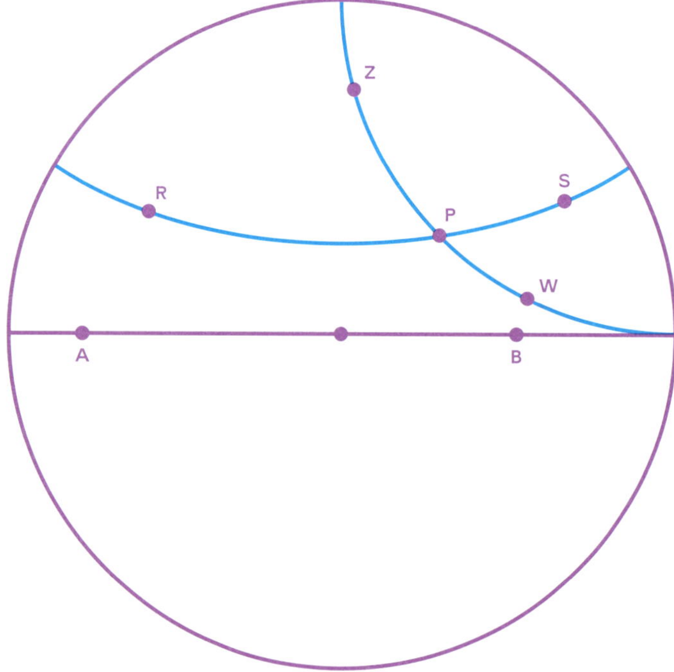

Daina Taimina, *Hyperbolic sketch*, 2006, crocheted microfiber

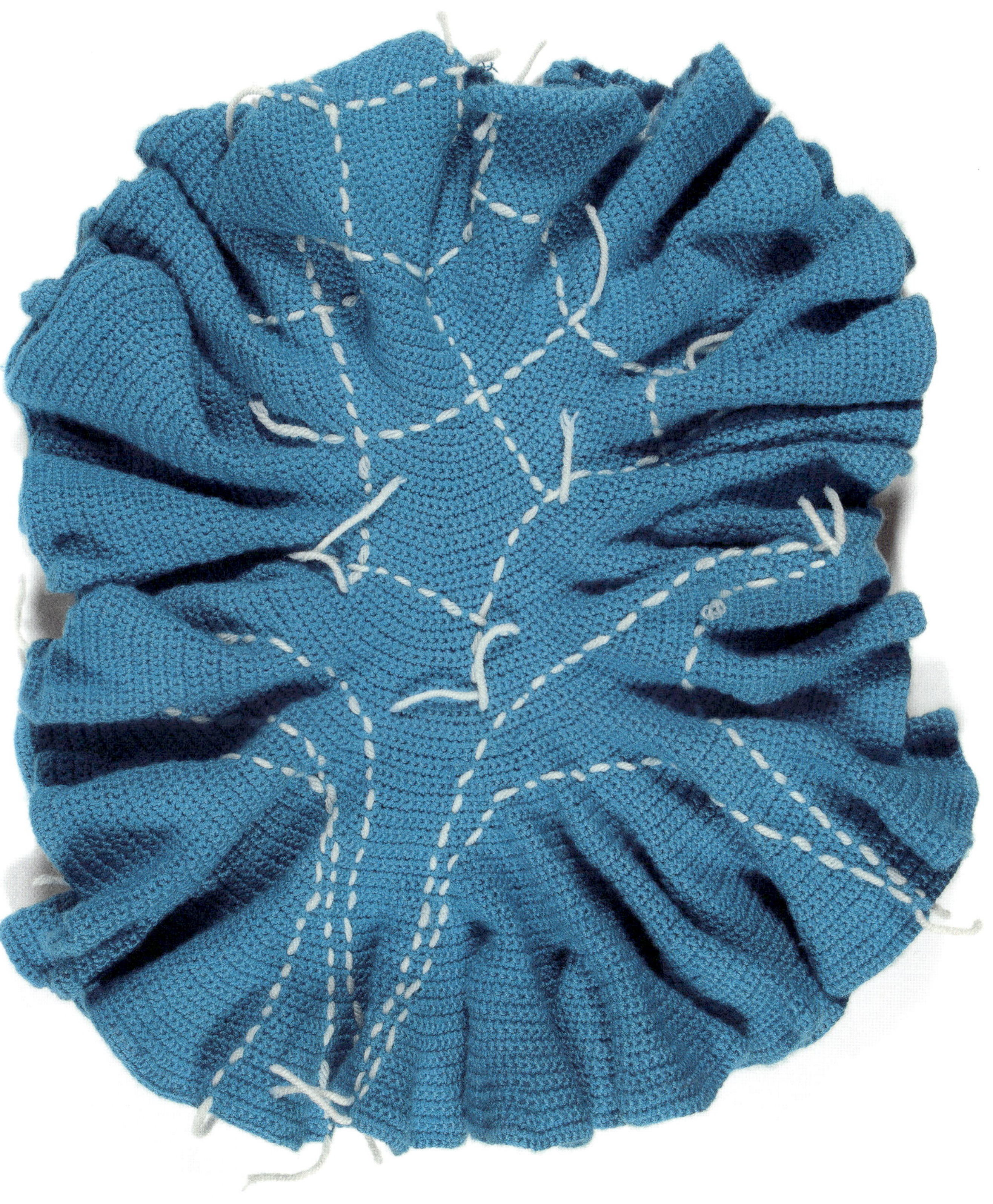

We can now connect this to the artwork by Daina Taimina. Suppose one wants a model in which the measuring rods do not shrink as we move toward the horizon. We could try to scale up the disk by making the radius of the boundary circle extend to infinity. However, hyperbolic distances do not scale up linearly, as they do in Euclidean geometry, but rather exponentially. Suppose we create a hyperbolic plane model out of a tablecloth and attempt to lay it flat on a (Euclidean) table. If we draw concentric circles around the origin of the tablecloth and the table with varying radii, the Euclidean and hyperbolic circumferences are (approximately) as follows:

Radius	2	3	4	5	6
Euclidean circumference	12.57	18.85	25.13	31.42	37.70
Hyperbolic circumference	22.79	62.94	171.47	466.23	1267.40

The circumferences of the hyperbolic circles on the tablecloth grow much faster than the corresponding Euclidean circles on the table. Because of this, it is impossible for the tablecloth to lie flat; it must instead contort itself into ever-wilder folds in an attempt to accommodate the rapidly increasing circumferences. Taimina did just that with her model *Hyperbolic sketch*: white threads illustrate various lines on the blue "tablecloth" (hyperbolic plane); one can see both asymptotic and divergent parallel lines as well as shapes, including a quadrilateral and a pentagon. The "triangle" that resembles a witch's hat illustrates an "ideal" triangle, which is the limit of a normal triangle as its vertices move toward infinity. The sum of the angles of the ideal triangle is 0, indicating that the sum of the angles of a hyperbolic triangle (which is always less than 180 degrees) decreases as the triangle is expanded. By picking up *Hyperbolic sketch* and suspending it at two points on any one of these lines, one can see that the lines really are paths having the shortest distance between the two points on the blue surface. Taimina's *Day and Night* beautifully illustrates one of the many ways a portion of a hyperbolic plane can appear when it is placed in Euclidean three-dimensional space. It demonstrates the *negative curvature* characteristic of hyperbolic planes, meaning that the surface resembles a saddle at every point of the surface. The piece suggests that such surfaces occur in nature, as, for example, in lettuce heads or corals. By using the craft of crochet, Taimina has given us objects of beauty and fascination that enable viewers to experience "bizarre" hyperbolic phenomena directly.

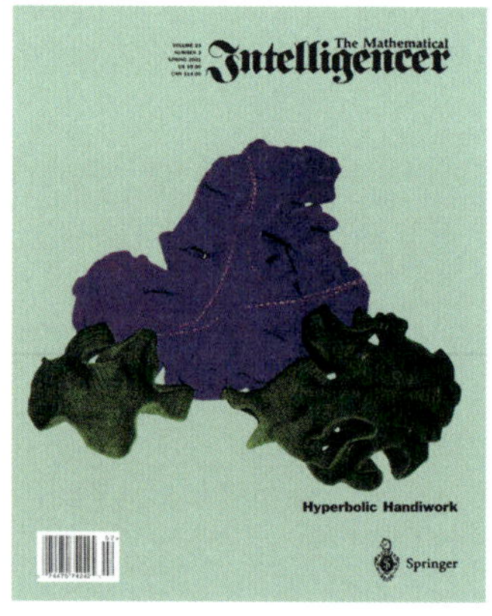

The Mathematical Intelligencer, vol. 23, no. 2 (spring 2001). The cover promotes the article "Crocheting the Hyperbolic Plane" by David Henderson and Daina Taimina and features Taimina's models.

1 János Bolyai quoted in Robin Hartshorne, *Geometry: Euclid and Beyond, Undergraduate Texts in Mathematics* (New York: Springer, 2000), 305.

Right and following spread: Daina Taimina, Day and Night (details), 2007,

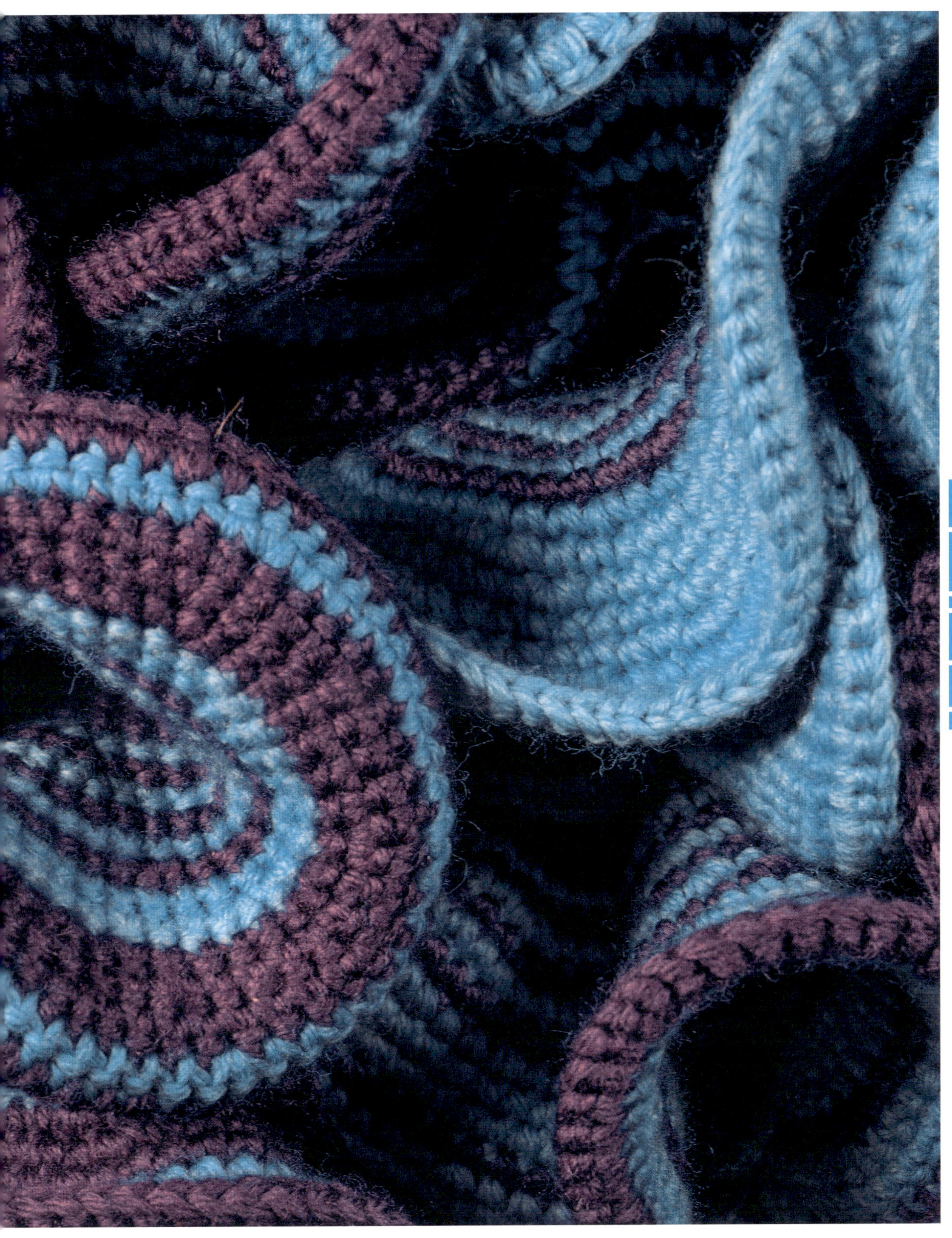

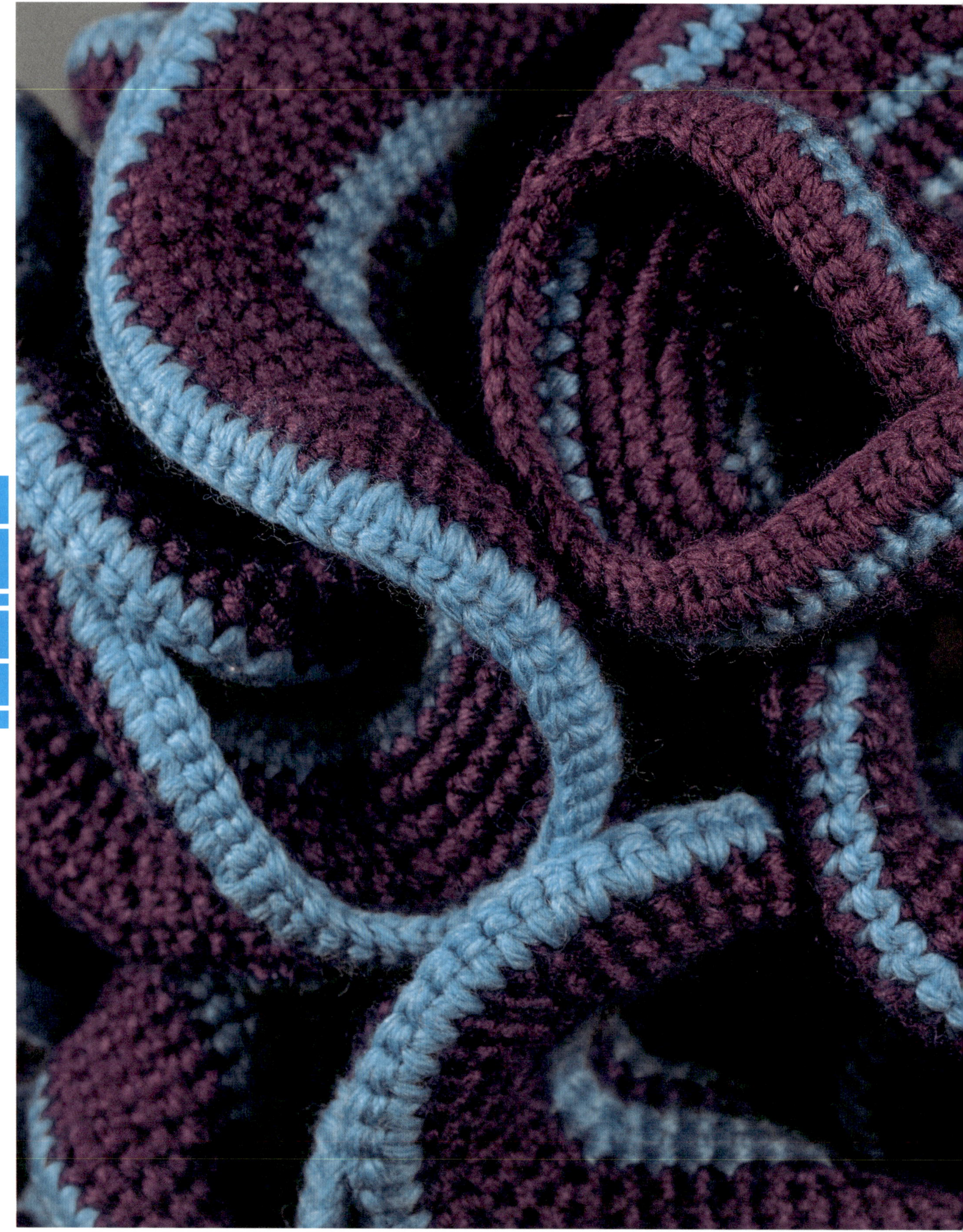

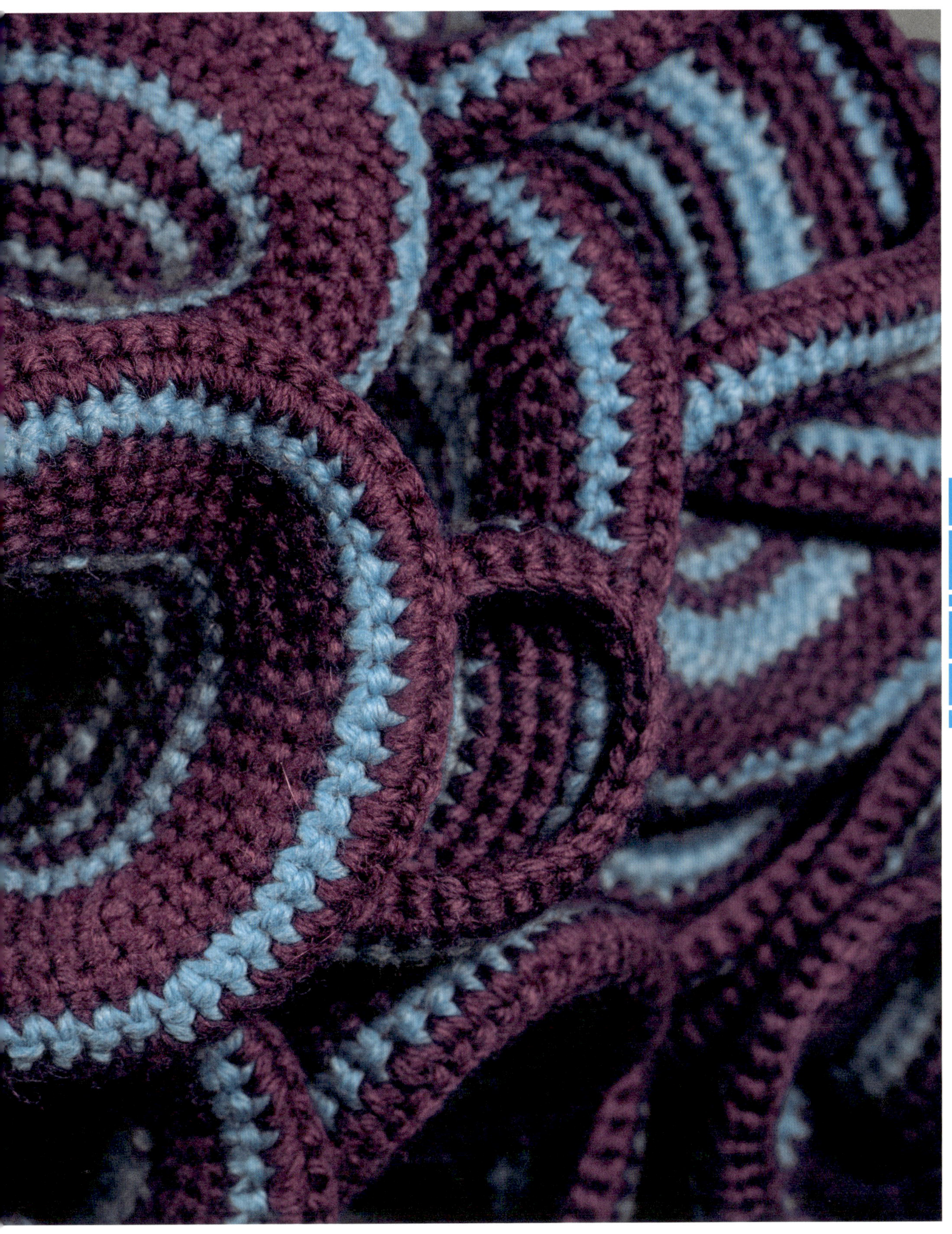

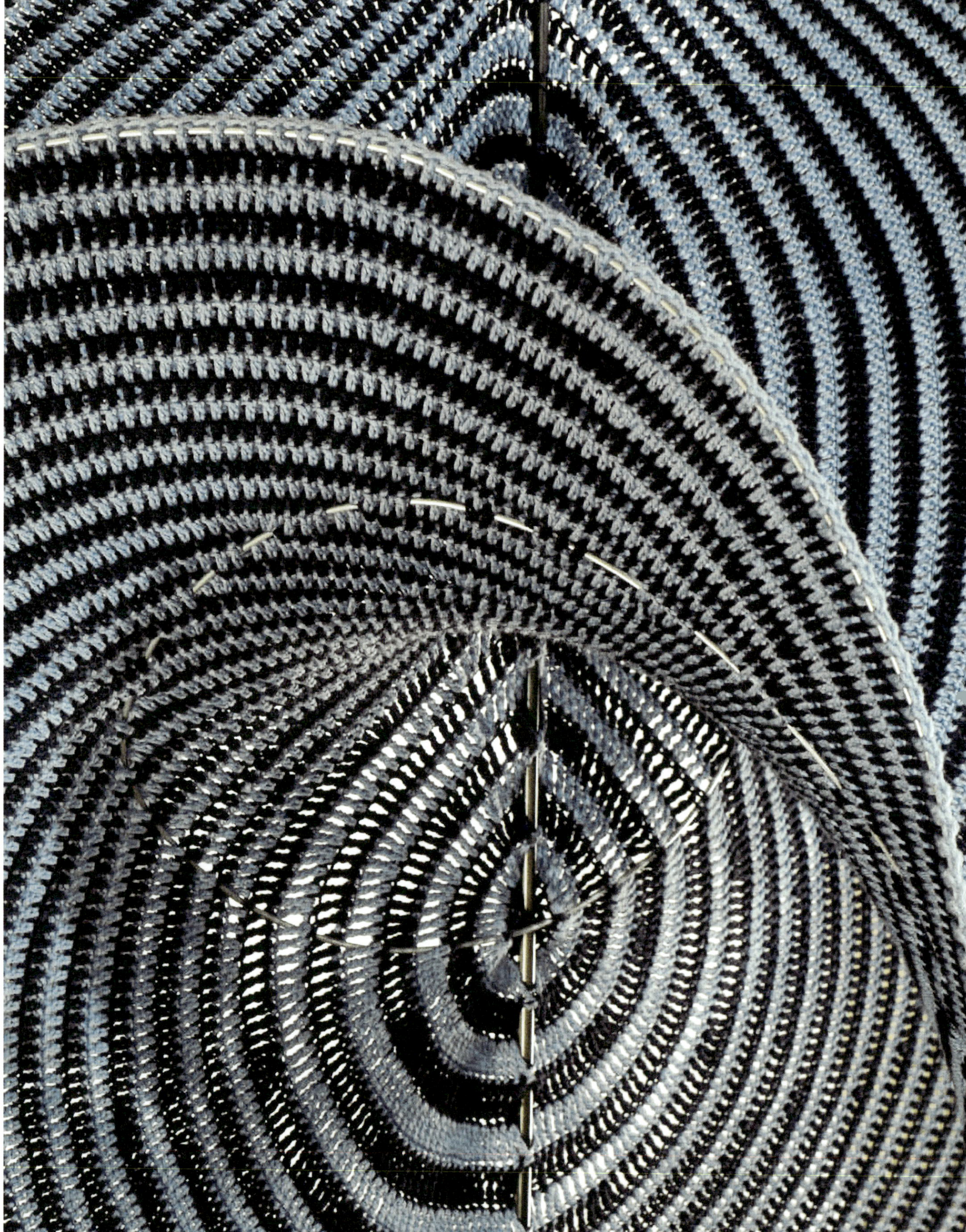

by Rachel Roe-Dale

STITCHING DYNAMICAL SYSTEMS

> "Imagine a leaf floating in a turbulent river and consider how it passes either to the left or to the right around a rock somewhere downstream. Those special leaves that end up clinging to the rock must have followed a very unique path in the water. Each stitch in the crochet pattern represents a single point (a leaf) that ends up at the rock."
> —*Hinke M. Osinga*

The motion of the leaf is an example of a dynamical system, one that evolves over time. In this example, leaves that start close together upstream may drift far apart. A single leaf may eventually cling to a rock, while its former neighbor is washed farther downstream, perhaps infinitely far away. This sensitive dependence on the initial state—where slight variations at the start can produce drastically different outcomes—is a hallmark of chaotic systems.

With undergraduate and graduate degrees in mathematics and a doctorate in meteorology, Edward Lorenz was one of the first scientists to notice this property. Working with a weather model in 1961, Lorenz observed that rounding the input value 0.506127 down to 0.506 drastically changed the weather his model predicted as the final outcome. This change suggested that, in some models, small perturbations or changes in input values can drastically alter the final outcomes or behaviors of a system. Lorenz's observations challenged the previous mindset, which predicted that small changes in input would result in only small changes in output, and that nature and mathematics would behave predictably if the equations used in the model did not depend on chance. Lorenz presented his groundbreaking results and observations in his paper "Deterministic Nonperiodic Flow," and then in 1972, he gave a presentation at the American Association for the Advancement of Science titled "Predictability: Does the Flap of a Butterfly's Wings in Brazil Set Off a Tornado in Texas?" This musing was subsequently popularized as the *butterfly effect.*

Four decades later, applied mathematicians Hinke M. Osinga and Bernd Krauskopf investigated the Lorenz system. Solutions to Lorenz's system

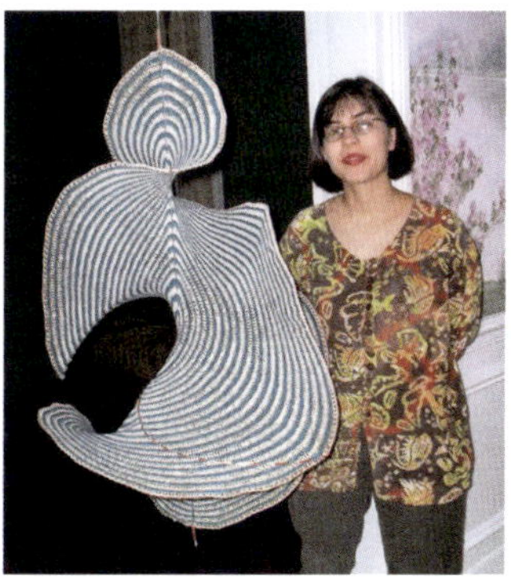

Carolyn Yackel with her crocheted model of the Lorenz Manifold. Hinke Osinga and Bernd Krauskopf offer the pattern at no charge so that anyone can gain the embodied knowledge of creating their own complicated surface.

Hinke Osinga and Bernd Krauskopf, Crocheted Lorenz Manifold (detail), 4-ply mercerized cotton yarn, wire, kiting rod, 35 ½ inches high (overall)

of equations form curves that typically converge to an object known as the *Lorenz attractor*—which coincidentally visually references a butterfly.

However, some solution curves converge to the origin of three-dimensional space ($x = 0$, $y = 0$, and $z = 0$) rather than onto the butterfly-shaped attractor. Osinga and Krauskopf were especially interested in these solutions and developed a computer algorithm to find this two-dimensional surface, which is known as the *Lorenz manifold*. While their code was accurate and effective, they found the resulting manifold renderings and visualizations insufficient to provide an intuitive understanding of the properties of this surface.

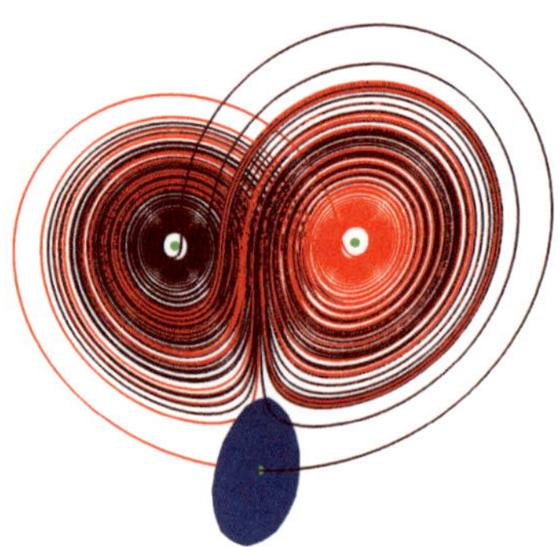

The Lorenz attractor (represented by brown and red solution curves) with an initial disk (blue) on the Lorenz manifold

During their academic winter break in 2002–3, Osinga and Krauskopf realized that the special way in which their computer algorithm generated the surface could be interpreted as crochet instructions. In the same spirit as Daina Taimina's crocheted models of hyperbolic planes, they saw the opportunity to create a tangible model of the Lorenz manifold. Osinga and Krauskopf mounted their crocheted creation, and they published photographs of the surface, the crochet pattern, and their mathematical analysis in *The Mathematical Intelligencer* in 2004. When looking at their creation, the middle of the smallest crocheted circle in the surface represents the origin of three-dimensional space (where $x = 0$, $y = 0$, and $z = 0$), and as time passes, solution trajectories that begin elsewhere on the crocheted surface will converge mathematically to this center point. The makers who have crocheted a Lorenz manifold based on the pattern published by Osinga and Krauskopf, or viewed and studied one in person, as Skidmore College mathematics students now do, are engaging with dynamical systems in a tactile way that would not be possible using only a computer screen.

See Hinke M. Osinga and Bernd Krauskopf, "Crocheting the Lorenz Manifold," *The Mathematical Intelligencer* 26 (2004): 25–37, doi.org/10.1007/BF02985416.

Hinke Osinga and Bernd Krauskopf, crocheted Lorenz manifold (three views),

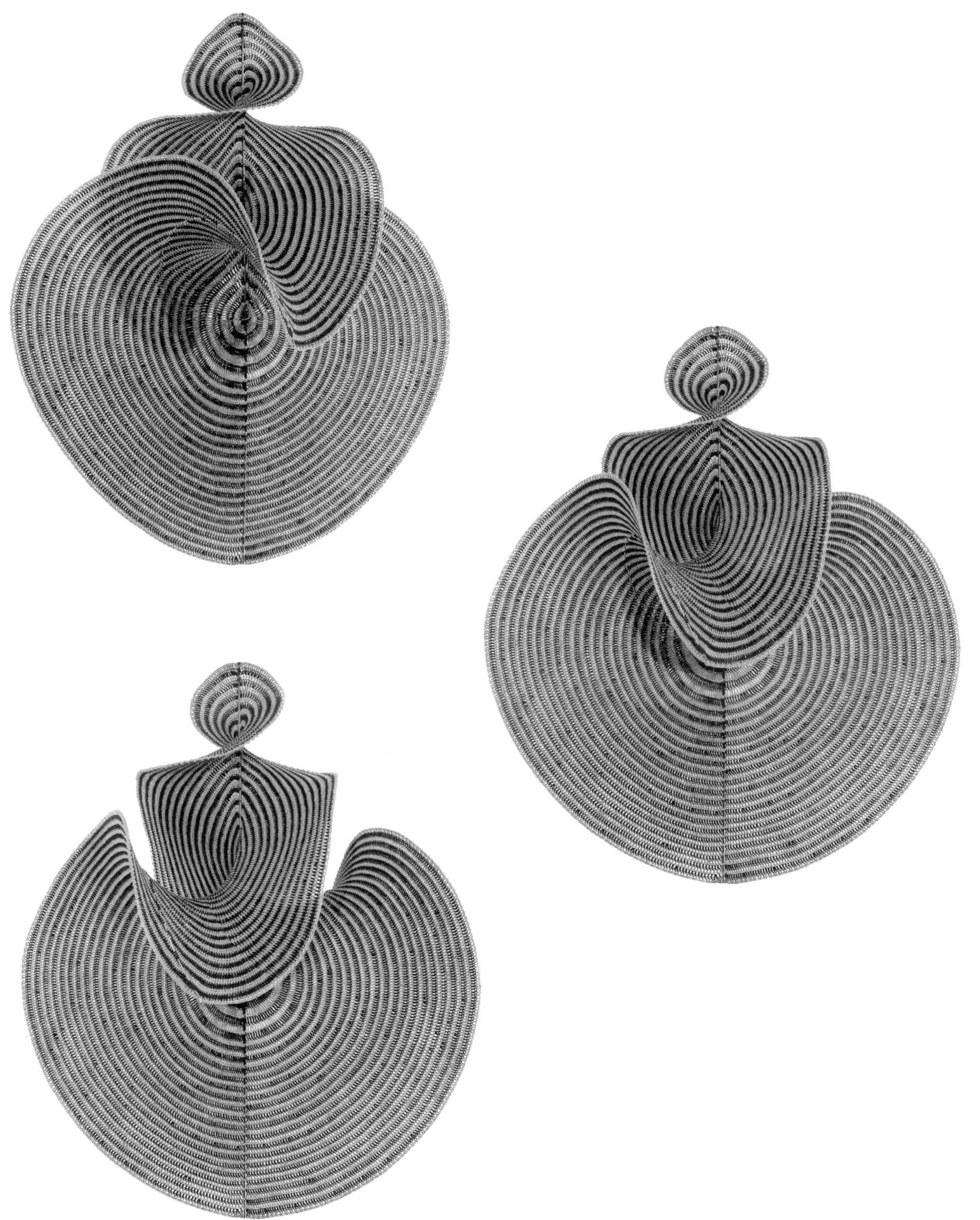

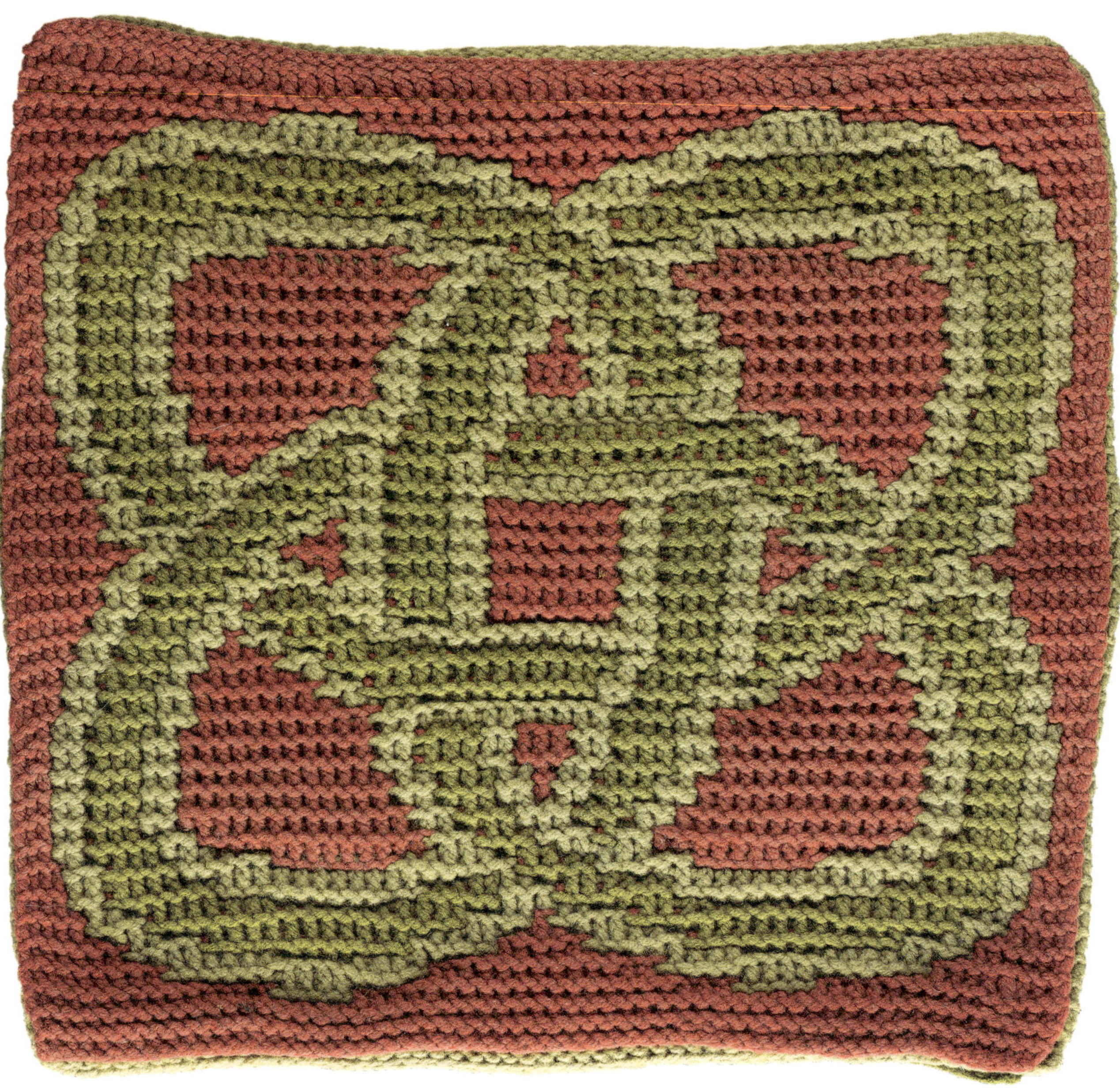

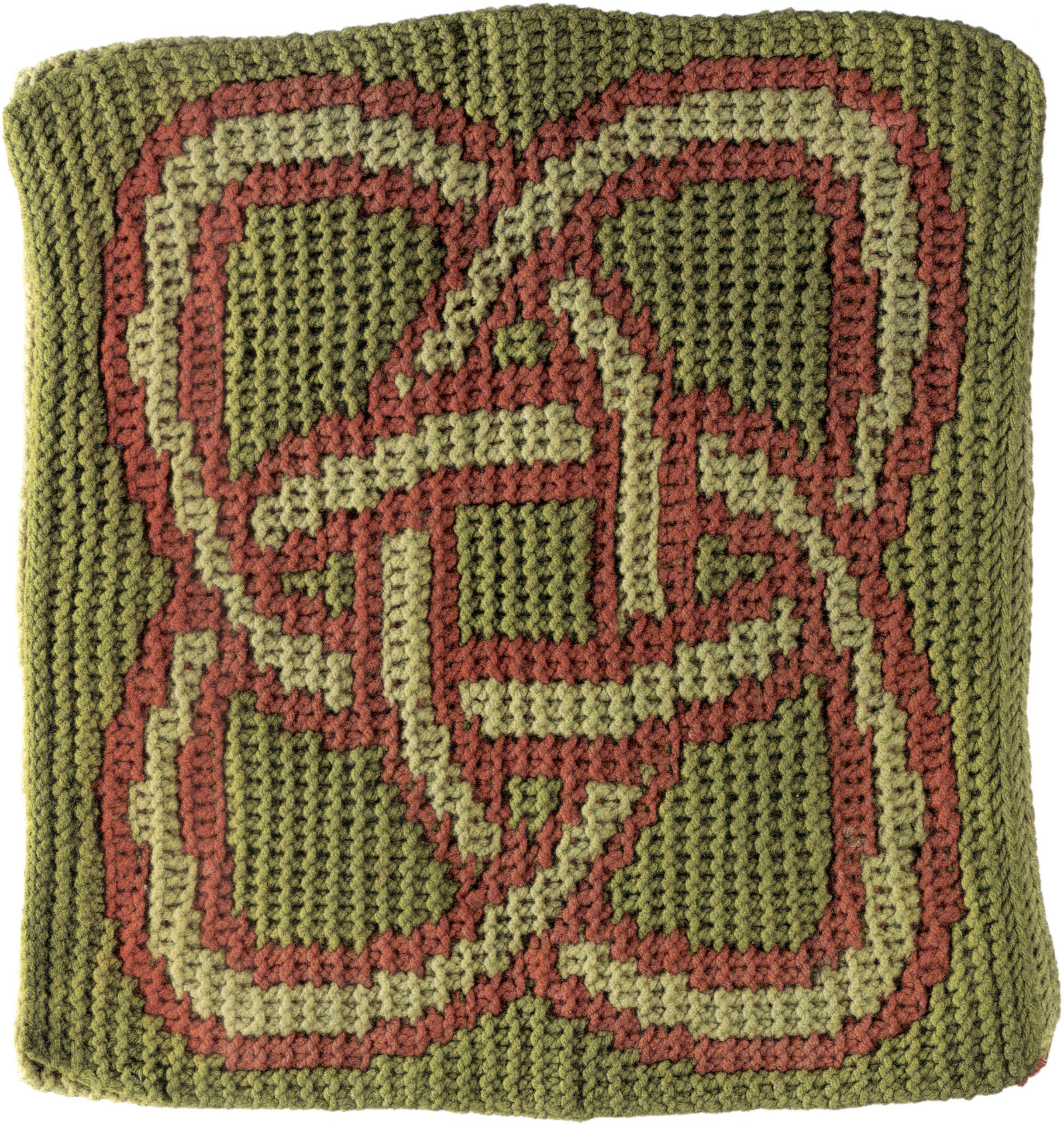

Alexander Crum Brown, triple-layer knitting sample, probably late 19th century, wool yarn, 13 × 13 ¾ × ½ inches; Scottish chemist and mathematician Crum Brown (1838–1922) created mathematical models through a variety of techniques and materials, including knitted wool, to illustrate his research. Here, a three-color, triple-layer knitted model, made with a complicated advanced knitting technique, displays a knot—and an early example of using fiber techniques to demonstrate mathematical principles.

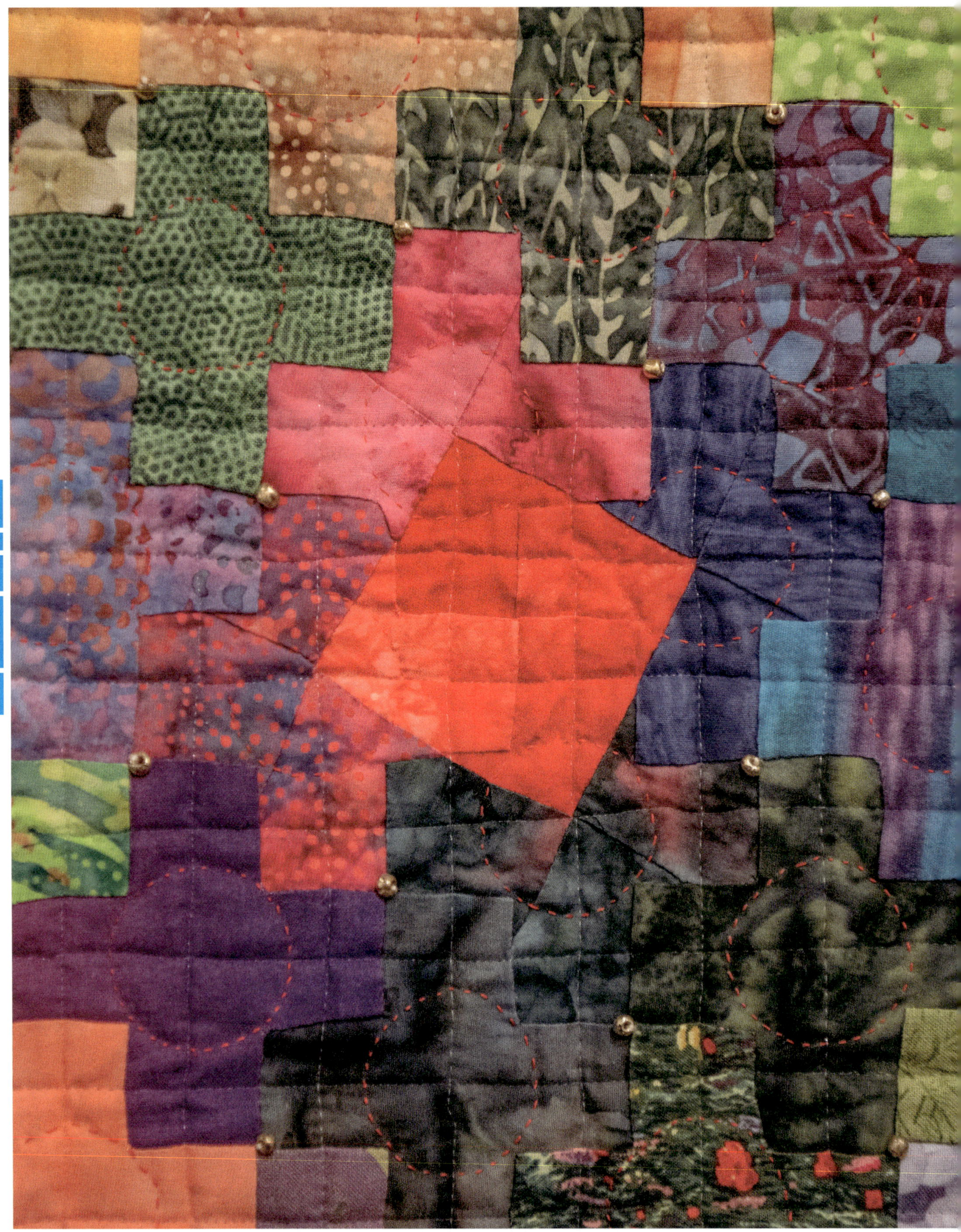

by Rachel Roe-Dale and
Rachel Seligman

TEXTILE MATHEMATICS: MODELS FOR PEDAGOGY

Fiber-based techniques such as quilting, crochet, knitting, and embroidery are particularly well-suited to making models for mathematical pedagogy.

In the 1980s, Elaine Krajenke Ellison, who spent her career teaching high school mathematics, began to create mathematical quilts and incorporate them into her teaching.[1] Ellison's quilts communicate the aesthetic aspects of mathematics and offer visual and physical form to an idea, theory, or data as a way to support mathematical comprehension and problem solving. Contemporary artist-mathematicians such as Ellison, Susan Goldstine, Hanne Kekkonen, Carolyn Yackel, and others similarly use fiber-based materials and techniques to model a wide range of ideas, helping them demonstrate and clarify abstract concepts in their pedagogy and in communication with nonspecialized audiences.

In *Greek Cross to Square Dissection*, Ellison explores geometric dissections: the dividing of a geometric shape into smaller identical units that can then be rearranged to form a different geometric shape. Here, the smaller unit is a six-sided, arrow-like shape that can be arranged to form a cross with arms of equal length or, with different orientation, to form a square. The units, or tiles, fit together with a swing-hinge dissection, meaning that the units are arranged around a center point to form the shape. For the cross, the four identical units fit together with the square ends out; for the square, the same four units are reversed, fitting together with the pointed ends out. In both cases, the cumulative shape can be used to completely fill planar space, as Ellison does for her quilted design. Because quilting involves the meticulous cutting and piecing of repeated shapes, it is an apt technique for understanding this type of geometry. The viewer sees that Ellison physically rotated the form without further manipulations, and in pairing the rotated form in different ways, created a new pattern.

Ellison has created over sixty quilts that explore myriad mathematical concepts, including fractals, orthic triangles, the Droste effect, the Fibonacci sequence, and more. Ellison's *Nature's Numbers*, for example, communicates both the aesthetic qualities and the frequency of occurrence of Fibonacci numbers in the natural world. The Fibonacci sequence is a list of numbers named in recognition of Leonardo Pisano, more commonly known as Fibonacci, who lived from approximately 1175 to 1250. In his text *Liber abaci* (Book of the abacus), published in 1202, Fibonacci posed and solved the famous "rabbit problem":

> Suppose a pair of rabbits is able to mate at the age of one month, and the gestation period for rabbits is one month. Assume that the rabbits never die and that the female will always give birth to a new male and

Elaine Krajenke Ellison, *Greek Cross to Square Dissection* (detail), 2013, pieced, appliquéd, quilted cotton, 53 × 53¼ inches (overall)

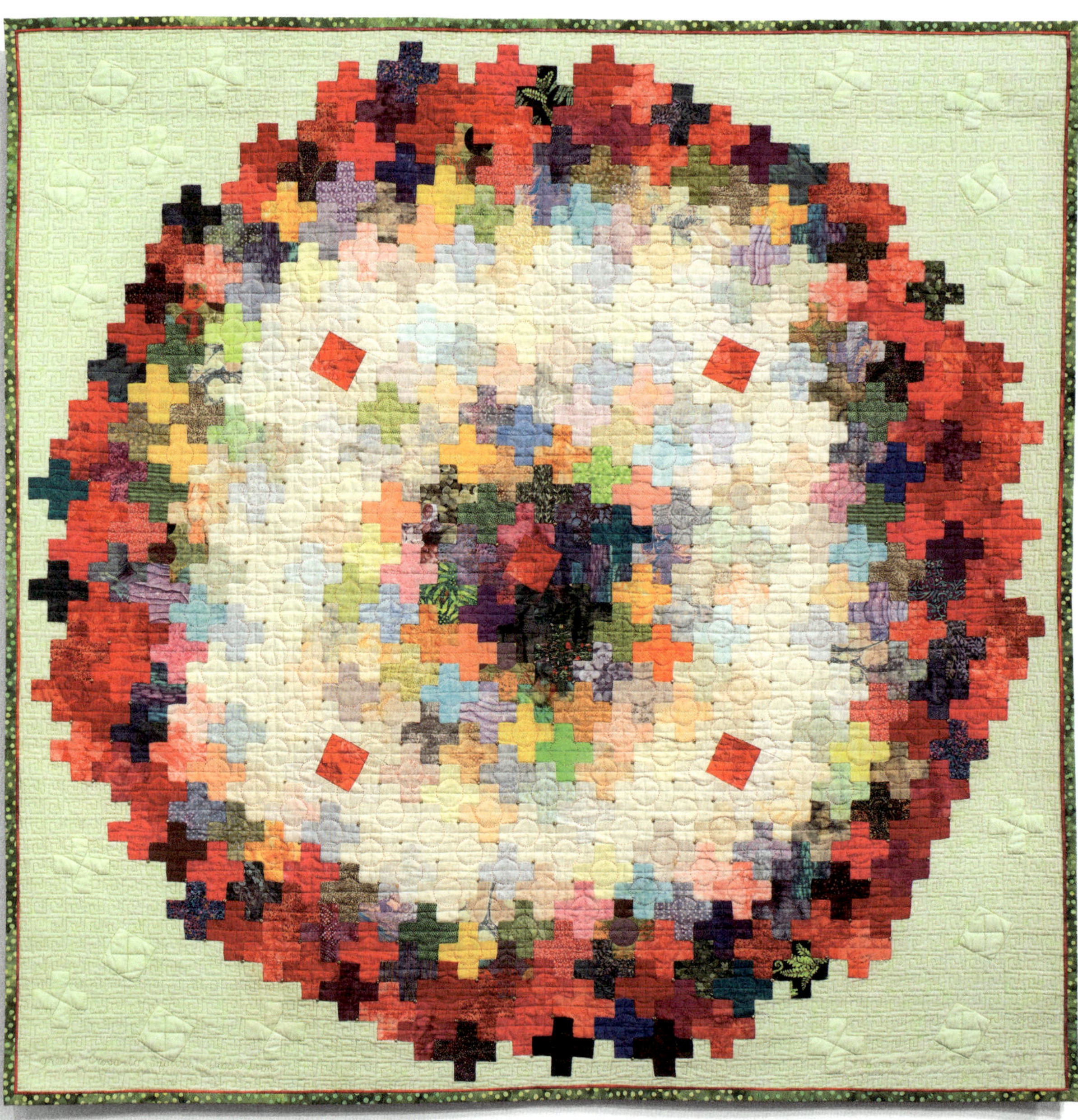

Left: Elaine Krajenke Ellison, *Greek Cross to Square Dissection*, 2013, pieced, appliquéd, quilted cotton, 53 × 53 ½ inches. *Right:* Elaine Krajenke Ellison, *Nature's Numbers*, 2008, pieced, appliquéd, quilted cotton, silk and cotton embroid

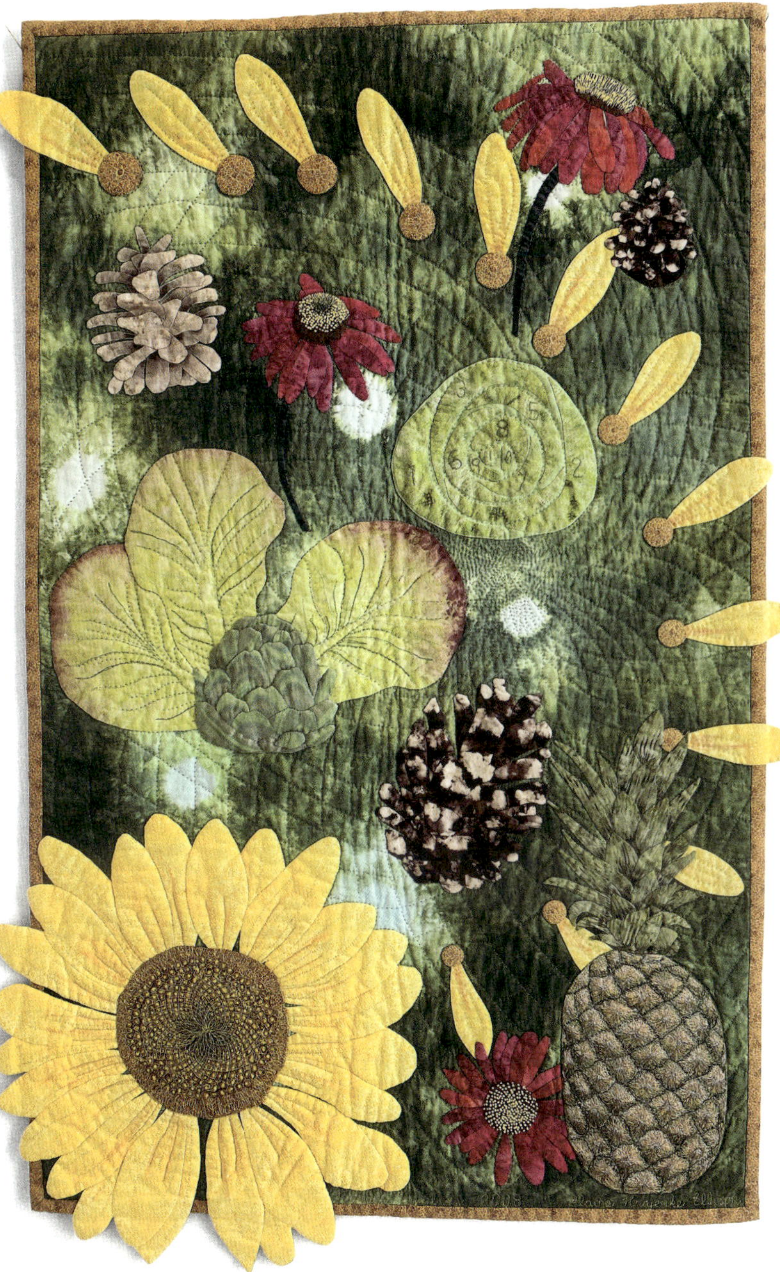

female rabbit every month. How many pairs of rabbits will there be after twelve months?

Solving this problem for the number of rabbit pairs after each month produces the mathematical sequence *1, 1, 2, 3, 5, 8, . . .*, where the next number in the sequence is equal to the sum of the previous two numbers. Numbers in this sequence are called Fibonacci numbers. In Ellison's quilt design, she includes plants such as a sunflower and an echinacea that have Fibonacci numbers of seeds in the spirals of the seed heads, and a pinecone and a pineapple, both of which have a Fibonacci number of scales—connecting mathematics to aspects of the natural world all around us.

Another way artists incorporate Fibonacci numbers is by using the golden ratio. This ratio, also known as the golden section or golden mean, is created by adjacent Fibonacci numbers. Two quantities are in the golden ratio if their ratio is the same as the ratio of their sum relative to the larger of the two quantities. Given $a > b$, we write: $(a + b) \div a \approx a \div b$. Mathematically we can show that this ratio is $(1 + \sqrt{5}) \div 2 \approx 1.618$, which is the golden mean. Artists and makers sometimes choose the height and width of their artwork to be in the golden ratio even if these measurements are not Fibonacci numbers. The ratio of the dimensions of *Nature's Numbers* is 42 inches $\div$ 26 inches ≈ 1.615. Ellison also showcases the golden ratio in this quilt by including a stalk of celery that spirals inward, with each stalk's radius in a golden ratio with the previous stalk's radius. This single quilt, then, provides numerous teaching opportunities for geometry classes, where students can see abstract concepts physically realized and can gain insight into how art making can deepen the creator's own understanding of the concepts.

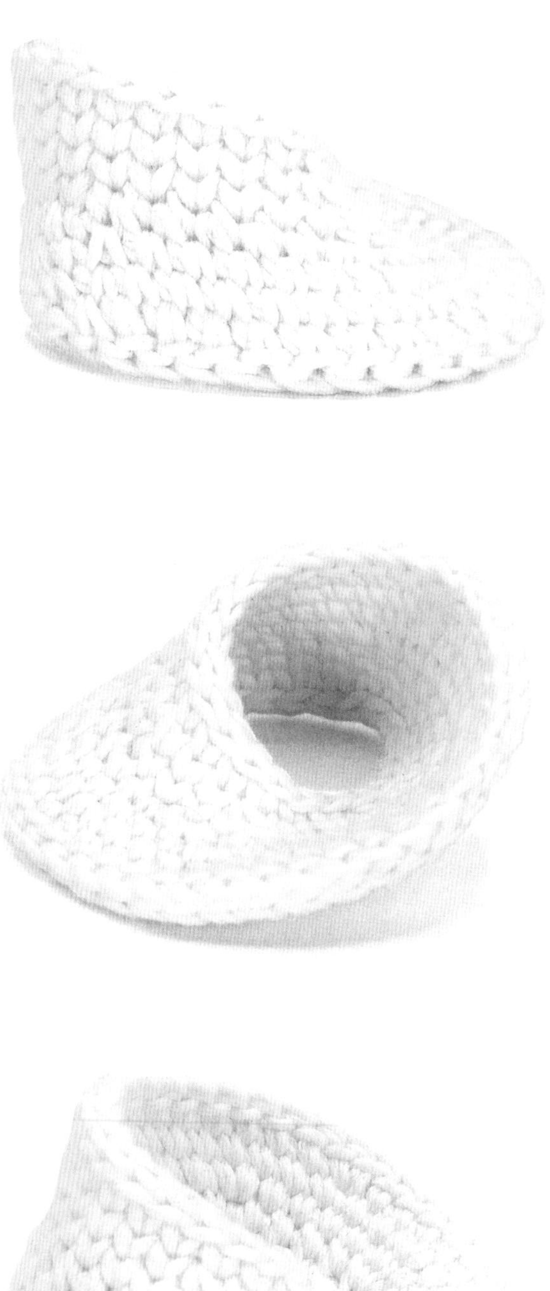

Hanne Kekkonen, Möbius band (three views, left), 2024, crocheted cotton yarn, metal ring, 2 × 4¾ × 4¼ inches; Klein bottle (two views, center), 2020, crocheted cotton yarn, 5 × 3½ × 3¼ inches; Möbius snail (two views, right), 2018–19, crocheted cotton yarn, metal ring, 2¾ × 4 × 3½ inches

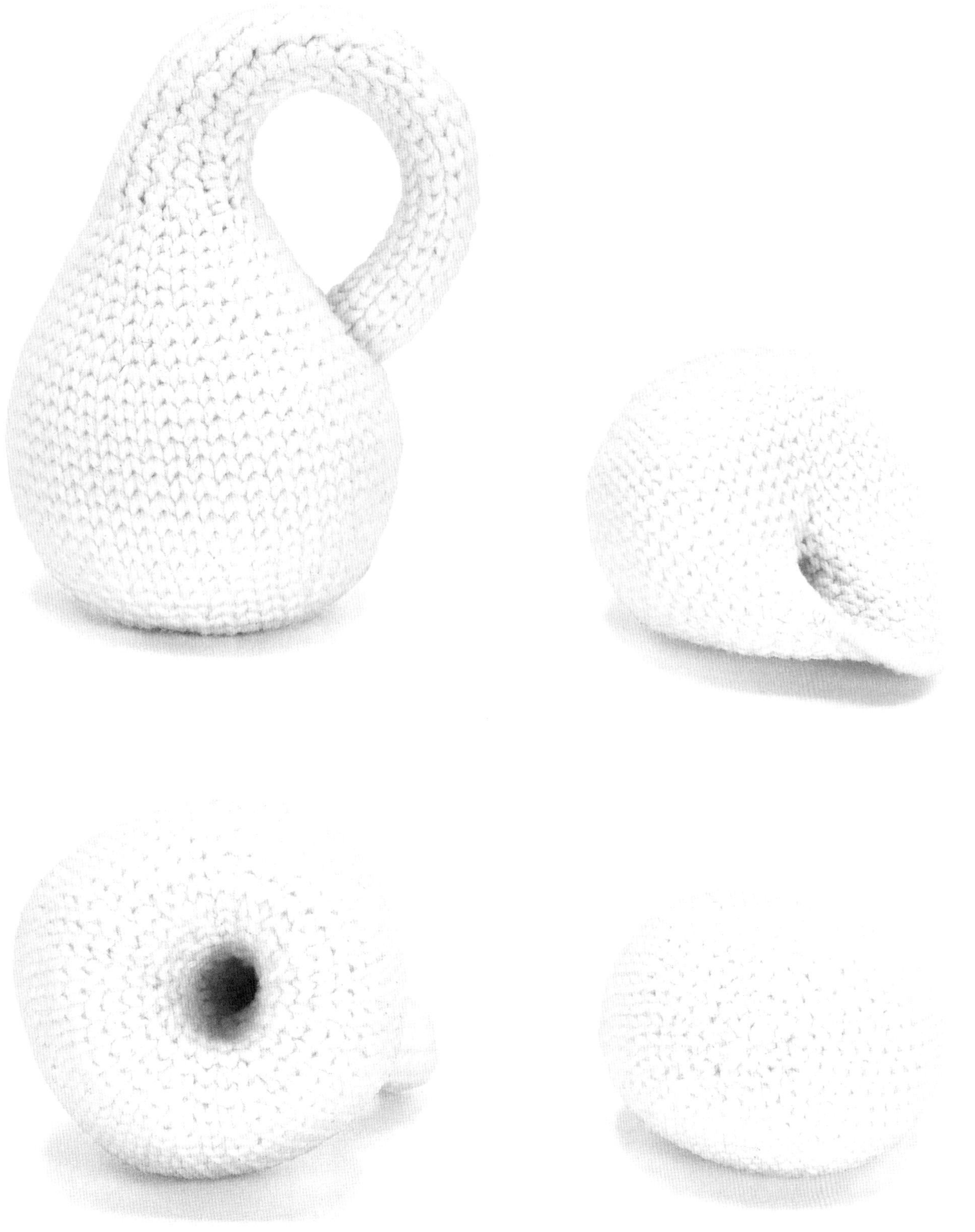

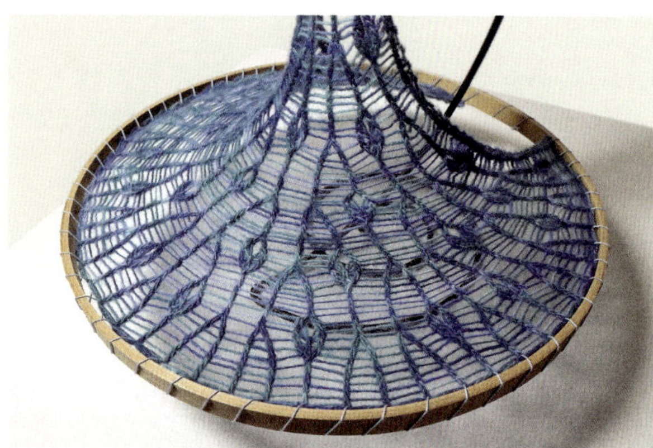

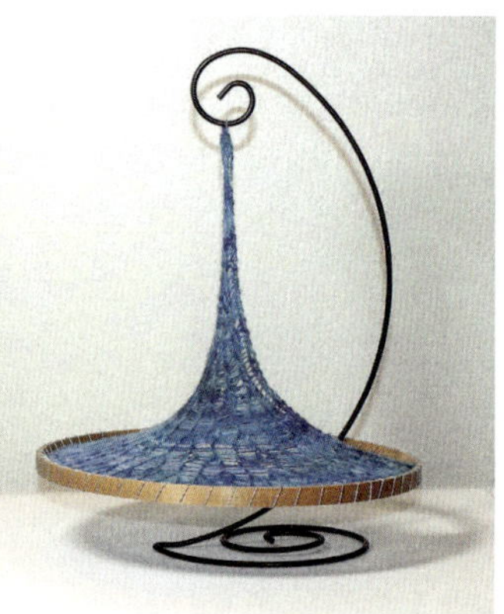

Susan Goldstine, *Fibonacci Downpour* (detail and overall), merino yarn, cotton thread, embroidery hoop, 13 ¼ × 10 × 10 inches

Carolyn Yackel, *Abundance* (top, two views), 2021; *Yearning* (lower left), 2021, *Interconnection* (lower right), 2021; embroidered thread

Susan Goldstine, meanwhile, visualizes the Fibonacci sequence using knitting. Because the sequence is defined in terms of itself (each number in the sequence is created by adding the previous two numbers) it is recursive in nature. In her *Fibonacci Downpour*, each row of knitting contains consecutive Fibonacci numbers in the branches and drop-like shapes. With each additional row, the sequence is manifested as an expanding pseudospherical form.

Models that make particular concepts tactile—offering embodied experiences to support the mathematical theories behind them—can be especially powerful teaching tools. In Hanne Kekkonen's crocheted models of Möbius strips, Klein bottles, and other nonorientable structures, each form comprises a single surface lacking an upward or downward orientation, meaning that they cannot be oriented in a single direction to define "up." Further, each surface is one-sided, a concept difficult to understand without a physical example. But by seeing or holding one of Kekkonen's crocheted objects, the viewer can observe directly that only one nonorientable surface comprises the shape.

Some techniques are more suitable than others for modeling certain principles. Carolyn Yackel harnesses properties of spherical geometry in her construction of Temari, stitched hand-size balls characterized by intricate geometric patterns. Yackel uses her designs to illustrate myriad geometric principles, including rotational symmetries on a spherical surface and how triangles on a sphere are able to have interior angles that do not add up to 180 degrees.

The accuracy that is inherently required for quilting, crochet, knitting, and embroidery makes them ideal mediums for communicating mathematical concepts. By creating fiber-based visual and physical forms, mathematicians see and understand math in new ways, as well as introduce complex concepts in familiar materials, making them more accessible to students and audiences of all ages. The opportunity for students to see and recognize—and, in some cases, touch or make—the patterns and structures that undergird mathematical concepts via the fiber arts makes them a powerful pedagogical tool.

1 For Ellison's patterns and teaching tools, see: Elaine Ellison and Diana Venters, *Mathematical Quilts: No Sewing Required!* (Emeryville, California: Key Curriculum Press, 1999); Ellison and Venters, *More Mathematical Quilts: No Sewing Required!* (Emeryville, California: Key Curriculum Press, 2003).

Answer: 233 pairs of rabbits

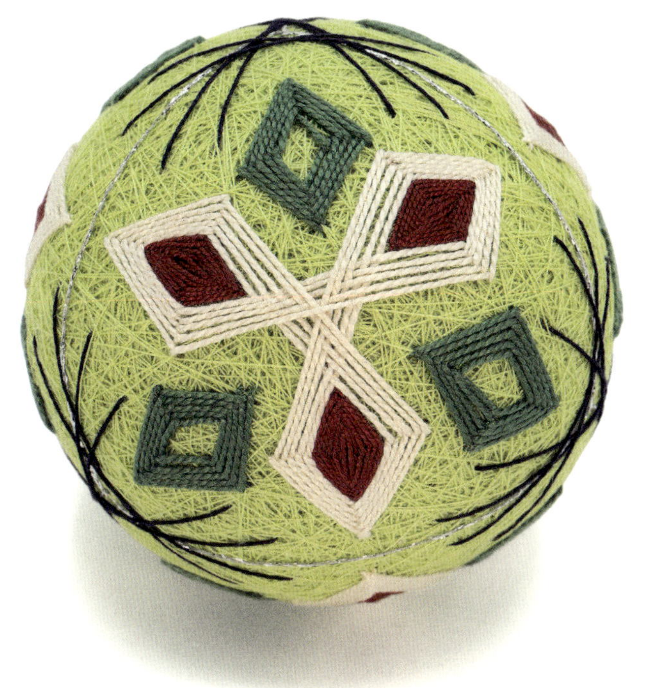
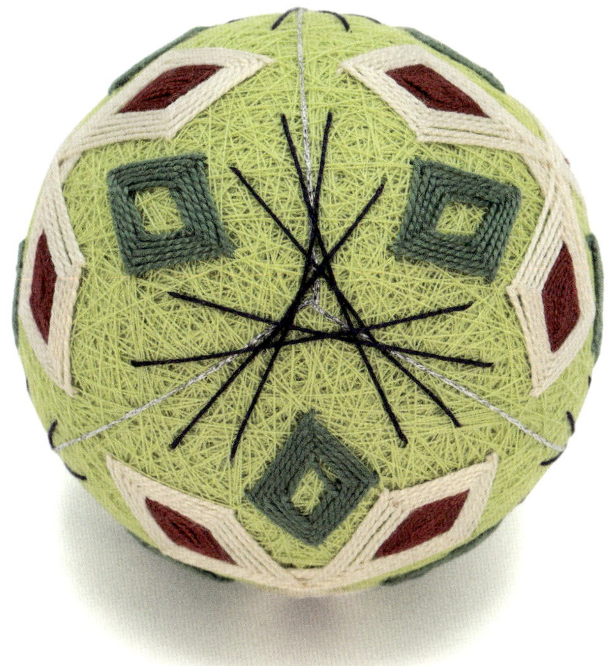
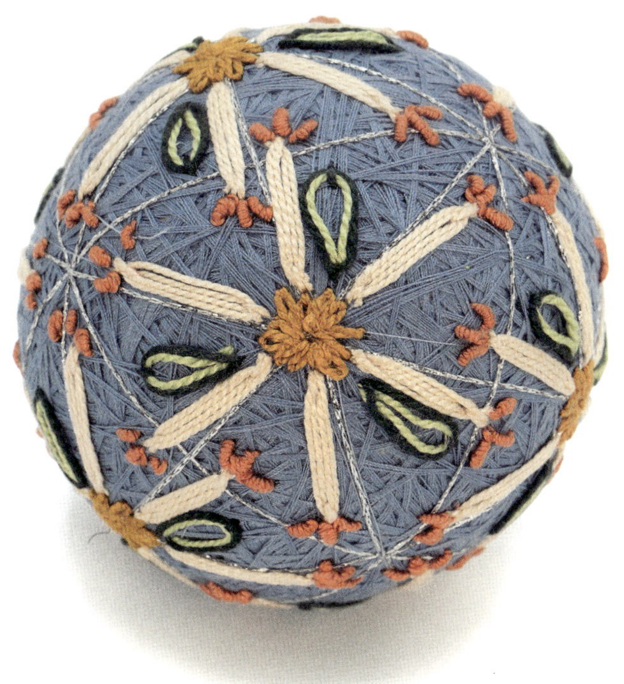
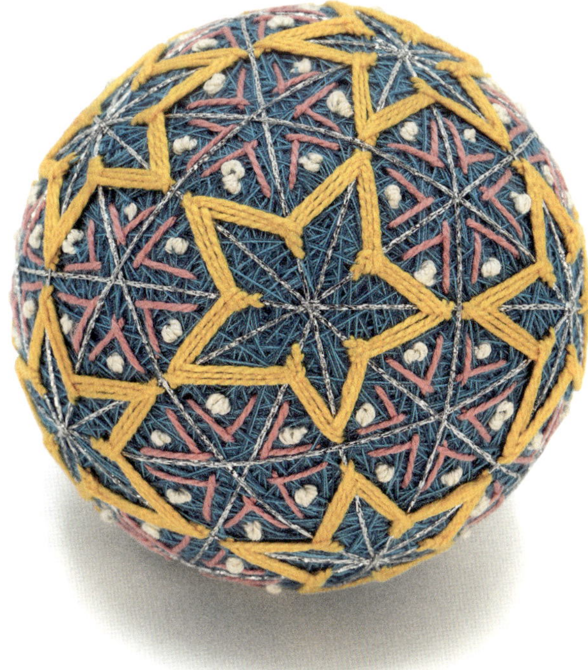

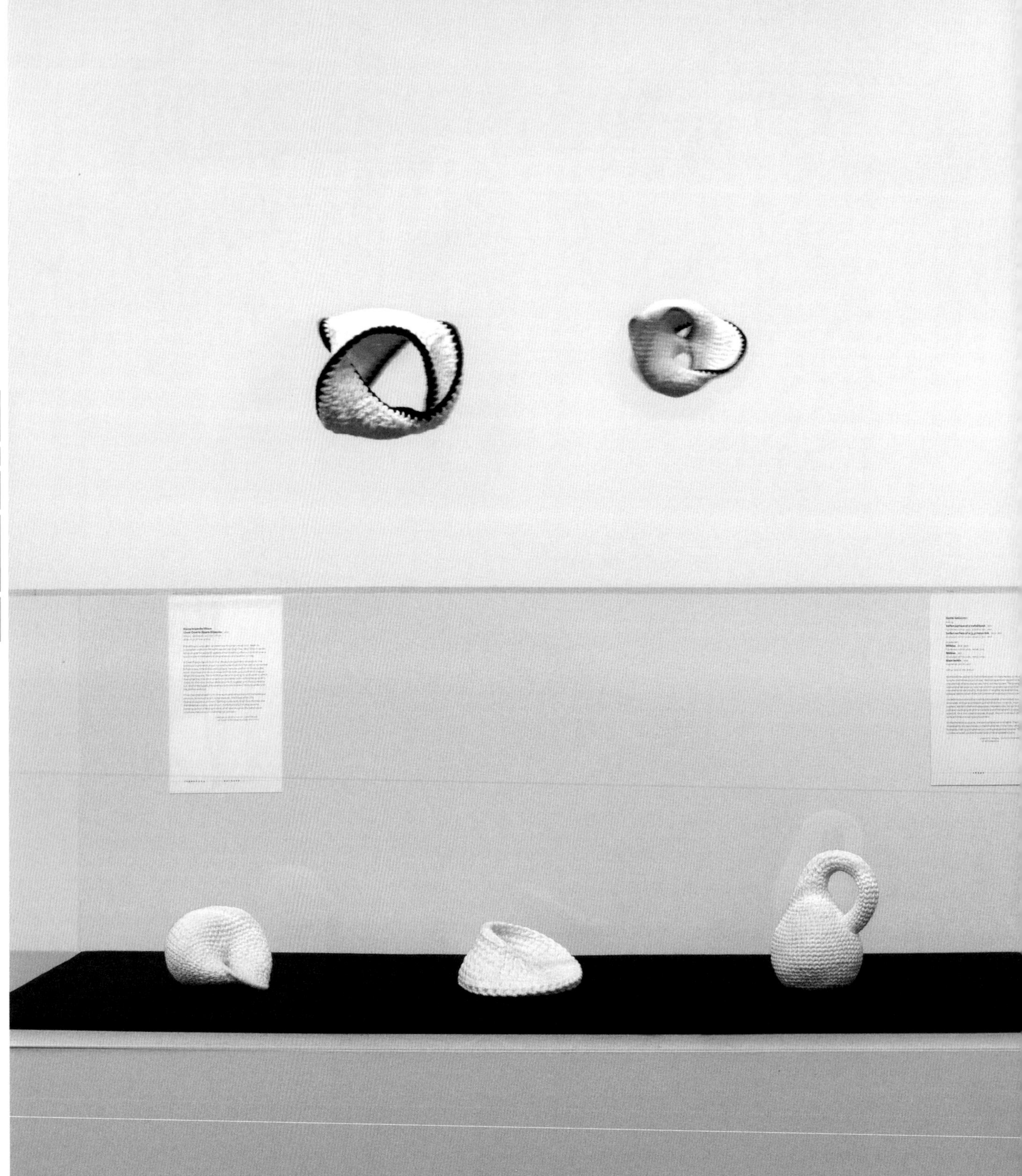

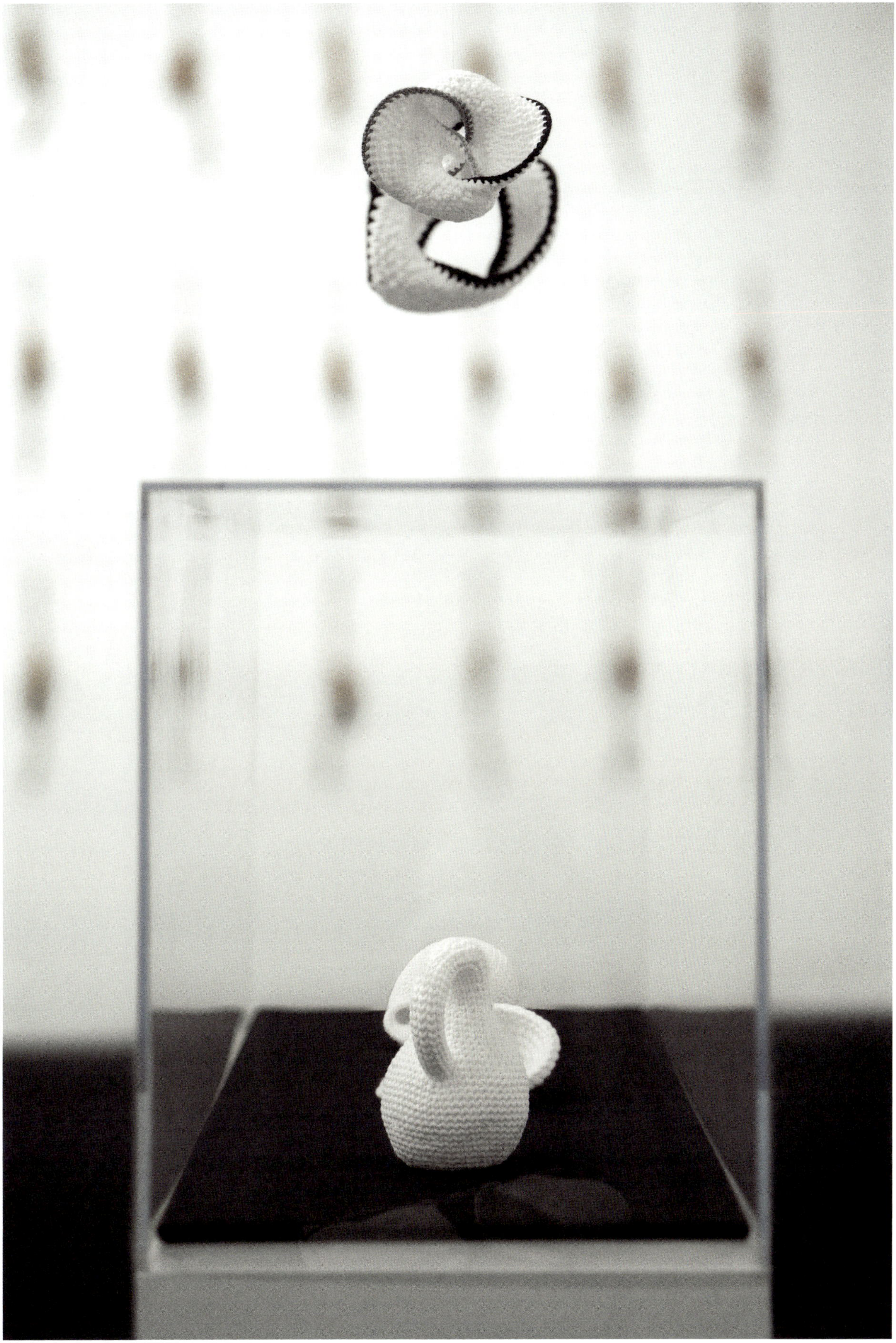

Installation views, Radical Fiber, featuring work by Hanne Kekkonen

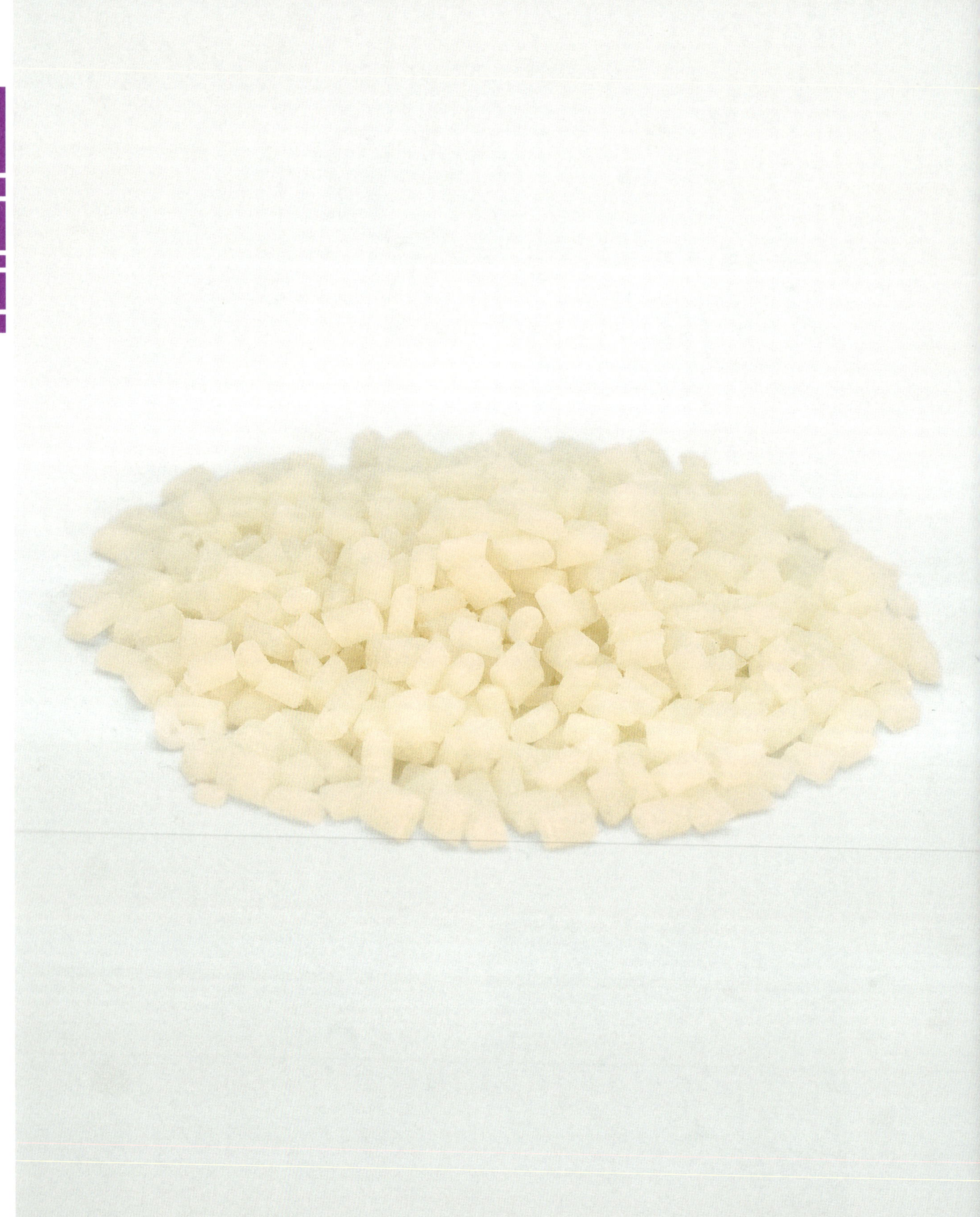

THE FUTURE OF TEXTILES AND SUSTAINABILITY

In Conversation: Preeti Arya, Alissa Baier-Lentz, *and* Juan Hinestroza *with* Nurcan Atalan-Helicke

NURCAN ATALAN-HELICKE: It's exciting to be in conversation with amazing inventors, entrepreneurs, and textile professionals who are making changes in the sustainable fashion industry. Could you each introduce yourselves?

PREETI ARYA: I am a professor at the Fashion Institute of Technology in New York, where I teach textile chemical finishing and performance textiles. I do research in textiles; I'm a textile scientist and a consultant. And I am a flag bearer for sustainability at every segment of the textile industry.

ALISSA BAIER-LENTZ: I'm the cofounder and COO of Kintra Fibers, which is a material science company on a mission to eliminate petrochemicals and microplastic pollution from the fashion industry. My role at Kintra is as the liaison between the science and the hand-feel and fashion components that brands are looking for.

JUAN HINESTROZA: I'm the Rebecca Q. Morgan Professor of Fiber Science and Apparel Design at Cornell University. I am a chemist by training, so I work at the interface of nanomaterials and textiles. In the last couple of years, we have been working on depolymerization and other strategies for eliminating contaminants in the textile industry.

NAH: Currently, polyester provides about half of the global fiber we use in clothing. Can you talk about the environmental footprint of our current textile production and why we need to think about sustainability in a different context? What are some alternative fibers that are already commercialized or are in the last stages of being commercialized?

ABL: When we, at Kintra, set out to design a material, we look at the full life cycle. We design toward the end of life or toward the next life of the material, right from the beginning. Today's textiles are problematic because they start from fossil fuel–based sources, and every time you wash garments made with polyester or nylon, which are the leading synthetic fibers today, millions of microplastic particles are released during the wash cycle, and they end up in the ocean. This is problematic. The equivalent of about fifty billion plastic bottles is released every single year in these microplastic forms, and they end up in our ecosystem. It is estimated that humans eat about a credit card worth of plastic every single week.

And then, if we look at the start of life, the production for polyester alone consumes more oil in one year than does the entire country of Spain. When we start putting these things into perspective, you can start to understand, on a human level, your impact. The fashion industry today is looking to recycle polyester as an alternative. But that process is kind of like taking a plastic water bottle, chopping it up into tiny pieces, and putting it back in the ocean; it disrupts the bottle-to-bottle recycling systems that are in place, effectively turning that plastic bottle into a one-way ticket to a landfill because the textile recycling systems aren't that advanced yet.

Our research at Kintra is designed to provide scalable solutions to address each of these problems. We start with 100 percent bio-based sources: for instance, we use sugar syrup or glucose for our inputs, rather than petrochemicals. That's the start-of-life component, and it provides a solution for that microplastic issue. The end-of-life component is important to consider as well, designing for that future circular system. We design for the garment usage but

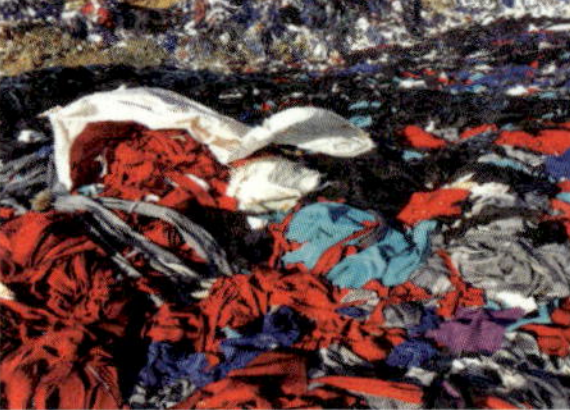

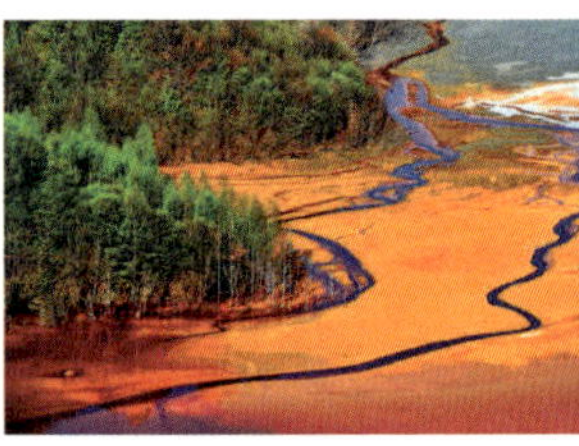

Textile waste in a landfill near Damascus, Syria (top); and a polluted river in China (bottom).

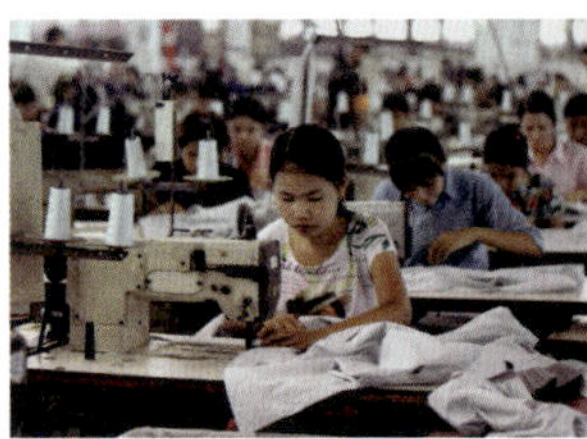

Worker at a garment factory in Yangon, Myanmar, 2012.

also consider issues like mono-material construction, which makes mechanical recycling easier. We design for compostability, for compatibility with nature.

PA: At FIT, because we are a leading design and technology college, we pay attention to the sustainability aspect as well. We know that fast fashion has done a lot of damage to the planet. However, consumers are not stopping their purchasing. Until recently, most people were not even aware of these issues.

What's going on in the manufacturing countries is a tough reality. **The textile industry is an oil-heavy industry, which doesn't offer biodegradable solutions.** These petroleum-based garments are produced so cheaply by so many fast-fashion brands that the picture is very ugly. Somebody pays a price because we are able to buy garments very cheaply—especially kids and people in other countries who are not being fairly paid. The person who makes a garment that we buy for fifty dollars might earn maybe sixty cents. Is that a fair deal? And how sustainable is this process? Does this mean we have been wearing tar all this time? Sadly, yes.

Sixty percent of the waste from textiles goes into landfills, where no biodegradation takes place. But there are solutions to revolutionize this throwaway society. We have sustainable fibers, but we have not paid attention to them because the revenue being generated by cheap, petroleum-based fibers has made some brands very, very rich. I hate to say that. Natural fibers typically cost more and the profit margin is less. But we have natural fibers—a lot of them. Hemp is growing in popularity. There are some issues with hemp, but once the red tape is removed, I think we will do well with the production of hemp. We also have cellulose-based and protein-based fibers. They are expensive and a little difficult to maintain, but they are biodegradable and last a long time. These are two categories of fibers we cannot ignore. And I think more scientists need to work on putting better textile-recycling systems in place. This idea about how we get rid of PET plastic or polyester is super important because it actually controls the entire economy for the worldwide textile industry.

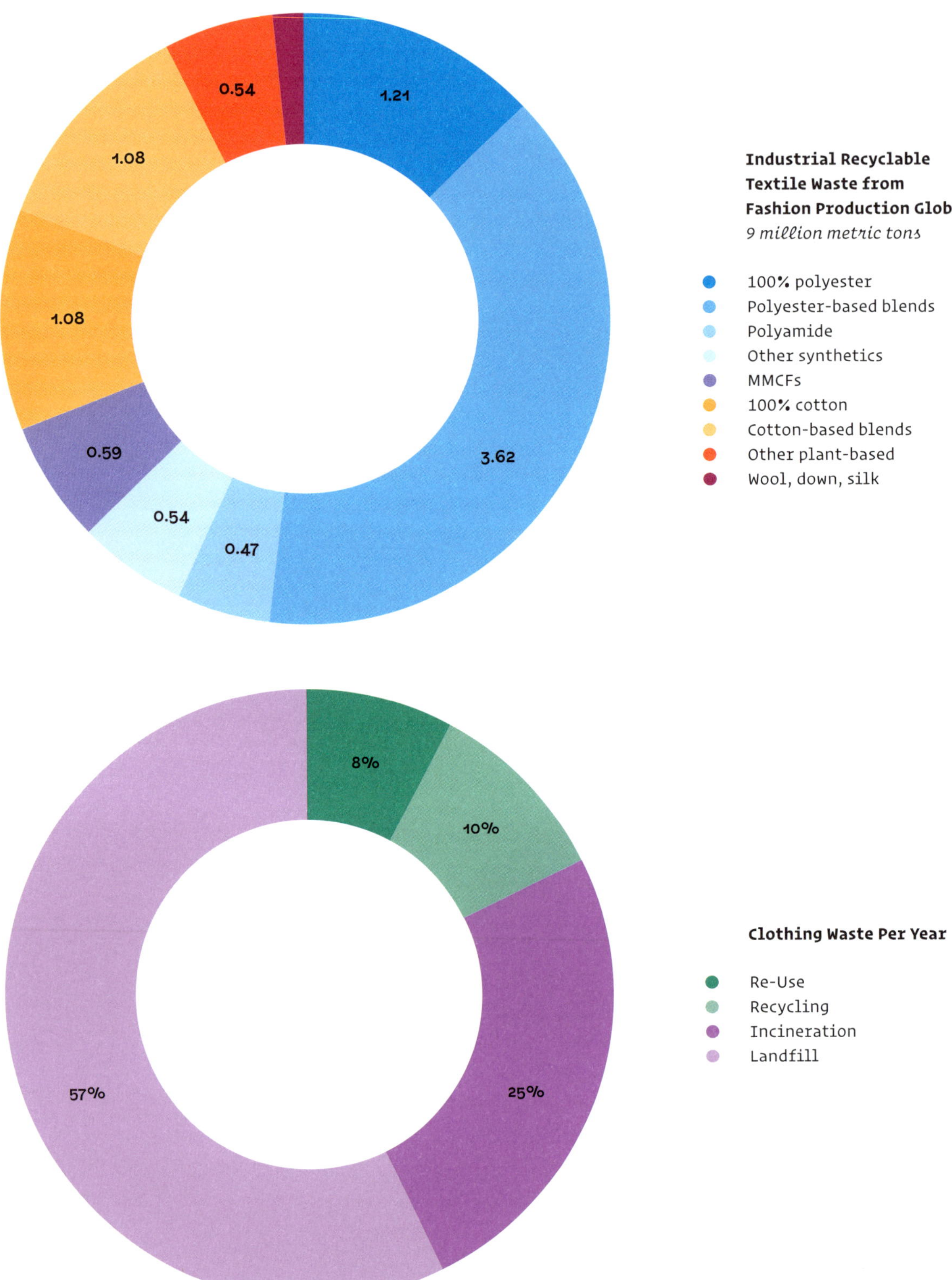

These diagrams, from Reverse Resources (top) and the 2017
Pulse Report (bottom), demonstrate the fashion industry's impact

We have some new terminologies in the textile industry: *biodegradable* and *bio-based* plastics. Biodegradable plastics can be broken down by microbes, and bio-based plastics are made partly from biological materials. Yesterday, I was speaking to a manufacturer who is focusing on biodegradable plastics. However, their product has not been approved because the current regulation requires degradation or decomposition to take place in 120 days. The polymer they have created—it's a nylon and polyester bicomponent yarn—degrades in something like 280 days. So, they still have some work to do, but the idea with biodegradable plastics is that, given a certain compost-friendly environment, they can be eaten away by enzymes. In terms of the bio-based, we have polylactic acid, or PLA, and we have polyhydroxyalkanoates, or PHAs. We also have thermoplastic starches, but they are not that popular in textile use for garments yet. And, as Alissa said, we have to take care with synthetic fabrics if we want to contain the waste, because when we are wearing or laundering or drying these fabrics, anytime there is friction, we are releasing microfibers into the air or the water system, and they are contaminating our environment. We shouldn't be surprised if we have microplastics in our bodies slowly cooking into a cancer.

More and more, we need to pay attention to certifications. *Sustainability* is a broad term, and people take advantage of the word. Production of PET can be made to look sustainable and production of cotton can be unsustainable—even though PET is nonbiodegradable and cotton is biodegradable.

> **JH:** This is not an easy problem; there is not a simple solution. It has to deal with chemistry, with psychology, with public policy, and with consumer behavior. I don't believe recycling is a solution because recycling only encourages more production and more consumption. Chemically, it's almost impossible to recycle or reprocess 100 percent of a single material. It is possible to use some of these raw materials in other processes that have more value. However, using them in the same process doesn't make sense: for polyester textiles to be recycled into more polyester textiles doesn't make sense. In our

lab, we work on decomposing, in this case, polyester with different chemical strategies, and then using them to create new lubricants, new dyes, or other new high-value compounds.

Some companies, unfortunately, communicate messages about sustainability through Instagram or similar platforms. They show beautiful pictures and beautiful messages. But when you look deeply into the material and understand the samples, you realize they are making empty promises.

This is a serious, serious issue because polyester molecules are incredibly abundant. If you think historically, we had mostly natural materials until 1938. We didn't have synthetic-based materials. So, this problem has been about eighty or ninety years in the making. And the effect has been cumulative: even if we stop the production of polyester today, the problem will not disappear.

Even if we replace all the textiles with bio-based materials, we will run out of sources for biomaterials. At the volume that we operate now, we will run out of trees. There is not a simple solution. We have to deal with economics, we have to deal with culture, and we have to deal with behavior. But, most importantly, we have to deal with public policy. Governments have to incentivize these changes—all the transitioning to new materials or managing the products from start of life to end of life—because most of the time the production system is invisible to the consumer. Consumers don't see what is actually happening when textiles are made in Sri Lanka, or in Bangladesh, or in India. And consumers don't see what happens when we throw away garbage in Malaysia, or in Chile, or in other countries.

The amount of waste we generate is insane. We produce about fifty million tons of polyester per year and less than 1 percent of that is actually repurposed—and usually on things that have less value and create more contamination. Chemical recycling and mechanical recycling require a lot of energy, which creates additional issues with sustainability. We have to find a comprehensive solution.

> **NAH:** Your summaries present a wicked problem that needs to be addressed it in all of its different aspects. Let's start by talking about the material design and production of alternative fibers. What are some of the challenges and concerns

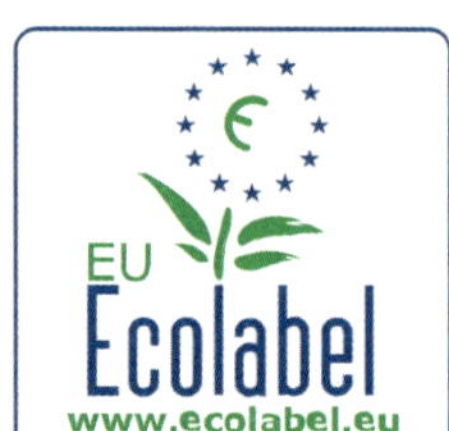

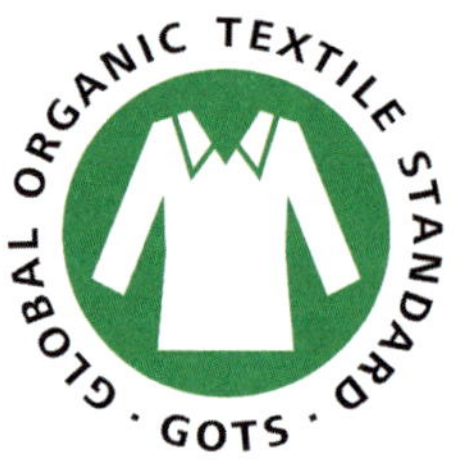

It is recommended that consumers look for specific certifications rather than the word *sustainable* (or similar) on product labels and to be aware of what each certification signifies.

166

around coming up with alternative fibers and commercializing them? Then we can talk about the psychology a bit more.

JH: When you mention *alternative fibers,* I believe you are referring to natural fibers, right?

NAH: *Alternative fibers* includes natural fibers, but it is also what you produce with nanotechnology, in the sense that you're changing the basic properties of the fibers so they are biodegradable, or can actually become more versatile.

JH: Fibers can be made, for example, from repurposed cellulose that comes from trees or from paper. Some companies have been trying to reprocess cellulose into another fiber for many, many years. The main issue with these companies is that they produce a very small volume of that biofiber, and they mix it with nonbiofiber—yet they generate a lot of publicity about the small volume of biofiber.

The other issue related to fibers is that they usually have a color—because, as humans, we are obsessed with color. But **color, especially gray, is one of the main sources of pollution.** If you are buying a fiber for a company, and you want, say, a specific blue, but the product is not that particular blue, you reject it, right? That blue fiber is actually sent back and it becomes black or another color. It's not only the raw material, but also the lubricants that are being used, the additives, and the dyes—they create this chemical mixture that is very difficult to separate out. It's a technical challenge from the chemical perspective.

ABL: This is everything we've been talking about in our Brooklyn lab since we set out to design a new material. My cofounder, Billy McCall, is a nanoengineer, so he has a background similar to yours, Juan. I love that you're bringing up the dyeing process. Synthetics are the biggest contributors to pollution; they are the world's most prolific fibers. The challenge of dyeing synthetics, at least PET, is that it requires high heat, and often chemical additives are utilized. And so, when we look at the value chain starting from oil extraction, we want to avoid these issues. Step one: design a material that can be bio-based. Step two: make sure the chemical composition is inherently compostable. Instead of being sent to a landfill, garments made with our material can go to industrial compost and return back to the earth from which they originated. Our compostable materials are specifically designed for aerobic digestion processes, meaning that our material, in microfiber form, can be digested to CO_2 and H_2O, which keeps the ocean clear of microfiber pollution. Next, when we think about the problems related to dyeing the material, we ask, can we do something in our material design component that can address some of these downstream challenges? For instance, can we design our material for lower-temperature dyeing processes, allowing it to take on dye much more easily than PET, and eliminating the need for some of these chemical additives that are currently used in today's dyeing processes for synthetics? When we set out to design a new material, we really look at each aspect of the value chain, look at the challenges, and try to address them at the level of their molecular structure.

  In terms of scalability, we've designed our material to be a direct fit into the same equipment—for resin production and fiber extrusion and textile yarn production—that is currently used for PET, which allows us to scale rapidly. We made about 100 kilograms of resin last year, which was our first year of business, and we have partners to scale up to 5 metric tons of resin this year. So, because our products have the ability to get out of the lab and into the hands of companies and our mill partners, we're starting to recognize additional components in the value chain that we need to address. And we've started to work with partners to achieve some of the pricing requirements that the companies need as well.

PA: I'm a neutral person; I am not taking any side or endorsing any brand or company. But we need to make consumers aware. Now, more and more brands should focus on sustainability-oriented advertisements in the sense of communicating what they do differently to make their process or their product generate a smaller carbon footprint. For instance, with rayon, people say, "It's rayon. It's like cotton." But no, it's not like cotton; it's bad. First, it is taking away all the trees from Canada and the United States. Second, the chemicals that

Installation view, *Radical Fiber*, featuring Kintra Fibers components and samples (wheat straw, lab vials containing resin and fiber, spool of yarn, dyed and undyed hand-knit fabric samples)

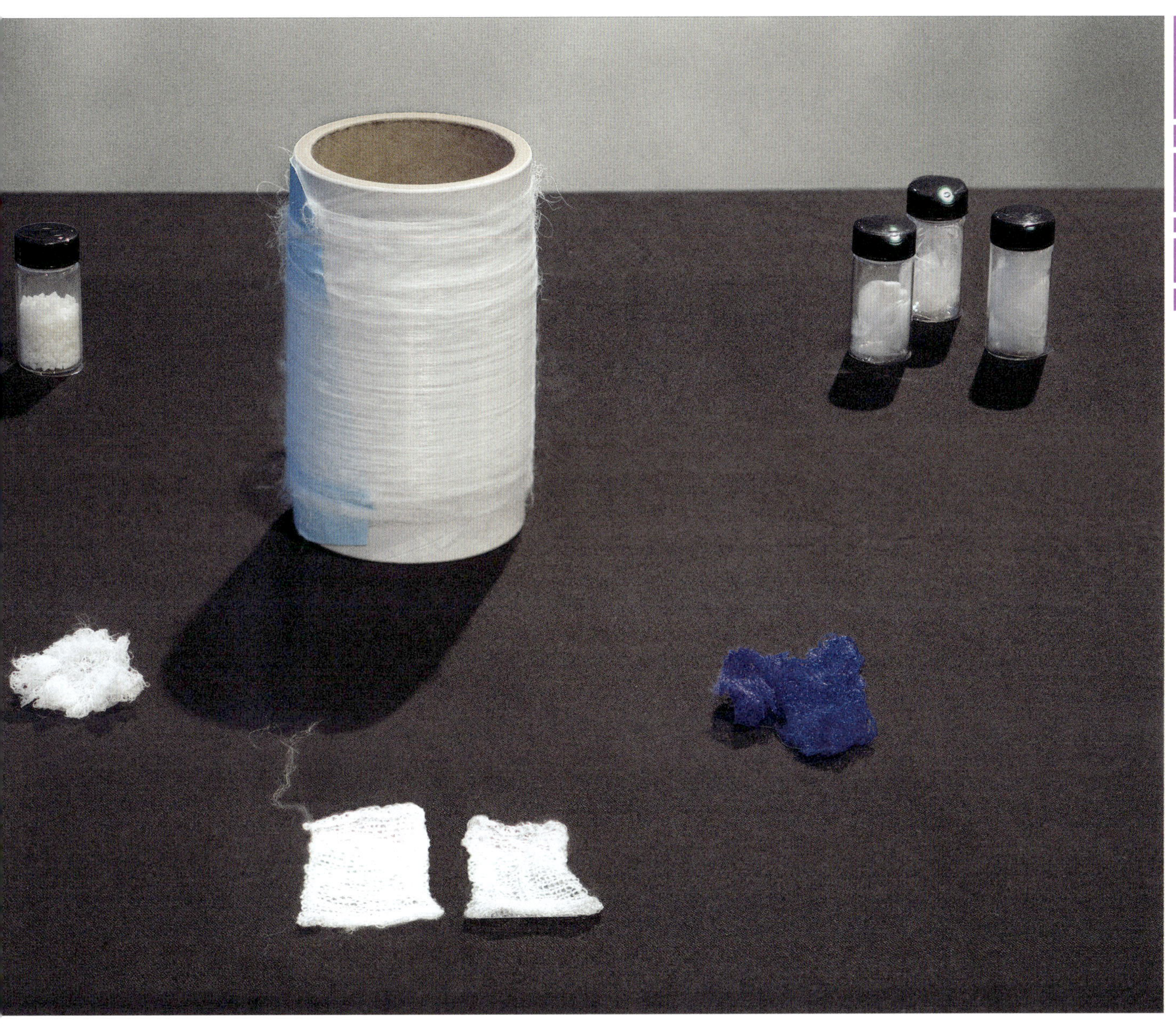

go into manufacturing rayon are bad. Third, the humans who work in those factories are getting diseases. Lyocell provided a little relief because its production is a closed-loop system. I tell my students, "If you really love rayon or regenerated cellulose, go for Lyocell instead of viscose." Making awareness-based advertisements will help. Media can help.

Alissa and Juan, please help me understand something. I know that polyester is cheap—around 90 cents a barrel. And the profit margin is high for polyester garments. Look at Nike, Adidas, Reebok, Under Armour . . . Their executives all became billionaires selling polyester. Natural fibers just don't have that kind of profit margin. As an entrepreneur, Alissa, how are you balancing sustainability and yet managing your profits to run a business? And, Juan, when you work with a company, what do they demand from you? Do they demand sustainability, or do they just demand a product that will sell and generate revenue?

JH: Well, I am very lucky. I'm a professor, so I don't sell anything other than my science, and I work with my students. But I do work with companies, especially investment banks that have a great interest in creating a sustainability portfolio. They want to give meaning to that particular expression. It's a complex problem because there are so many solutions, but generally they are too small in terms of scalability and volume. I'm not in a position to defend the polyester industry, but something positive is coming from them: they have the ability to produce garments in large numbers and at very low costs because they use materials that are easily available around the world. Remember that cotton and other natural fibers don't grow everywhere. Some places have no trees, so what are they going to do? And there are not enough trees to replace all those synthetic materials, at least not at the moment.

A few years ago, there was a company that claimed that they were producing a garment that was 100 percent recycled polyester. The company sent us samples. When we analyzed them, we found that they were not 100 percent recycled polyester. But this company knows that if they make that claim, they can charge 30 percent more. We went to the factories and found that they labeled the polyester as "recycled" without it really being recycled—because, chemically, it's impossible to identify that type of material. Basically, once one of these organizations provides a company with a certification, then the brand can hide behind the certification and say, "We have a recycled polyester."

I have seen instances where the factory for bottles is next to the factory for yarns. They take the material from bottles and transfer it to yarns. So, technically, the fibers are made from plastic bottle material, but it doesn't make any sense. Governments have to start with better regulatory processes and understand the volume of the issues. **Even if we replaced all synthetic textiles today, we would still have to deal with what we have now.** And there is no incentive at this moment to solve that problem. There is some incentive to reduce a little bit of the consumption, but there is no incentive to clean existing pollution. And that's a wicked public policy problem. I'm so glad that this generation cares more about these issues than mine did when I was a student. We were always looking to make better materials, to increase and speed up production. I wish there was a silver bullet that said, "Replace this for these." But there isn't one.

PA: We need to unite our forces as academicians and make our students more aware not to sell out to these big companies and brands and to continue to push for awareness.

JH: At the same time, I don't want to trash the brands and the big companies—because they have an important role to play, too. If something is going to change, we are going to need those big companies. We are also going to need these big plans to help us change. And we are going to need the big retailers to force their suppliers to make those changes. It's happening, but I don't know if it's happening fast enough. The rate at which we are consuming is insane. And everybody wants other people to change, but not themselves. They think, "I can keep my twenty pairs of shoes and my forty pairs of jeans, but you can only have two." A public policy may need to be implemented, like a shop plan. I hope that a merge between big brands, educational systems, governments, and NGOs can occur to beat this wicked problem.

ABL: I share that perspective as well. We're not setting out to compete with virgin PET. That's impossible for us to do in this day and age. However, with cooperation from governments, from brands, from suppliers, from consumers demanding change across different industries...If we take into account the externalized cost of virgin PET and factor that in, whether it's through a tax or a different component, policy can create a more equalized playing field. According to the research that we've done, our product is actually going to be price-competitive with recycled PET as we scale and grow. That's an attainable category. And remember, recycled PET still starts as oil. But our product has that higher price point that consumers look for, and the benefits that it delivers, again, are reducing microplastics, moving toward a bio-based source, and providing that end-of-life component. It delivers on that while still being price-competitive because we use the same manufacturing equipment and systems that are already in place. We can actually work together with the brands, with those industry stakeholders, to shift in a new direction and make it really easy from the material standpoint.

NAH: *Sustainable fashion* has become a buzzword, and there is a lot of greenwashing going on. Any company can make claims. As Juan was saying, there is no accountability, there is no monitoring, and there are no punishments or sanctions if a company makes false claims.

JH: It's even worse: there actually are incentives for them to make false claims.

NAH: And with social media influencers, this issue has actually been taken to another level. But, like you said, with government regulations; more international standards; more collaboration between governments, businesses, and NGOs; more cooperation between professionals, entrepreneurs, and academicians, in terms of research; and more monitoring in place, positive change can happen.

Let's talk about consumer psychology or psychology in general. What are some of the general business trends that we need to be thinking about? Can we look at consumer trends,

in terms of their expectations and willingness to pay, and see
how they match up?

PA: So far, fast fashion has taken the lead. Earlier, I discussed some nasty truths about the fast-fashion industry. So now, very slowly, some industry leaders are moving toward slow fashion and mass customization. At FIT, we organize a workshop every January with the Massachusetts Institute of Technology: one company sponsors the workshop, and members of the two faculties mentor the students. We come up with a product that is a little high-end but has desirable qualities and that will last longer. Through education and awareness, we are trying to break this fast-fashion cycle.

But that's a slow method. It would be more impactful if the big brands adopted the slow-fashion concept: instead of selling five T-shirts for $5 each, sell one for $100, but give it some e-textiles, some chemical finishes, make it high-performance and smart in a few ways. The point is that if we cannot get rid of PET completely, we can at least change the product's life span so it doesn't hit the landfill in five years or six months; it can be used for at least ten, twenty years. And mass customization is the opposite of mass production, where there are sweatshops and people are dying and the earth is dying. No, none of that. Mass customization is creating whenever there is a demand. So, we are working on it that way as well as promoting more biodegradable plastics.

JH: At Cornell, we are trying to use tools to incentivize behavior. For example, we use virtual reality that allows students go into the VR world of these factories and see how things are made. And we try to create messages that buying so many things is not as important as the real consequences of our actions. For many of us, a product's life ends when we throw it into a garbage can. Very few of us know what happens to that garbage after it leaves our home. And maybe we don't want to know. But when we travel enough, we find piles with our logos in other countries.

Things do not disappear; the Principle of Mass Conservation still exists. It just doesn't exist in the same country, but it exists in other places. The majority of the world doesn't live in the United States, so a big population in developing countries is not being served. At some point, the governments will have to act. And I think it's possible. The same thing that happened with cigarettes is happening with oil, with climate change, but there must be some coordinated action because shipping the problems—the contamination, the garbage—to other countries is not sustainable. **We only have one planet. And we need many, many actors to find solutions and create change.** As an educator, I think bringing awareness and knowledge and understanding is our tool. Other businesses will have other tools, governments will have policies, and, hopefully, this trend will change.

ABL: I'm going to share one anecdote about microbeads—those plastic beads that were put into face wash that went right down our drains. They were banned in the United States in 2015, and that gives me hope. There have been some initiatives to ban microfibers and microplastic-causing materials from entering the fashion industry, which would include polyester, nylon, and variations of polyester and nylon that are bio-based but that still have that biodegradability problem; as Juan has illustrated for us, it has to do with the chemical compositions of those materials. Some policies are moving in the direction of looking at the fashion industry's contribution to microfibers and working toward eliminating them. One of them is The Microfibre Consortium. Kintra is part of that community. So, there are some industry movements in the works, but looking to other industries that have taken action against these components can serve as case studies for us in the textile industry.

NAH: It is great that we're talking about the globalization of the problem—both in terms of the production chains and the waste chains. I come from Turkey, and in the 1980s, when I was growing up, Turkey was a textile production center. Now, the production centers have shifted to other countries where labor is cheaper, where there are fewer environmental standards in terms of chemicals. One of the things Juan mentioned is about government and how structural solutions, in terms of regulations, can really work. We have discussed awareness a lot, in terms of what consumers can choose. But I want to ask two questions. First, what or who should consumers trust? There are so many claims, like "eco-friendly," "responsible," and "recycled," and unless I'm willing to read a lot about each of them, I may just buy the rayon or recycled wool and be happy with it. And the second question relates to access and justice: What is the cost of these materials? Can everybody have access to more sustainable materials?

PA: I'll start with the second question—access to materials. A lot of countries don't even have access to firsthand clothing. These days, what we are doing is dumping our clothing waste into different countries—Kenya, for example. If we are a developed nation, and we have all the big brands, it is our corporate social responsibility, our ethical responsibility, to actually make an effort. That's why I agree with Juan. The government has to step in. But, it still falls on our shoulders, as educators, because we care. We'll continue to knock at the door of government and we'll continue to push, but we have to resolve these issues with a lot of discussions, with a lot of initiatives, with a lot of petitions. It won't be an easy path. We have to keep touting the benefits of sustainability and biodegradability.

And the first question you asked was about confusion. Yes, it's very confusing as a consumer, even for me. I have to get onto a company's website and read the details—and sometimes even that is misleading. A product will say it is 100 percent cotton. Then if you read the label carefully, it will say 96 percent cotton and 4 percent elastane, which is nonbiodegradable. It even happened with a very big label. We are social media aware and active, but we need to become a little more product aware and active as well.

JH: I agree with you. Our students bring in products and the labels claim to be something, and I say, "I'm a chemist. Let's go to the chemistry lab." And then we prove that the label does not correspond to the product. I've been to many factories. Sometimes, the labels are produced in one factory and the product is produced in another factory. Sometimes the mismatch is not intentional, but many times it *is* intentionally misleading, and that is scary. But this is the power of knowledge—having access to information, having access to ways to verify the composition of the material. We are working on a portable device that could scan, for example, and tell you if a product has nylon, or has a particular dye. If we have spectroscopy technologies that allow us to capture that information in real-time and educate consumers, they may make better decisions.

But there is also a consumer problem to solve. We can recycle, we can replace one material for another, but if the volume doesn't change, the problem will not disappear. And that's where psychologists, people who work on behavior of big numbers, on public policy messaging, on social media can act. You cannot keep producing these misleading messages that everybody retweets or reposts and that claim to be sustainable on Instagram. Ask: What does it mean? How many kilowatt-hours are you saving? How much water are you consuming? When you ask these questions, most people don't have answers. And the role of professors is to educate students to be critical. They can read these magazines that portray beautiful tales, but ask them the tough questions.

ABL: At Kintra, we're scaling. We're really small right now. We're a team of three. But the three of us talk frequently about how we envision this future that we are discussing today: What can we do as a producer to ensure we're looking at things like land and water usage? What kinds of farming practices are our partners doing on the farm? What kinds of feedstock can we use? Glucose can come from a number of different feedstocks. So, what is the life-cycle assessment of each individual type?

And then we compare our product to the global-warming potential for nylon, for PET, for recycled PET, and we make sure that we actually offer benefits throughout the entire system. We take that very seriously. We look at the data that's publicly available, read life-cycle analyses, and let that guide our decision making, so that when we grow, we know we're partnering with people who are doing the right types of farming, working efficiently with the feedstock so it doesn't take as many steps and release as many emissions—from the cradle-to-gate standpoint.

Some companies give us real hope. If we can set up the right partnerships, the cradle-to-gate calculation will potentially be carbon negative. We let the data lead us. And then, we can help brands communicate with specific scientific terminology about the benefit that our material provides when they seek to market their products. And we will make sure that our material isn't blended with spandex or another component and that the integrity of our material remains through the entire value chain. It's a great point that you, Preeti and Juan, raised about the life-cycle responsibility from the material's maker, all the way through to the brand communications.

NAH: Juan has talked about the industry taking measures, but also about how research can be integrated into these false claims. I think the industry itself can start policing or punishing companies that make these false claims. If a business feels that its interests are being jeopardized by these false claims, then there could be bigger changes as well providing better information to consumers.

I'm a food studies scholar, and I also teach environmental sustainability. One of the exercises my students do is take a processed food item and start tracing the ingredients. Usually, they get stuck at certain places because some of these items are patented or trademarked and the companies do not necessarily share the information. The students feel frustrated, and the assignment shows that our consumer knowledge is limited. Unless we can have more transparency in the industry itself, the consumer knowledge or awareness will be a weak component of the solution.

AUDIENCE: For fiber artists looking for practical guidance, is there a yarn on the market right now that isn't going to kill the environment? For me, the more I try to be good about what I buy, the harder it is. I see something that says "cotton yarn" and I flip it over and read the ingredients list and it's 15 percent acrylic. It is really difficult to buy yarn without plastic.

JH: If you go to some countries, this information is readily available. For example, in Finland or Denmark, you can get more information about where things are coming from and what the materials are. But some countries have regulations that hide access to the knowledge. The best way is to actually test the material yourself. **We have students in fashion and management, and we teach them chemistry. Testing is how to know if a product is real or not.**

The real difference for climate change will come from individuals, and the only advice I can give at this moment is to consume less. Replacing materials one for one is not going to happen; creating biodegradable materials doesn't have the volume or the capacity. Knowledge about what is actually in the supply chain is important, but at the end of the day, most people make decisions based on their pocket, and we need to create the proper incentives so the pocket will decide to move toward solar energy, or toward a particular material, because otherwise it's not going to happen. We created this problem eighty years ago and we cannot solve it.

PA: We deal with organic chemistry and polymer material science. What we do—if we can, if the situation permits—is we burn the fiber and smell it. If it smells like paper or burnt leaves, then it is cellulose or 100 percent cotton. If it smells like burnt hair, then it is 100 percent wool or silk. Petroleum-based fibers create a black bead at the end and a very sooty flame.

I have been interviewed so many times by journalists and attorneys about false claims and misleading information. Don't put your heart and soul into a brand. They are running a business. Some of them are very good, but some of them are not so good. Have an open mind for any brand, new or old, that gives you a quality product and gives you what it claims.

ABL: On the manufacturing supply chain side, we're really excited about some of the mills that we've had a chance to interact with—people who are incorporating more and more natural dyeing processes, natural dyes, and innovative new fibers. Brands are emerging—smaller, midsized, some that are actually growing very fast—that do look stringently at their supply chain.

When I look at a product, for example, skincare, I'll look at each individual ingredient and do a quick search. Typically, I can find information relatively quickly—actual data that's accessible, rather than just the top-line score that they give. It's the same for the fashion industry. Step one: look at the label. Step two: look at each individual component and understand what you are really purchasing. It takes a little bit of effort, but I find that I then keep the product longer and really take care of the material and become a more responsible shopper.

And I love thrift. I try to live in natural materials and natural dyes, then go out into the world for limited times in thrifted traditional clothing.

PA: I support thrifting so much. Go for garage sales.

NAH: I have been working with small farmers in Turkey who are reviving the wheat varieties that were domesticated ten thousand years ago. And several years ago, I met a Pakistani entrepreneur who is reviving colored cotton. There is a farmer in California who has been experimenting with these colored cottons, which means they're not dyed, and they have been producing and selling these yarns. I always suggest working with local farmers. There are a lot of fiber farms around us, near Saratoga Springs, New York, with llamas and sheep. Some of them come to the farmer's market, but some of them also organize fiber tours and other programs. Once you connect with them, you will be amazed—they can put you in touch with a larger network of more sustainable yarn people.

AUDIENCE: Normally when we discuss sustainability, we are looking forward and considering new technologies and future impacts. Are there historical models for sustainability or specific movements across time and cultures that could create a framework for our present efforts?

JH: I'm not a historian, but because of my work, I have read a lot about the history of fibers, and **most of the processes involved with making textiles have never been sustainable**. In terms of producing pollution or using materials that contaminate, some cultures actually isolate the people who, for example, deal with leather or produce the color for fiber. They have to be a different class, and often are far away from the town. As an engineer, I think thermodynamically—you cannot produce something out of nothing. What you can do is minimize the impact, but the impact will be there. People in the past had some incredible ways of thinking and doing things, but because their volume was not so big, the impact was not so noted. Now, we have a humongous volume. The impact will be noted. The ecosystem, the earth, is not able to dampen those effects fast enough to be avoided, like it did in the past. Also, the population has dramatically increased, and as we consume more, we need more energy, we need more products. I have not found any examples of enterprise that are completely sustainable for the long term.

PA: You are completely right, Juan. The most unsustainable trend is our exploding population. Regarding textiles, India is a warehouse of traditional textiles. It has lots of growth in natural fibers, hand-spinning, and hand-weaving, so that is very sustainable. And they are trying to preserve these traditions as much as they can. Is every company following that model? No. There are limited companies, and they follow the certification that they get from the government of India and say, "Okay, this is handpicked, handspun, handwoven." It is a sustainable practice because the spinning communities are run by nonprofit organizations, so are the weaving communities, the artisans. But those are very limited in numbers and their marketing is not so good. So, are there such companies? Yes, but it is a cottage industry.

AUDIENCE: For Kintra Fibers, what is the source of your glucose? How sustainable is it to divert foodstuffs to fiber production?

ABL: The entire production of bioplastics, even incorporating growth, is about 0.017 percent of the global agricultural market. We start with a bio-based plastic that is then melted into a fiber; it is the same production as PET. This is not to say that it's not important to consider food systems. It absolutely is. But today, we make enough food to feed the world. The fact that we aren't actually feeding the world is a supply chain failure. We need to fix our food-distribution systems rather than say we should never pursue bioplastics that are based from sugar because doing so can interrupt the food-production system. We get asked quite frequently whether we are using a first-generation feedstock. The answer is yes, because it is what is commercially viable and available. We could use a second-generation or even a third-generation feedstock. And just to clarify: first-generation feedstock is something like a corn or wheat; second-generation would be something like a corn stover, so a leftover biomass; third-generation would be something like an algae.

Kintra's industry partners can take first-, second-, or third-generation feedstock for the production of our inputs. The inputs that we use are sold to various industries, including food and beauty. When you compare the glucose from corn to the glucose from corn stover, first-generation actually has fewer carbon emissions. So, we really need to look at the whole system. And when we compare our fiber to PET nylon and recycled PET, ours is drastically less, in terms of its global-warming potential. And the corn, the first-generation feedstock, is actually carbon negative at that cradle-to-gate stage.

It is important to consider that question about food systems, but more importantly, we need to consider the emissions profile. We look at the agriculture practices of the people with whom we are partnering. What type of agriculture are they doing and how does that contribute to global warming? At Kintra, we are responsible for building a supply chain, looking all the way down to our farm partners, and that is something we will continue to do as we scale and grow.

JH: Again, there are no easy solutions. Not all all-natural materials are sustainable; not all synthetics are evil. It is a big, wicked problem that requires a lot of different approaches to be solved—and it is urgent. We cannot postpone this for another generation, as we have done for the last eighty years. **We don't have the excuse that we don't know the consequences—because we see them and we drink them.**

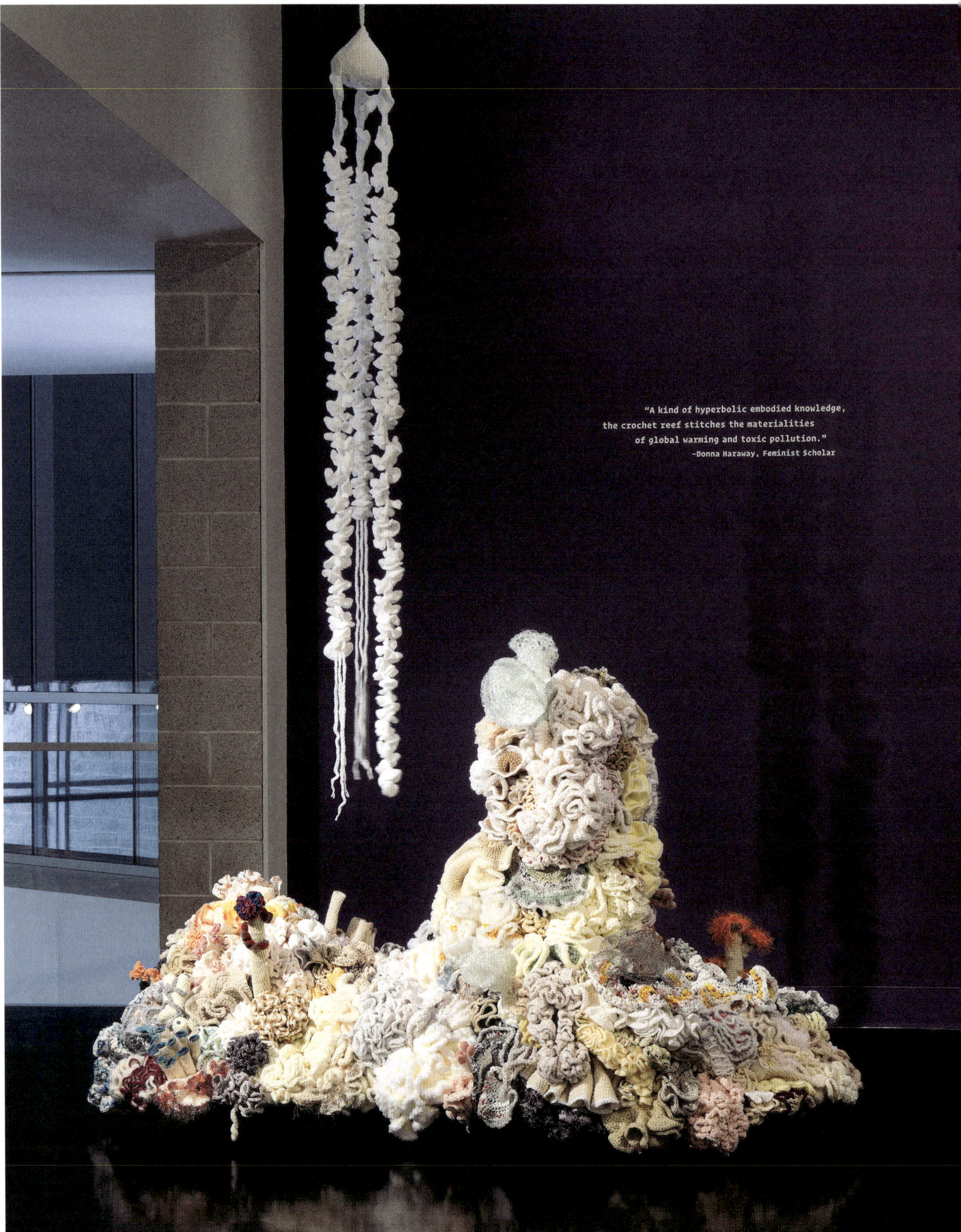
"A kind of hyperbolic embodied knowledge,
the crochet reef stitches the materialities
of global warming and toxic pollution."
—Donna Haraway, Feminist Scholar

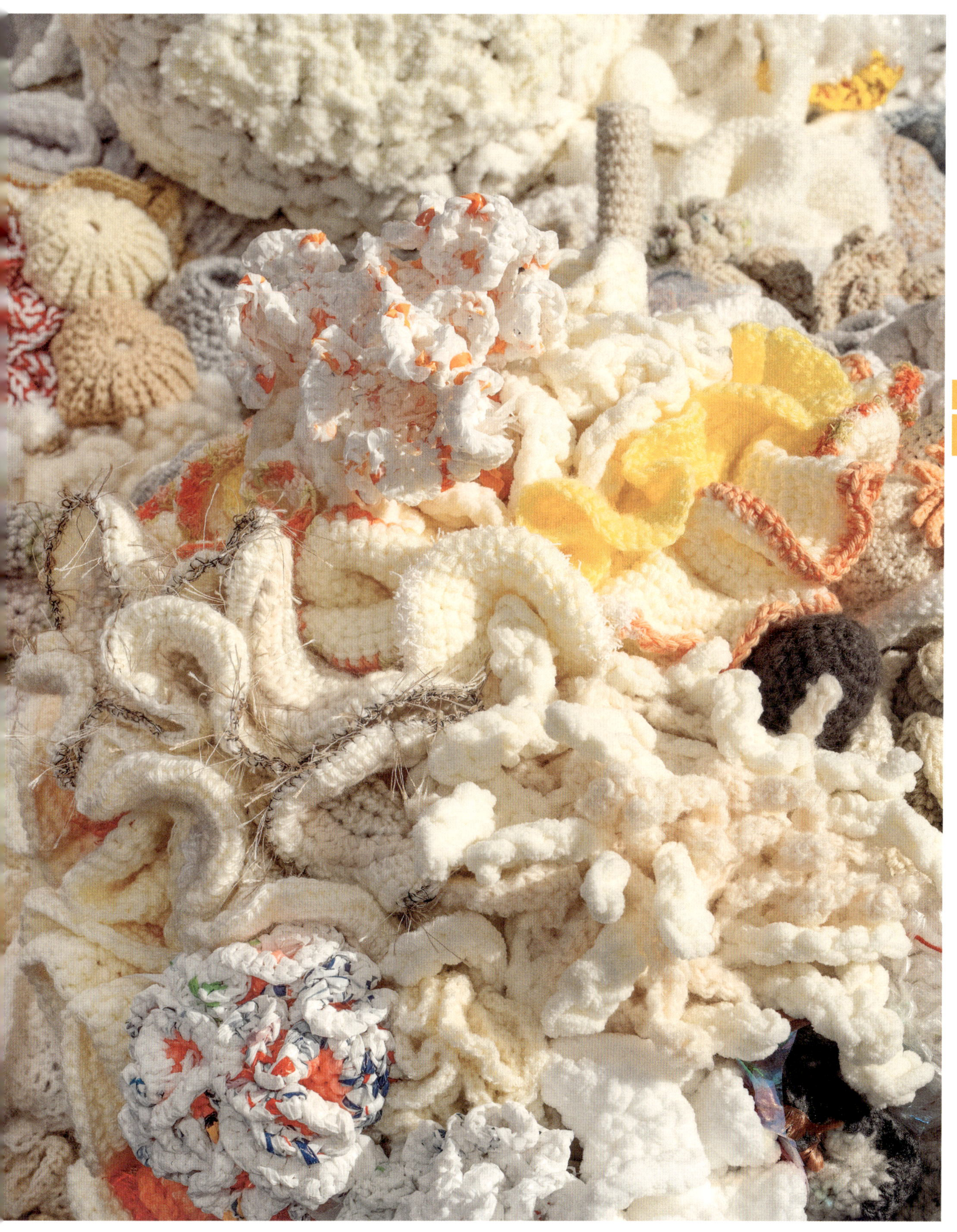

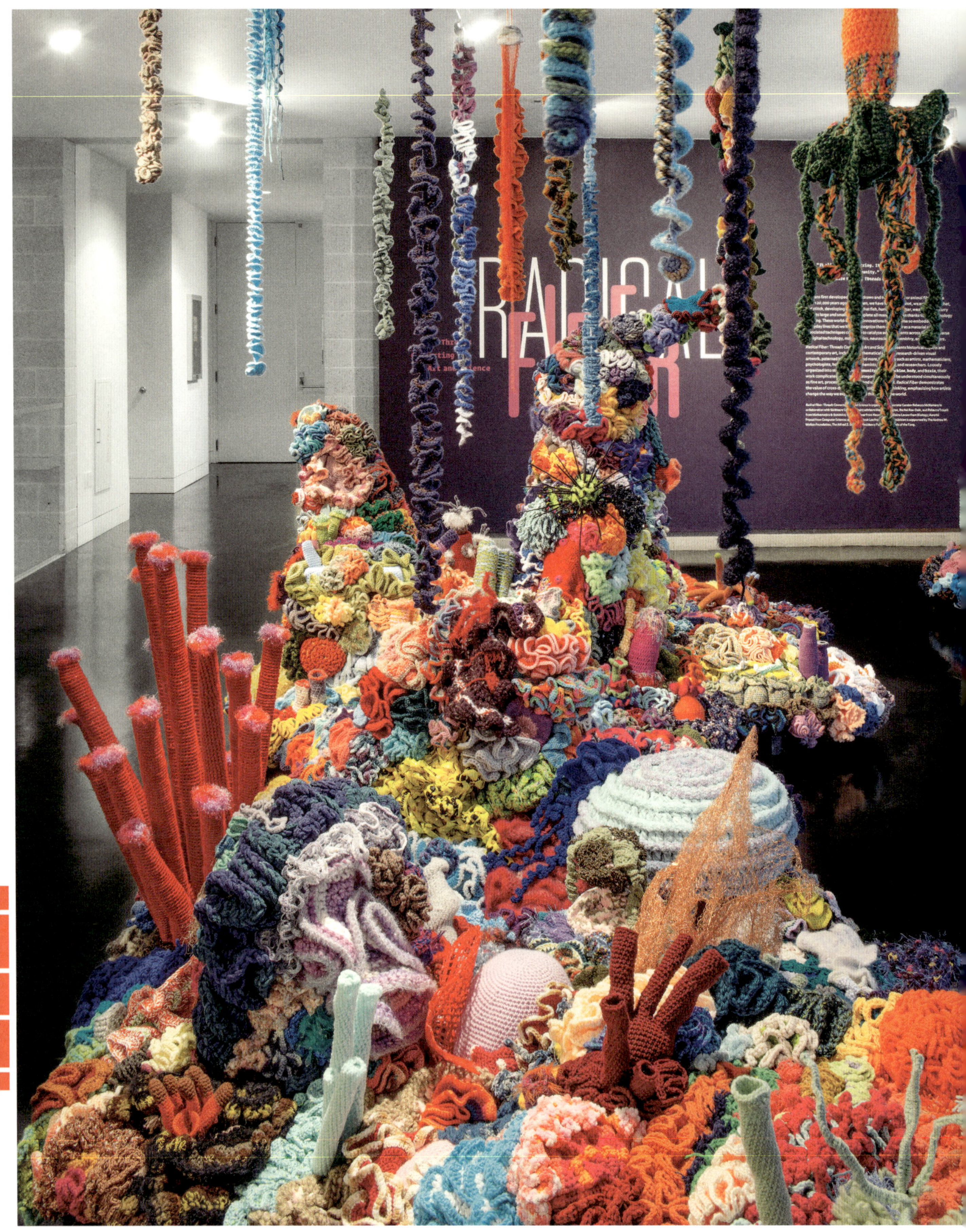

by Rebecca McNamara

THE SARATOGA SPRINGS SATELLITE REEF: CROCHETING ECO-CONSCIOUSNESS IN COMMUNITY

The *Saratoga Springs Satellite Reef* was created with more than two hundred participants from the Tang Teaching Museum's global community. Since 2007, makers around the world have created fifty *Satellite Reefs*—all part of the worldwide *Crochet Coral Reef* project by Christine and Margaret Wertheim and the Institute For Figuring. This massive social-practice artwork accomplishes myriad objectives: bringing attention to the devastation of coral reefs, due in great part to a warming climate; interrogating and challenging "women's work," textile histories, and the ways in which fiber-based art is perceived and valued; building communities through craft; providing embodied knowledge of mathematics; and more.[1]

Math, Science, and Yarn

In 1997, mathematician and artist Daina Taimina successfully modeled the first three-dimensional, interactive hyperbolic plane using crochet. Hyperbolic geometry was discovered in the early nineteenth century and challenged existing Euclidean principles.[2] It is a complicated area of mathematics, so when Taimina was teaching it to Cornell University undergraduates, she wanted to offer a model that students could physically interact with—a feat that, up until that point, had been deemed impossible; a paper model had been constructed, but it was fragile and couldn't withstand touching or manipulation. After mapping hyperbolic geometry's exponential growth on paper, Taimina recognized it as a crochet pattern. And she crocheted the seminal hyperbolic plane model. Using the oft-underappreciated tools of feminine handicraft—yarn and a hook—and her knowledge of both mathematics and craft, she defied the expectations of the mostly male-dominated field of mathematics.[3]

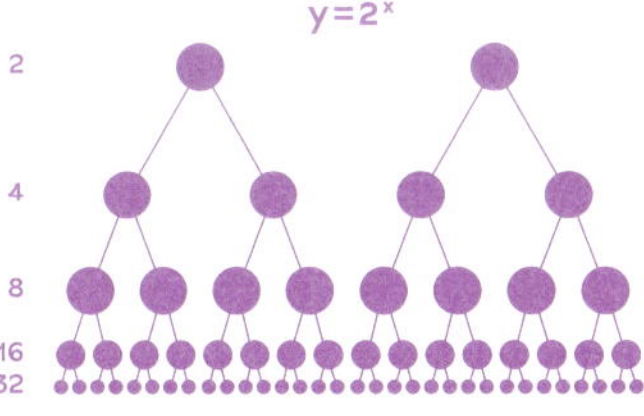

Mathematician Daina Taimina mapped out exponential growth and recognized its likeness to a crochet pattern. From there, she began crocheting a model of a hyperbolic plane.

Inspired by this work, the Wertheims, and subsequently tens of thousands of *Satellite Reef* participants, realized they could adjust rates of increase and other variables to create ever-more enchanting and unique forms. These slight variations—a *queering* of the code, in their parlance—forms the basis of the *Crochet Coral Reef* project. As Margaret Wertheim writes, "As Earthly organisms accrete variations, so our crochet reef community has accreted a library of pattern formations, bringing into being a wooly taxonomy of crochet coral 'species.'"[4] The Wertheims recognized yarn's potential for bringing awareness to coral bleaching and the deaths of corals around the world, including near the coastlines of their native Australia—enticing viewers into the conversation with extraordinary creations made from modest, everyday materials and the centuries-old technique of crochet.

The Saratoga Springs Project

The "rules" to participate in the *Saratoga Springs Satellite Reef* project were simple: there were hardly any. We prioritized openness and access and made a commitment to include every coral submitted to the Tang by the deadline. We invited participants to use yarn of any material, size, color, and hook size, and to follow one of six patterns provided by the Wertheims or to create their own underwater-inspired creations with no pattern at all. To create a coral-like form, the crocheter increases the number of stitches in each row. So, while a rectangle (think: scarf) has even sides, with the same number of stitches per row, for a coral—or *queered* hyperbolic plane—a crocheter may add a new stitch every two stitches, or every six stitches, or every random number of stitches. These choices, along with yarn material, yarn weight, hook size, and the tightness or looseness of one's stiches, affect the outcome.[5]

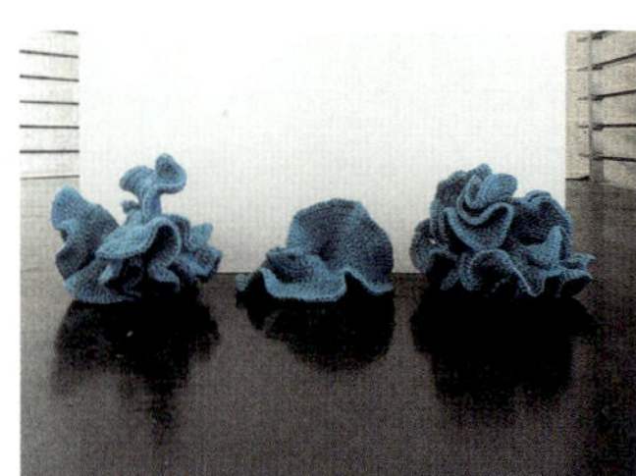

The corals shown here use the same yarn and hook size and were all created by the author with an effort at consistent tension. However, they look completely different from one another because the rate of increase varies in each.

For many of us, crocheting corals was a novel type of *meditation by making*: there were no feet to fit the socks, no sleeves through which arms must go, no blanket edges needing to be straight; there was no right or wrong. Crocheters who had previously followed stringent patterns now exercised their creativity in new ways, making art for viewing rather than for function. New and experienced crocheters alike embraced the inevitable mistakes of learning or expanding a skill: the "mistakes" could stay. Nature, after all, is imperfect.

Community Building

Every *Satellite Reef* is based on the same premise established by the Wertheims, but each one evolves experientially and physically in distinct ways. For the *Saratoga Springs Satellite Reef*, variables included not only our distinctive Skidmore College campus community, but also the early years of the global COVID-19 pandemic and the safety regulations it generated. On February 9, 2021, we held the first of what would come to define our project: free, online public crochet programs, open to all. Throughout that year, we hosted six Tuesday evening online workshops, forty-four Wednesday lunchtime online craft circles, five in-person Saturday afternoon

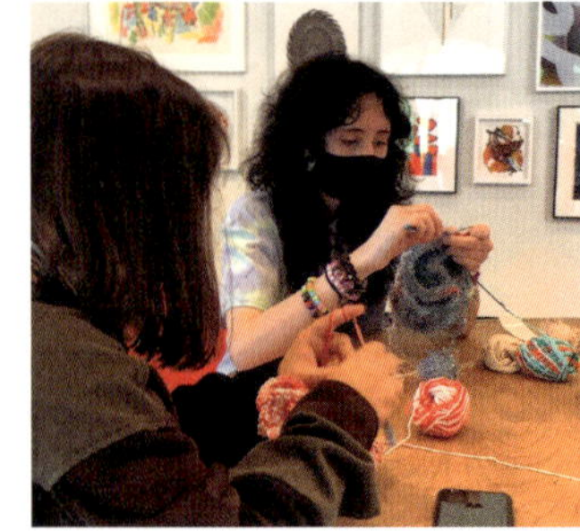

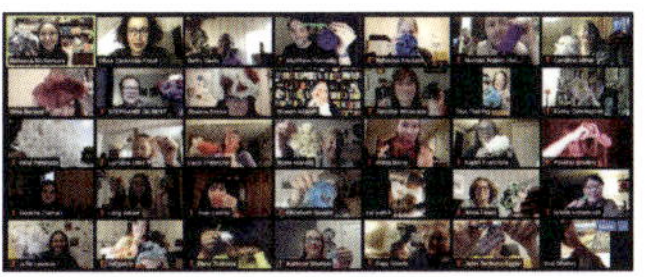

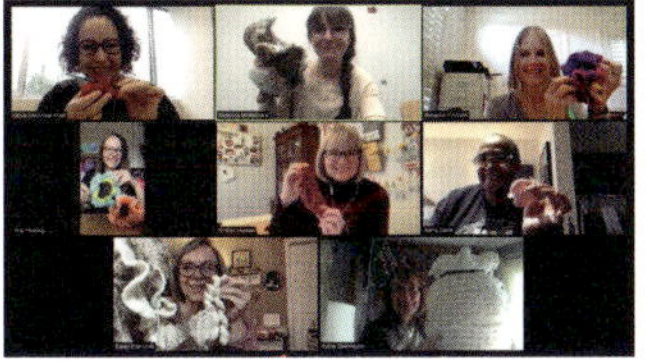

craft circles, a Family Saturday program, a student-run crochet workshop, and a "Learn to Crochet with Plastic" workshop—a total of fifty-eight public programs, plus additional private events, where we gathered and talked and crocheted.[6]

Workshops allowed a wide variety of people at different stages in their crocheting journeys—sometimes more than a hundred participants—to share a digital space. I talked about the project and its themes; seasoned crocheters offered ideas and tips while newer crocheters received instruction from fiber artists and teachers Lucy Beizer (Skidmore class of 2019), Victoria Manganiello (Skidmore class of 2012), and Cal Patch; and we all shared stories about what brought us there and what craft means to us.

The Wednesday lunchtime craft circles were more casual and intimate, with a core group of half a dozen "Wednesday regulars" joining me on Zoom each week, other crocheters coming in and out as their schedules allowed. This intergenerational group of (usually) women from all over the country—Austin, Texas; Dunn Loring, Virginia; New York City; locally in Saratoga Springs, New York; and elsewhere—became a point of connection during an ongoing pandemic that disconnected many of us. We talked about our families and the hardships some faced, what brought us to the places we now call home. We talked about vacations and gardening and access to vaccines. We shared stories about gendered experiences with fiber: splicing is "masculine" on the sailing docks while macrame is a "feminine" domestic craft. We shared our work, either timidly or proudly, and we motivated, empowered, and inspired one another. If one person showed

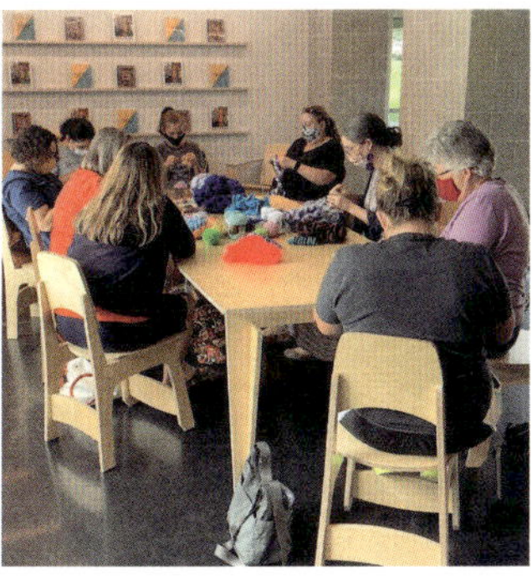

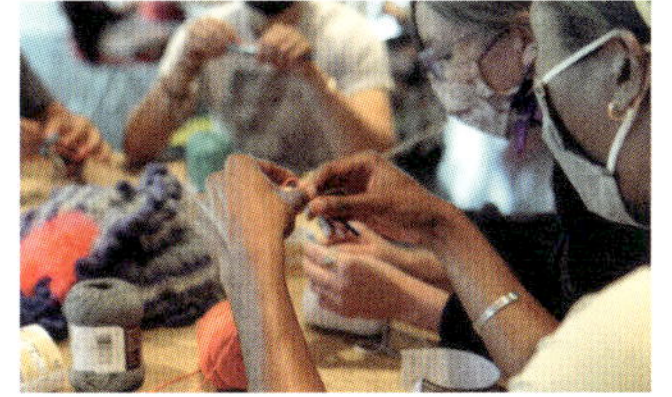

a long, winding coral filled with yarn scraps, it was likely that someone else would make a long, winding scrap-yarn coral the following week— each as unique as a signature. We learned new skills. We complained about weaving in our ends (the finishing touch of any yarn project). Although our weekly craft circle was scheduled for thirty minutes, we often crocheted and chatted for a full hour, sorry for the time to end. Later in 2021, Saturday afternoon in-person craft circles offered similar experiences for regional participants; what a joy it was finally to share our work through tactile investigation and to offer one another hands-on teaching.

When museum professionals speak of *community*, the word is, frustratingly, too rarely defined. Who makes a community? Is it a group of people with shared interests and backgrounds

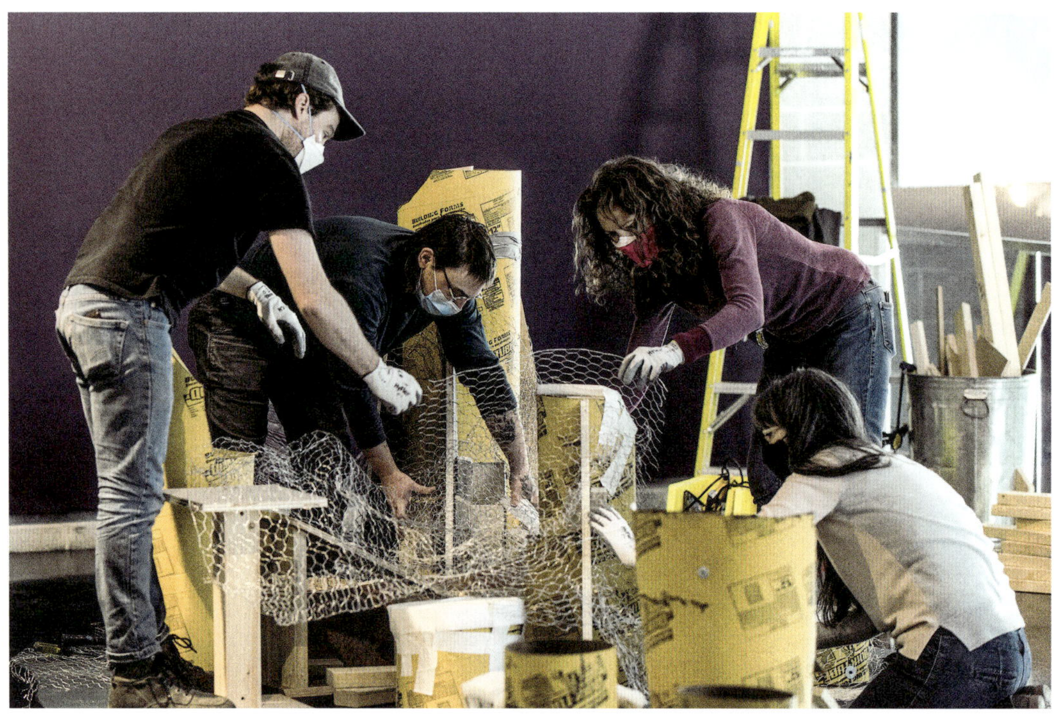

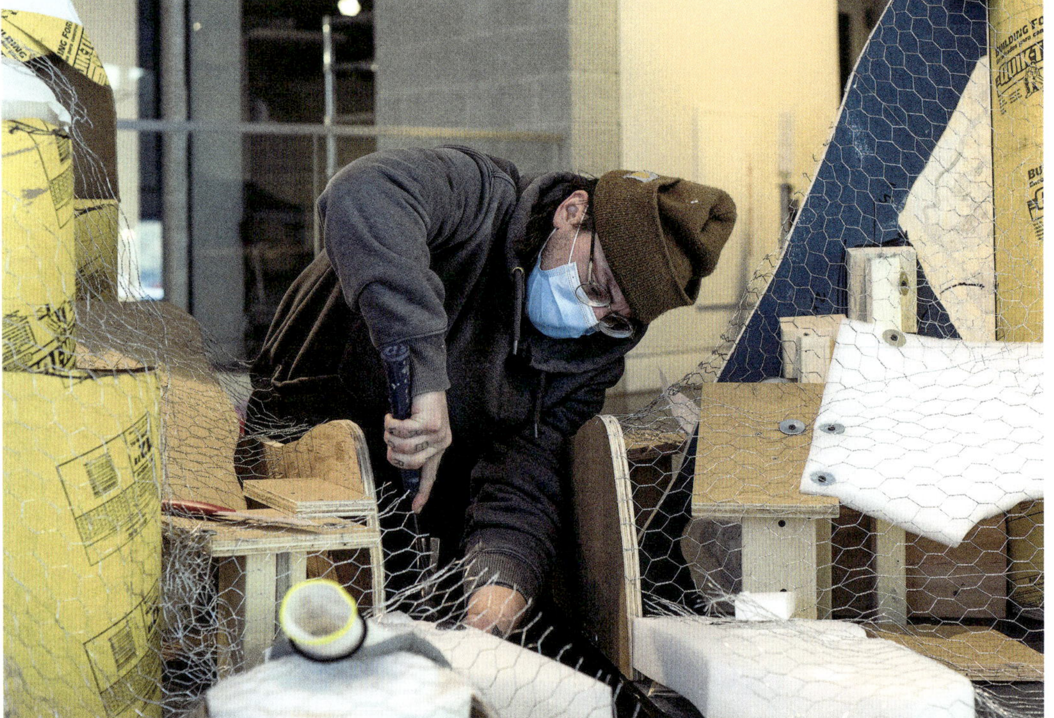

Tang staff and crew members assemble the understructure for one of the *Saratoga Springs Satellite Reef* islands in the gallery. Following this phase, the chicken wire was covered with large pieces of felt, and each coral was individually stitched onto the structure by a team of local volunteers.

spending time together? Maybe sometimes, but as the *Reef* project demonstrates, not necessarily. We, the *Saratoga Springs Satellite Reef* makers, were (are) a community. Throughout the week, we crocheted apart. Some of us met on Zoom or in person; some Zoom participants only listened, keeping their video and audio turned off. One participant, a retired lawyer and army veteran, left me a voice mail to say she "wasn't ready" for sharing herself on Zoom, but found "kindred spirits" in listening to others' stories. Some people never joined a program at all. Clearly, it was not physicality that brought us together, nor was it shared backgrounds. We were intergenerational, interracial, from all over the country and outside the United States; in school, working, and retired; married, single, dating, with and without kids and grandkids; from different educational backgrounds and different socioeconomic classes. Our *Reef* community was defined by positive energy, a shared goal, and our collaborative artistic process.

Individual Stories, United Voice

More than 200 people submitted more than 1,500 corals to the *Saratoga Springs Satellite Reef*. To create the sculpture, I mapped out the *Reef*'s footprint and modulations with cardboard, and the Tang's installation crew constructed a wood-and-Sonotube base and covered it with chicken wire and felt, and a team of volunteers stitched each individual coral on; we also hung some from the ceiling. Together, the corals formed an enchanting, colorful, awe-inducing sculpture.

Individually, each coral chronicles the moments, days, or months of its maker's life. At workshops and craft circles and through letters sent with their submissions, participants shared some of what their corals hold. Steve Medwin wrote that he began crocheting the day his wife had a stroke, on October 7, 2021. In the North Carolina mountains with their son and his family, they lacked cell service and drove forty-five minutes to a hospital, where COVID-19 prohibited Steve from staying with his wife. He used material from his grandchildren's textile project kits to busy his hands and distract his mind from worry. He soon learned about the *Saratoga Springs Satellite Reef*, and throughout his wife's recovery, which involved much waiting around in doctor's offices, Steve busily crocheted corals. Stories like this one fill the *Reef*, some told, others untold. Each individual coral contains so much more than we will ever know.

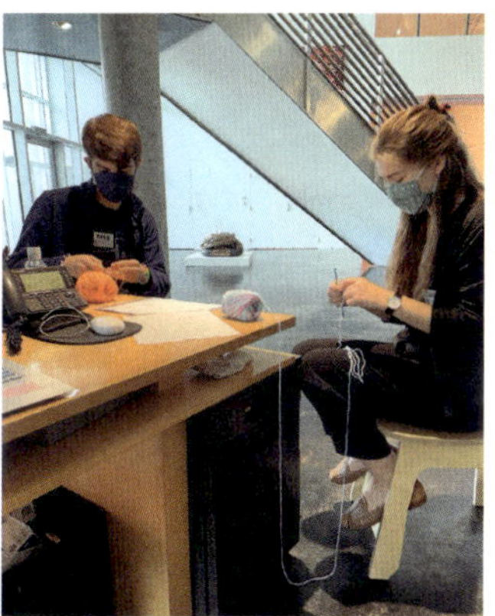

The Tang provided free crochet kits with cotton or secondhand yarn, a crochet hook, pattern instructions, and a take-away card about the project to encourage Skidmore students to try crochet, no strings attached.

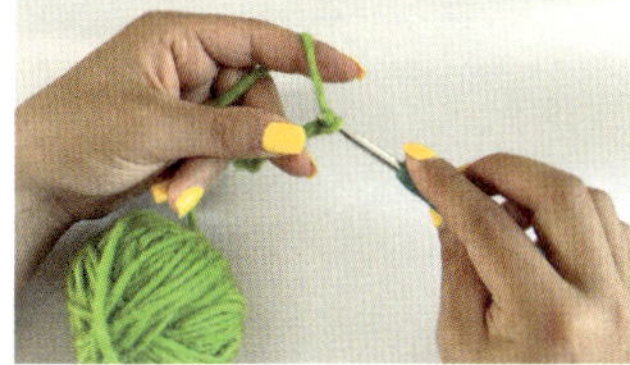

An additional offering for anyone who wanted to learn to make corals is the video "How to Crochet a Hyperbolic Plane," led by Lucy Beizer.[7]

Many participant letters simply reflected an eagerness to make and be part of a museum in a new way. They offered humbled variations of "I hope this contribution will be suitable," while frequent contributors, determined to execute all of their ideas before their yarn and time ran out, wrote "I will send more soon." And, importantly, we often saw the rewarding phrase, "I hope to visit." Indeed, the Tang welcomed first-time visitors eager to see their and their loved ones' contributions in the *Reef*, pilgrimaging to the museum from near and far.

Plastics and Corals

By prioritizing community building, introducing crochet to new people, and encouraging long-time makers of functional pieces to innovate and express themselves as artists—challenging the stereotype of fiber-based making as a strictly domestic pursuit—we allowed for a material-conceptual incongruity. *Saratoga Springs Satellite Reef* is filled with (alongside wool and cotton) acrylic, nylon, polyester, and synthetically dyed yarns—some of which had been buried in stashes for decades, but also some purchased new for the project. These materials and the fossil fuel–producing processes used in their manufacturing are devastating to the earth.[8] Troubled by this incongruity, I became increasingly invested in understanding my own yarn purchases and soon learned the difficulty of finding truly eco-friendly or sustainable yarns, especially affordable ones. Packages claiming "wool" on the front of the label may reveal a significant percentage of acrylic in small type on the back, for instance. Material restrictions would have rendered the project financially inaccessible or simply unappealing to many potential participants.

This lack of parameters generated some creative thinking around sustainability, however, as some "Reefers" began to examine their consumption more closely. Participants cut up plastic bags that would have ended up in curbside pickup bins, transforming them into corals. And longtime fiber crafters dove into their stashes: the leftover fluffy pink yarn that made a friend's baby blanket, the forgotten sweater that was too ambitious to complete, the small but irresistible skein purchased on clearance, a couple feet of blue sport-weight here, a small ball of chunky yarn there, tossed into baskets, jammed into Rubbermaid containers, tucked into corners of closets, forgotten in tote bags. Not small enough to discard, but rarely enough for a new functional piece, these bits and balls were given new life as corals.

Tang staff and local volunteers organized more than 1,500 coral creations submitted by makers from around the world by size and type in preparation

Participant Rebecca Knutson's list of plastics used in her creations demonstrates just how much plastic is in our everyday lives: Bulk wrapping plastic from cases of Diet Coke, Halloween-sized Nestlé candy, quilt padding, paper towel (Bounty, Kirkland Signature) and toilet paper packaging; construction site "Caution" tape; plastic Christmas tree netting; Dartek packaging film; plastic bags from home, friends, and family, including grocery bags from supermarkets in Maryland, Virginia, and Colorado (Giant, Wegmans, Safeway, Harris Teeter, King Soopers, Trader Joe's), a Disneyland souvenir bag, shopping bags from retail stores (Talbots, Walmart, Target, CVS, Rite Aid, Norm's Beer & Wine, Staples, Marshalls, Jo-Ann, Michaels, The Home Depot, Lowe's), newspaper delivery bags, take-out food bags, dentist "Smile" bags.

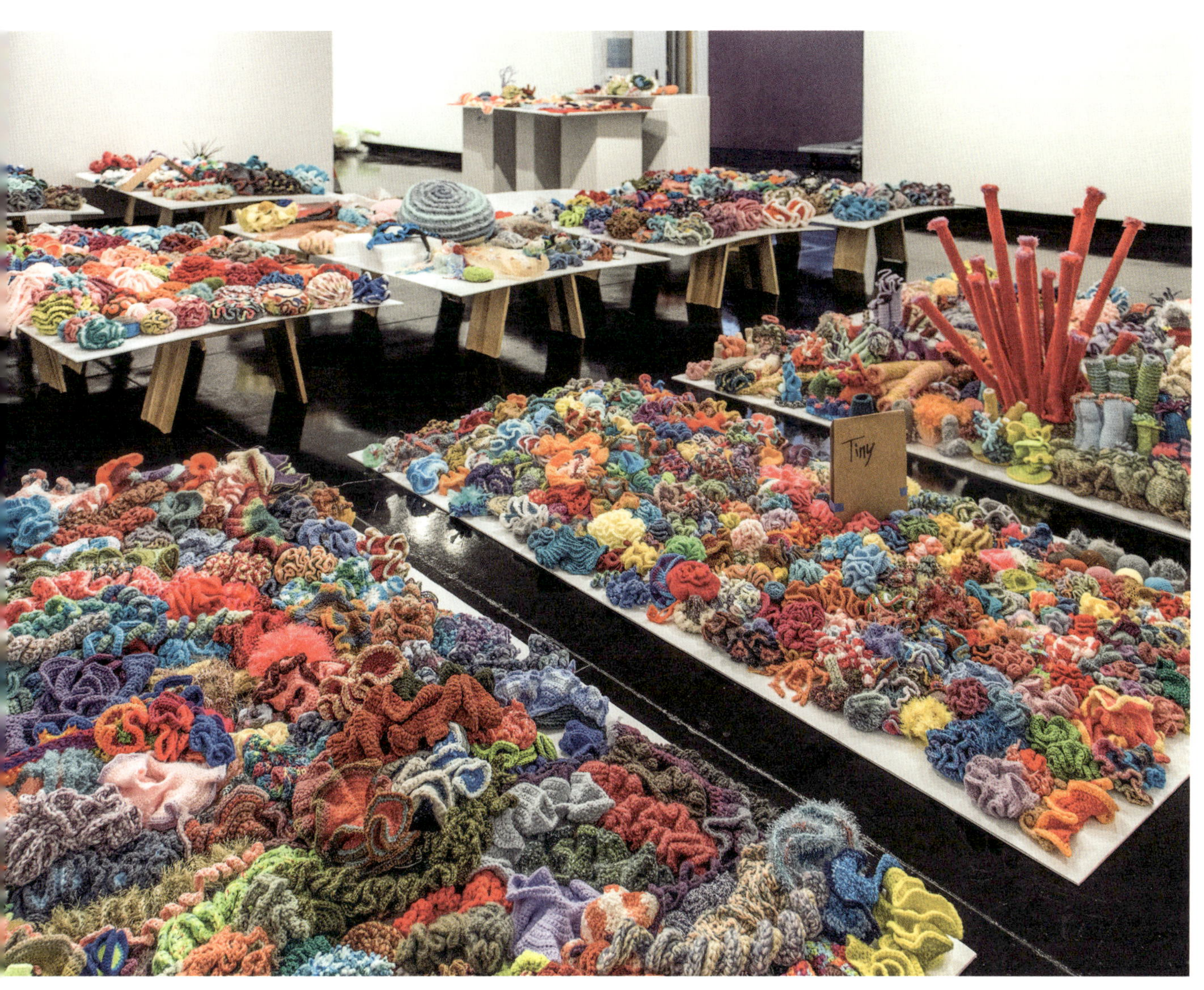

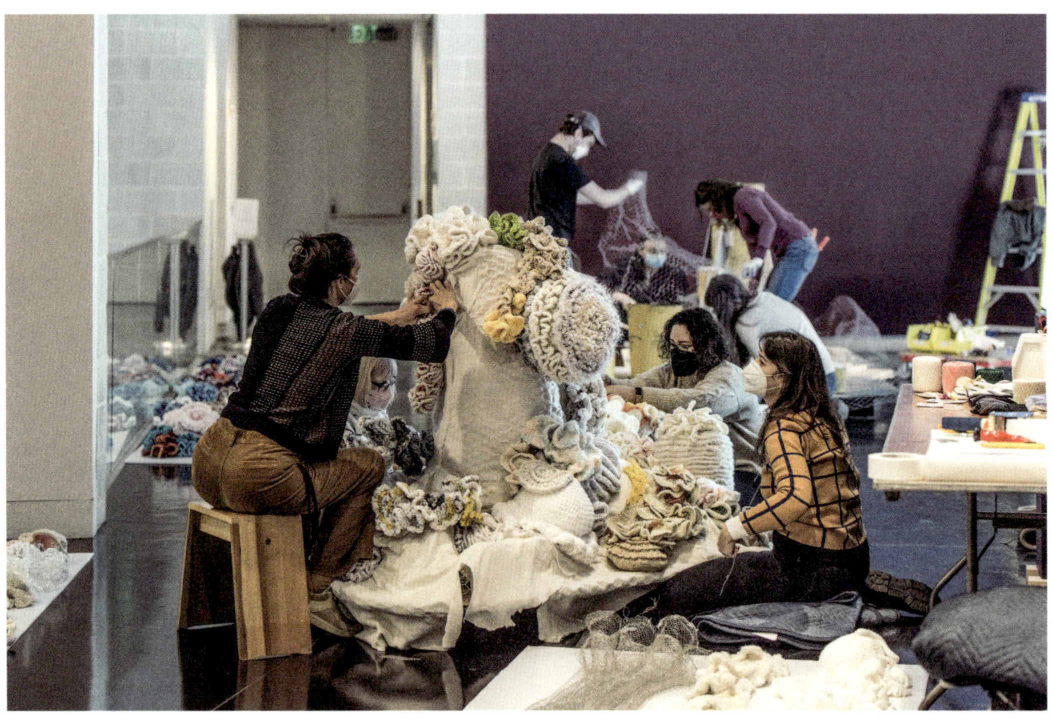

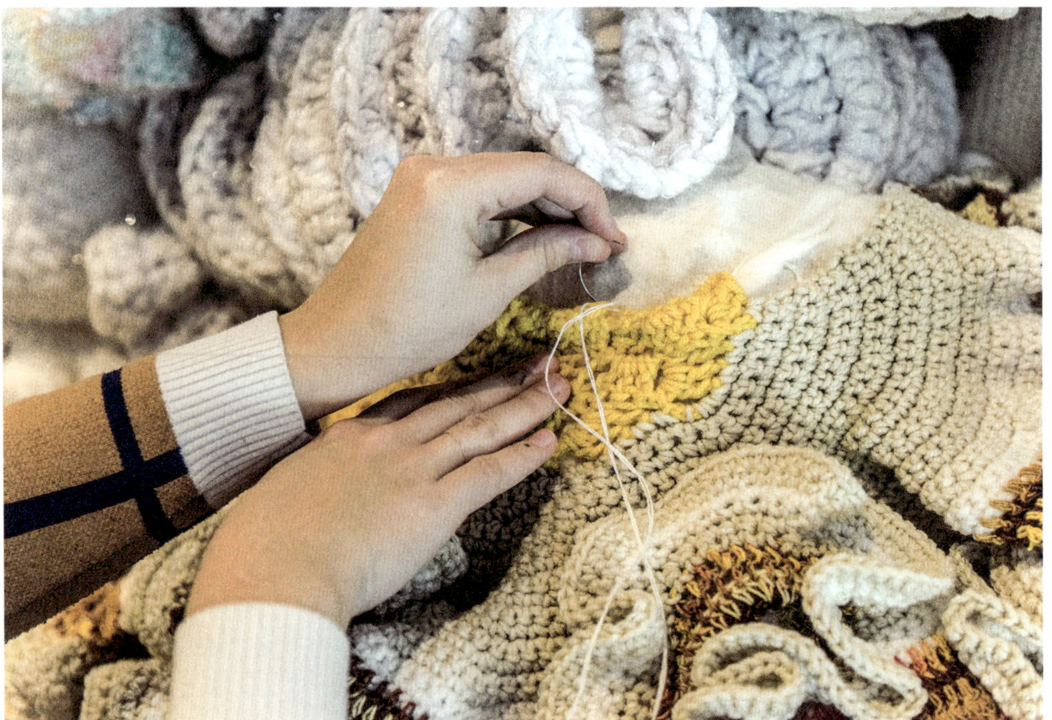

Tang staff and local volunteers begin to place and stitch coral creations onto a *Reef island* that draws attention to coral bleaching.

The *Reef* serves as a salient metaphor: a single person cannot accomplish what two hundred people can. To what extent can we use this logic to help mitigate the climate crisis? Even if we are not all sitting in a room together, we can individually contribute to improving our ecosystem in small ways and find community in the knowledge that others are also composting, purchasing eco-friendly sunscreens, avoiding—not merely recycling or reusing—plastic whenever possible, buying second-hand clothing rather than new, and so on. In our actions, we can expand the existing worldwide community that cares about our Earth and its creatures. In colorful and "bleached" crocheted corals, the *Saratoga Springs Satellite Reef* reflects both the wonder of our natural world and the human-wrought devastation we must contend with. And perhaps, it also offers hope that remote, seemingly dissimilar people can come together toward a common goal.[9]

1   Margaret Wertheim and Christine Wertheim, *Crochet Coral Reef: A Project by the Institute For Figuring* (Los Angeles: Institute For Figuring, 2015); Udo Kittelmann, Christine Wertheim, and Margaret Wertheim, eds., *Margaret and Christine Wertheim: Value and Transformation of Corals*, ed. (Baden-Baden, Germany: Museum Frieder Burda, 2022).

2   See essay by Mark Huibregtse in this book for a detailed history and explanation of hyperbolic geometry.

3   Daina Taimina, "Crocheting Hyperbolic Planes," filmed June 14, 2023, at TEDxRiga, Riga, Latvia, video, 17:14, www.youtube.com/watch?v=w1TB-Zhd-sN0.

4   Margaret Wertheim, "Crochet Codes and a Crafty DNA," in *Margaret and Christine Wertheim*, 77.

5   For crocheters contributing many corals, including myself, we gained invaluable embodied knowledge about our materials. As Christine Wertheim wrote, "One of the pleasures of participating in the *Crochet Coral Reef* project is to viscerally experience how much *matter* matters, how much the materials with which one works contribute to the qualities of the final object … [T]his understanding is acquired, not through [reading] books, but through the activity of *doing* or making itself, a process known as 'acquiring a skill.'" Wertheim, "Matter, Form, and Technology: Materiality Counts," in *Margaret and Christine Wertheim*, 63.

6   I am especially grateful to Tang Special Events and Publications Manager Olivia Cammisa-Frost for smoothly and skillfully running the behind-the-scenes technology for these online programs as well as for being a key collaborator throughout the project.

7   "How to Crochet a Hyperbolic Plane," *Radical Fiber: Threads Connecting Art and Science*, The Frances Young Tang Teaching Museum and Art Gallery, tang.skidmore.edu/exhibitions/286-radical-fiber-threads-connecting-art-and-science.

8   "Polyester," Materials Index, CFDA (Council of Fashion Designers of America, Inc.), accessed January 31, 2023, cfda.com/resources/materials/detail/polyester; "Preferred Fiber and Materials Market Report," Textile Exchange, October 2022, textileexchange.org/app/uploads/2022/10/Textile-Exchange_PFMR_2022.pdf.

9   After the close of the exhibition, the largest *Saratoga Springs Satellite Reef* island was moved to Skidmore's Scribner Library. The three smaller islands remained on view in various public areas of the Tang through 2022, and for the spring 2023 semester, they were moved to Skidmore's Billie Tisch Center for Integrated Sciences, where they continue to bring joy, wonder, and knowledge to passersby.

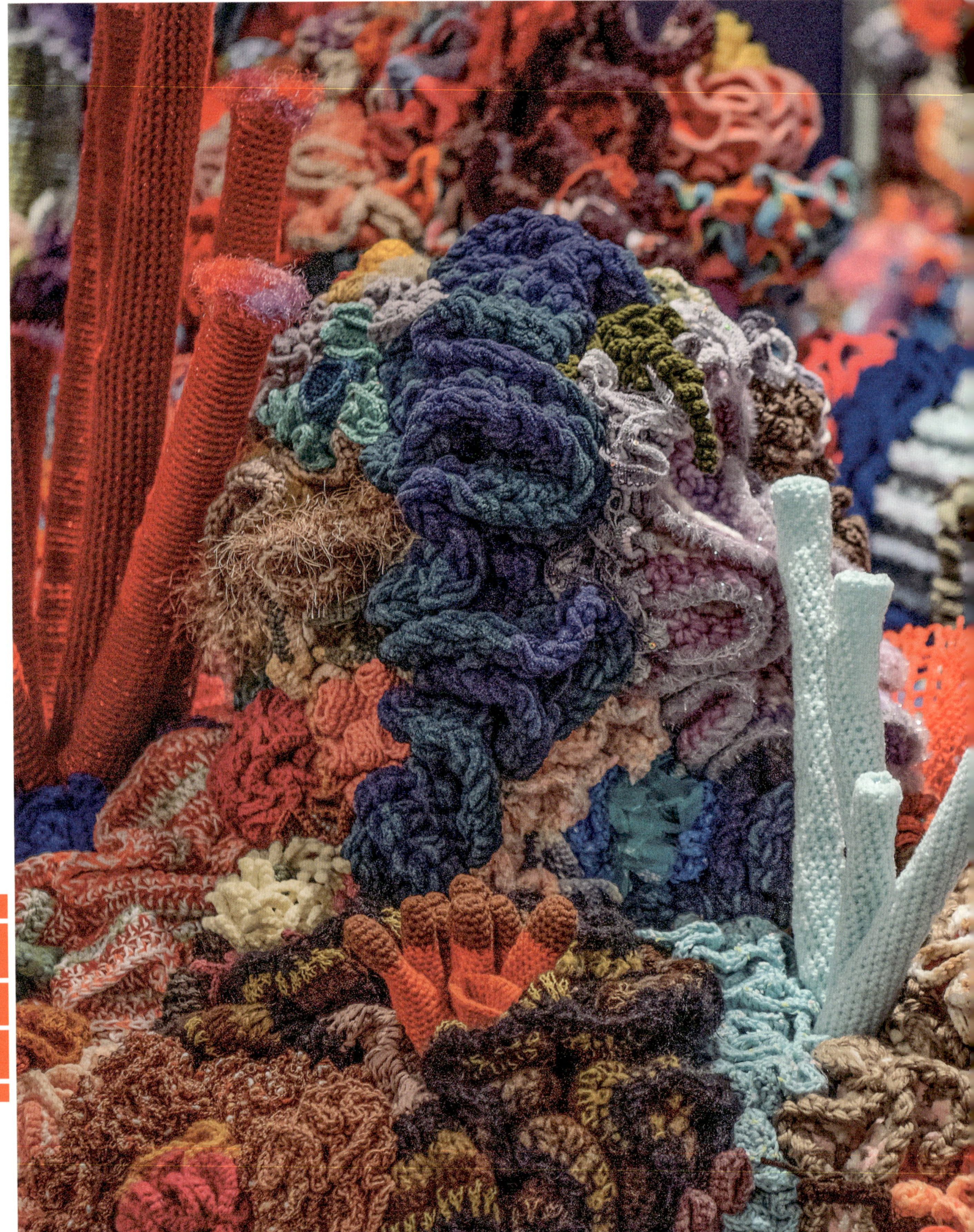

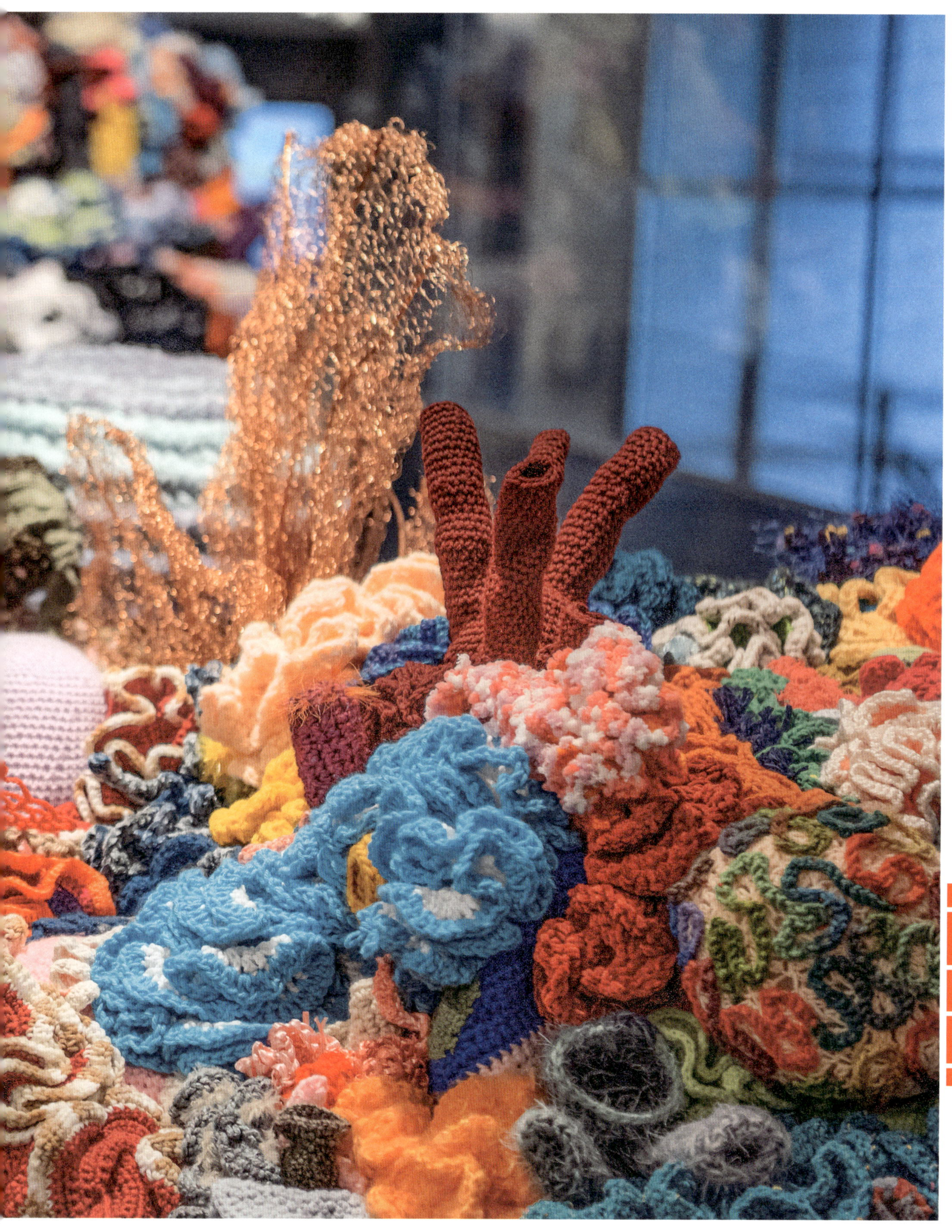

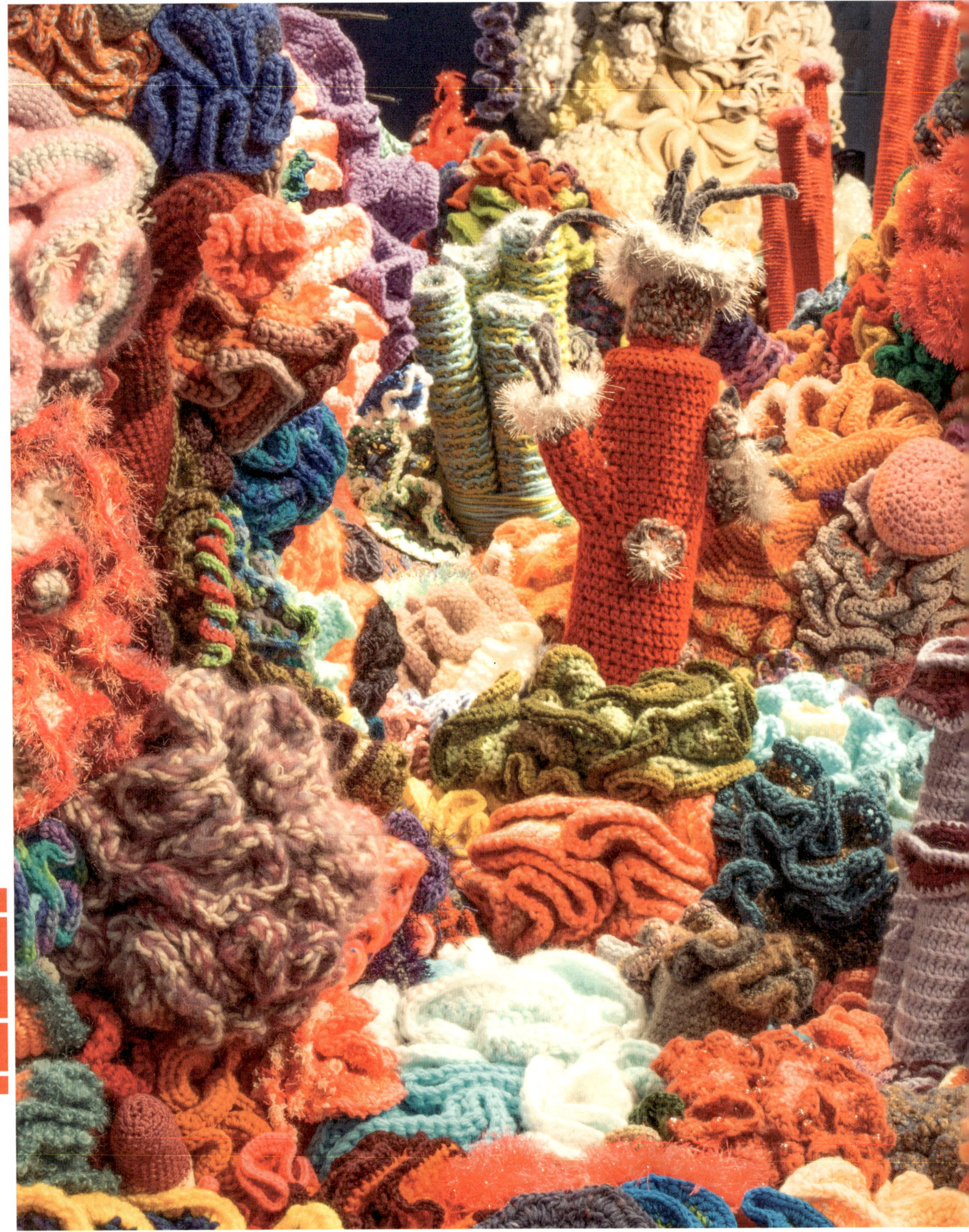

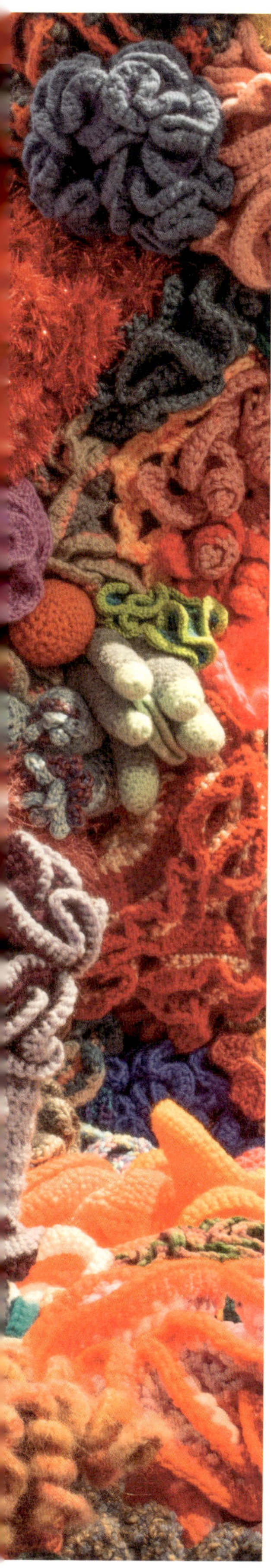

Left, previous, and following spreads: Installation views, Radical Fiber, featuring the Saratoga Springs Satellite Reef, part of the Crochet Coral Reef project by Christine and Margaret Wertheim and the Institute For Figuring

PROTECT
CORAL REEFS

Corals are animals that first developed about 500 million years ago. They are a vital resource for all living things on Earth, including humans. **Each of the five mass extinctions on our planet was preceded by the death of a coral reef.**

Coral reefs have the greatest biodiversity of any ecosystem globally, and the fish that rely on corals provide a significant food source for more than a billion people worldwide. The reefs drive coastal economies in areas such as tourism and fisheries, and they protect coasts from erosion and flooding by absorbing an astonishing 97 percent of the energy from waves heading toward the shore.

But warming ocean waters, increased ocean acidity, overfishing, and pollution stress the corals, causing them to expel the algae living in their tissues and turn white, known as a "bleaching event." Corals can survive bleaching if it is identified and counteracted quickly, but, as with any other animal, once a coral dies, it is dead forever. Today, coral reefs are 50 percent less able to provide food, jobs, and climate protection than they were in the 1950s. From 2008 to 2019, about 14 percent of the world's coral reefs were lost due to rising ocean temperatures; during the next twenty years, it is estimated that 70 to 90 percent of all coral reefs will disappear.

What can we do? As with any sustainability effort, the answer is complicated and frustrating. Individually, we can restrict our use of lawn and garden chemicals, reduce plastic waste, use coral-friendly sunscreens, compost food scraps rather than throw them in the trash bin, avoid purchasing plastic-based synthetic yarns and garments, buy secondhand rather than new clothing. Perhaps most vitally, we can demand that political representatives support positive climate initiatives, such as the reduction of greenhouse gas emissions, and to legislate greater transparency in labeling of synthetic materials.

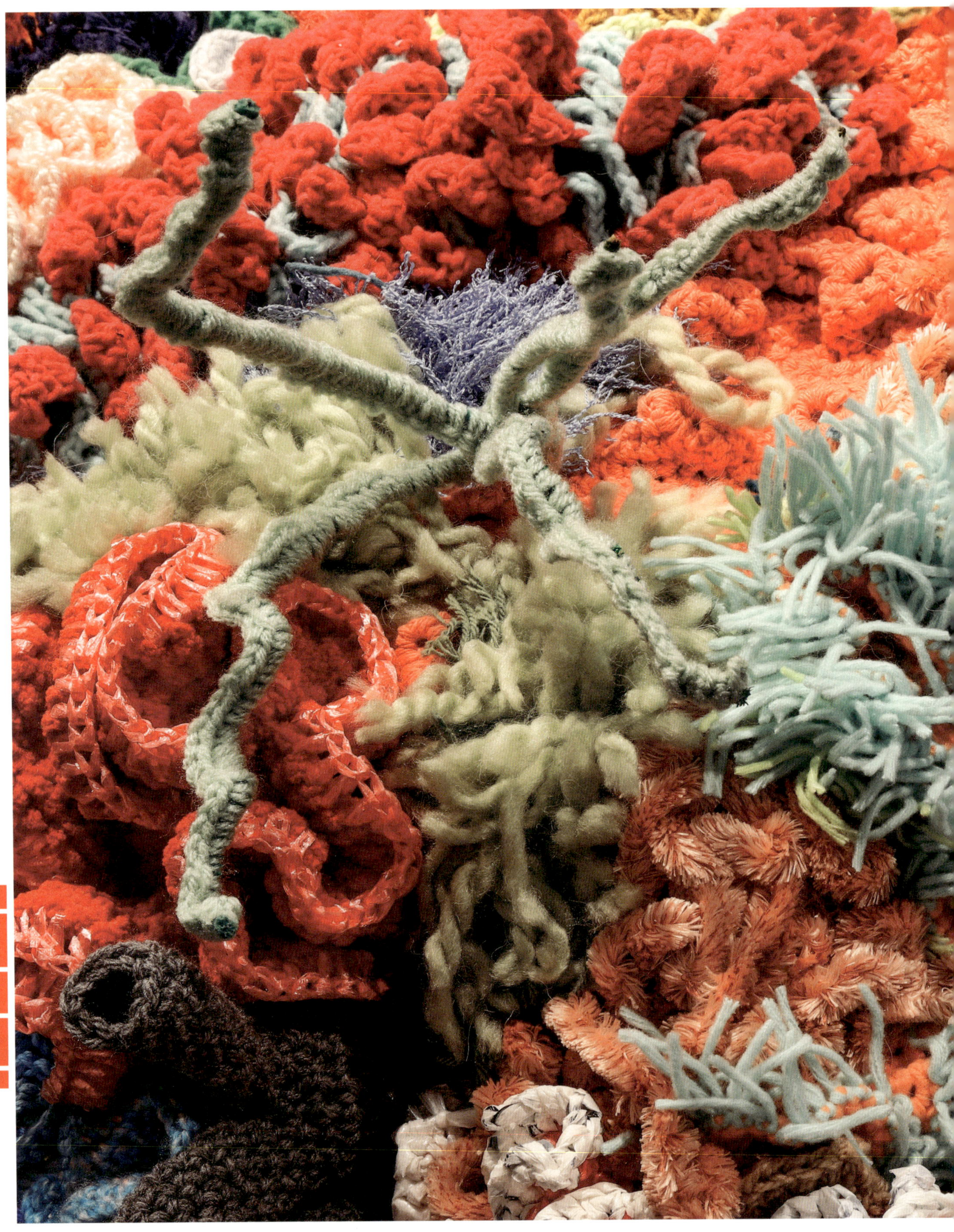

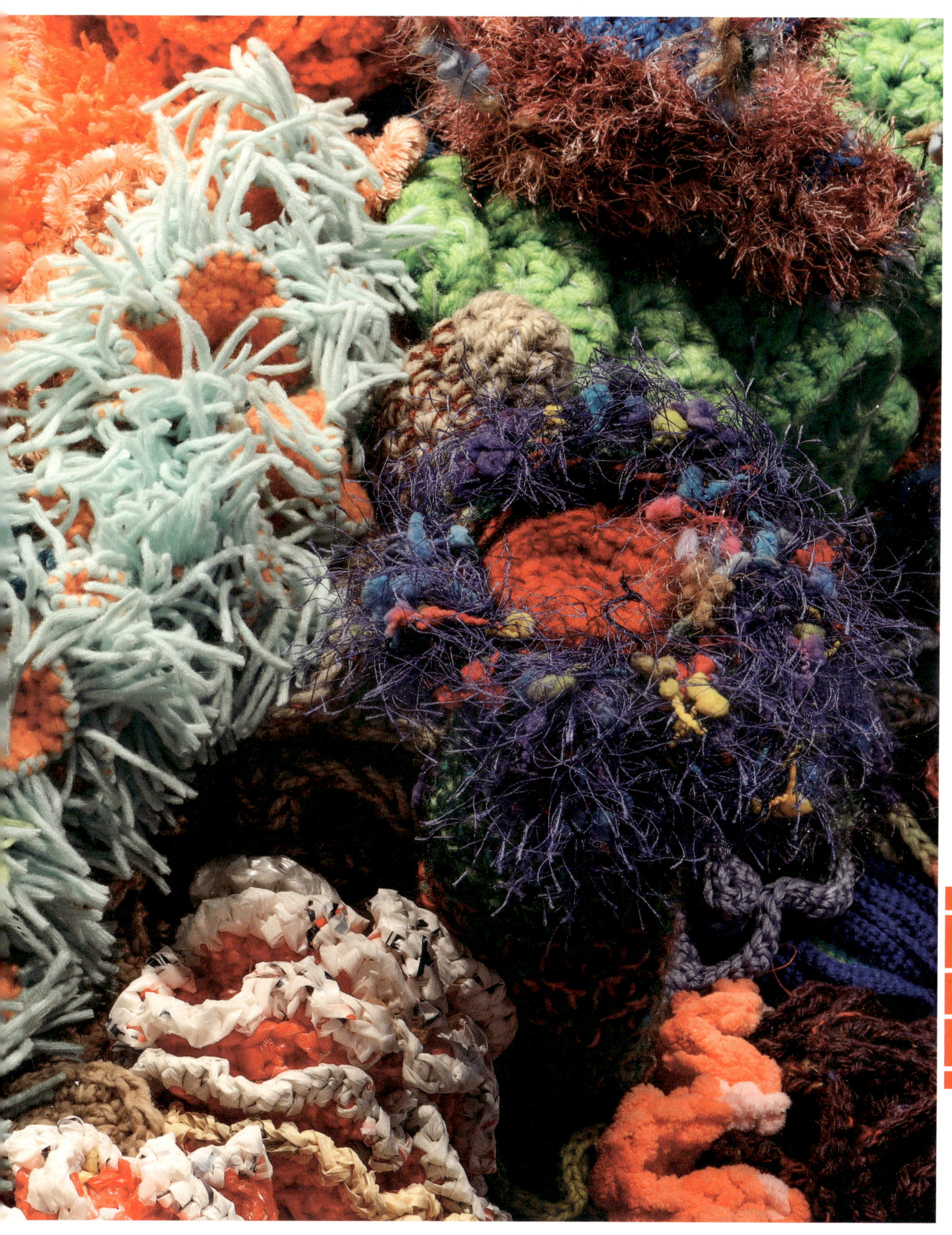

Saratoga Springs Satellite Reef
Contributors

Jackie Allaire-Macdonald
Liz Archer
Karen Arciero
Sharon Arpey
Nurcan Atalan-Helicke
Sonia Athalye
Amy Sue Axen
Katie Backaus
Adelaide Barkhorn
Judith E. Barlow
Margaret Barry
Rachel Baum
Fiona Berry
Mary Berry
Laura Bill
Sarah Boiwz
Clare Boland
Cassie Bond
Tamar Bordwin
Angie Bowles
Carrie Boyd
John Boyd
Indira Brechko
Kaitlyn Burch
Emily Cabral
Meghan Cabral
Kelly Callahan
Elaine Cammisa
Olivia Cammisa-Frost
Ricki Carroll
Elyssa Carson
Louise Carson
Cece's Wool craft circle
Kathy Ceceri
Molly Channon
Kathy Chen
Nadja Choroszylow
Eliana Colzani
Claudia Comay
Elizabeth Custer
Betty A. Davis
Caroline Declercq
Candace Deisley
Jaydelis Deleon
Cynthia DiDonna-Nethaway
Fatou Diop
Stacey Downey
Karen Drozdyk
Zoe Drozdyk
Kate Dudding
Kyle Dzurica

Jessica Eisenthal
Cathy Eliseo
Sheena Emma
Jordan Epstein
Naomi Epstein
Kendra Farstad
Lorena Ferreira
Carol A. Firestone
Natalie Fischer
Kara Fisher
Kerry Flynn
Karen Franciola
Jaimie Frank
Jutta Frankie
Amy Frappier
Leah French
Madelyn Friend
Donna Galloway
Chelsey Giammarino
Stephanie Gilbert
Emma Gill
Michaela Glinsky
Kathy Gordon
Daisy Graham
Kylie Green
Hanhui Guo
Mary Hague
Brook Heston
Kirsten Hogenson
Betty Holck
Sonia Holmer-Herrmann
Shiloh Hurley
Kristen Hyman
Taylor Jaskula
Brianna Johnson
Kelly Johnson
Patty Johnston
Eileen Kaiser
Annelise Kelly
Erica Kemp
Matthew Kennedy
Anna Knapp
Rebecca Knutson
Kelli Kozak
Sujin Lapinski-Barker
Elaine Larsen
Julia Lawless
Gabriela Lebron
Linda Leskovitz
Joan Levine

Anna Lewis
Maya Litton
Angel Lockhart
Caroline Love Miller
Jessica Lubniewski
Genesis Maldonado
Victoria Manganiello
Rosie Manley
Bettina Marlow
Emma Martell
Sasha Mathrani
Donna McAlpin
Darlene McNamara
Kat McNamara
Rebecca McNamara
Bailey McShane
Samantha McShane
Steve Medwin
Sophia Mehta
Alice Mensching
Elysse Meredith
Michele Merges Martens
Lorraine Mignone
Jenny Miller
Laura Miller
Sarah Miller
Sharon Milzoff
Frances Moore
Joanne Morrison
Kitty Muntzel
Tommy Myhill
Nancy Nelson
Diane Nolan
Max Nolan
Christiana Nuzzi
Michiko Okaya
Calen Olegnowicz
Barbara Oliver
Lorraine Oller
Kathy Overington
Sharon Paddock
Susan Palmer
Gina Peterson
Renee Phaneuf
Melisa Ramirez
Barbara Raymond
Angela Rella
Marjorie Renko
Avery Roberts
Rachel Roe-Dale

Gabriella Root
Stephanie Ryall
Amy Samson
Katie Saunders
Hans Schepker
Marcy Schepker
Jen Schildge
Paul Seggev
Rachel Seligman
Kathryn Shelton
Smith College students and staff
Madeline Alexa Smith
Libby Smith-Holmes
Deborah Soffel
Susan Sofia-McIntire
Gina Soressi
Elizabeth Sovern
Rica Spector
Max Spencer
Cathy Splinter
Maria Staack
Megan Stasi
Linda Stern
Barbara Sullivan Sicke
Csilla Szabo
Ilona Szabo
Emily Theisen
Tina Thering
Luce Threads
Cyndee Tovsen
Rebecca Trousil
Peris Tushabe
Bryn Valenti
Victoria Van Der Laan
Lorelei Von Stackelberg
Amy Wagner
Holly Walsh
Riva Weinstein
Kaliyah Wilson
Cynthia Winika
Dianne Winter
Maurice Wisdom
Sophia Wittemyer
Cheryl Wood
Betty Young
Tricia Young
Cynthia Zellner
Isabel Zellner-Coe
Lisa Zimmermann Jordan
Nancy Zimmermann

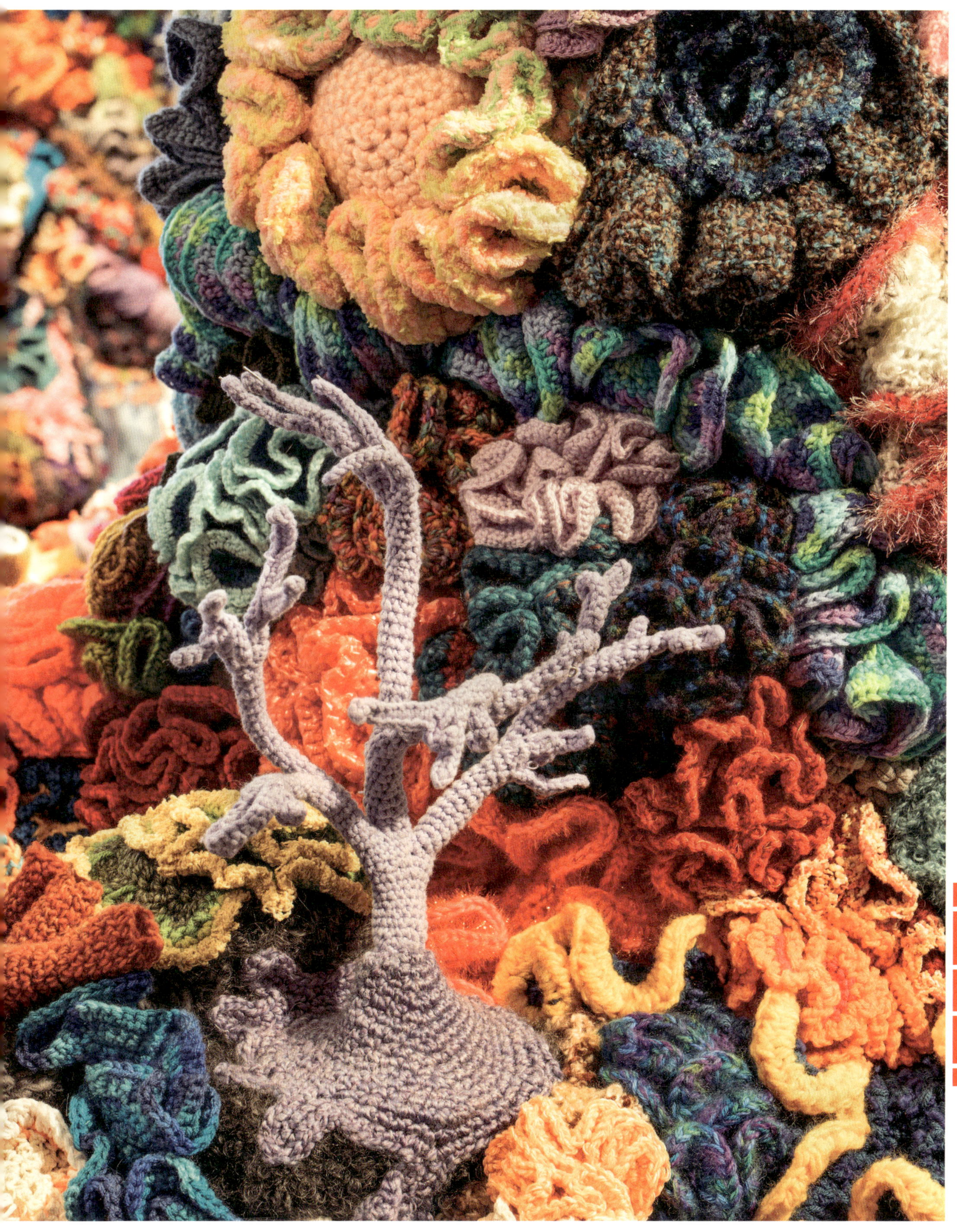

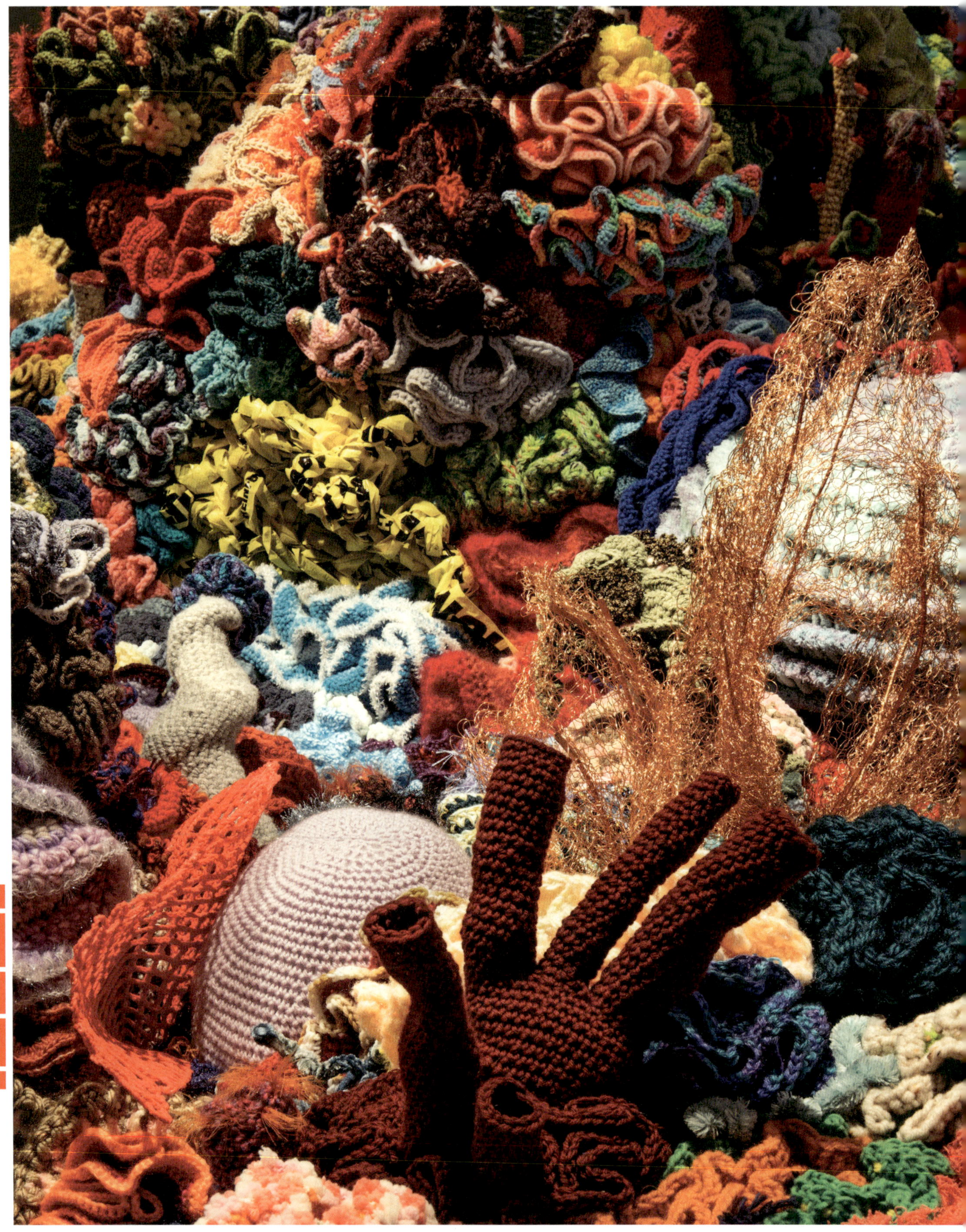

"It amazed me to see what can be created out of a
simple piece of string, when both friends and
strangers in a community can come together to
create something extraordinary." —*Erica*

"I was happy to be able to contribute my work to
raise awareness about climate change and its
impacts. Art has a way of reaching and bringing
light to many difficult subjects and more so when
done in unconventional ways. Something as
tactile as a work of crochet in yarn has a surprise
element and in that surprise is an opportunity
to communicate and perhaps enlighten. We will all
be impacted by the effects of climate change, some
in small ways and others in measurably large ways
and we need to come to terms with that and start
looking for solutions, not only for us humans, but
also for all of Earth's creatures." —*Claudia Comay*

"I'd never tried freeform or hyperbolic crochet
before! This really taught me new skills and
pushed me to look at crochet in a sculptural way."
—*Elyssa Carson*

"I swelled with pride in seeing the magnificent
final product. It was a treat to see my modest
contribution and other pieces the other crocheters
had shared on the check-in webinars while they
were in progress." —*Stacey Downey*

"As a remote Skidmore student during the 2020–2021
academic year, this project provided a critical
lifeline, connecting me to the Skidmore and Tang
communities." —*Sophia Mehta*

"I was the first person in my family to graduate
from college and have gone on to work as an
educator since 1996. I grew up in a family of
modest means, in a log cabin without electricity.
My parents both grew up in poverty and could
not provide any financial help for me to attend
college . . . Fortunately, Skidmore offered generous
aid that made it possible for me to attend and
graduate in four years . . . During the time I was a
student, my mother and grandmother and great-
grandmother supported me in the ways that they
could. They sent me letters, handmade gifts, an
occasional $5 or $10, and constant and abiding love.
Both my grandmother and great-grandmother have
passed away, but I inherited their sewing tins.
As I made the ivory brain coral, I used my mother's
hand-spun wool. I also used silk thread from
those inherited sewing tins to attach beads.
The thread is over a half-century old, but still
strong . . . an apt metaphor for the resiliency
and self-sufficiency that those women bequeathed
to me. As I attached each bead, I allowed myself to
sit with the many memories I have of both my
Gram and Nana, since I was lucky enough to reach my
thirties before I lost either of them . . . My mother
is sending her own corals from New Hampshire.
We have talked about these corals, crocheted via
Zoom, and we feel that we are sending the spirit
of five generations of women to campus. It helps
us to feel connected to my daughter, Lyra Flinn
[Skidmore class of 2025], who is beginning to find
her own path on campus." —*Amy Samson*

"I was excited about the environmental science
component but also the possibilities that you could
produce with yarn . . . Over time, the projects
I did were scientifically accurate representations
of real coral reefs . . . I wanted to make them
something important, so one of them is an invasive
species." —*Nurcan Atalan-Helicke*

"Since learning how to crochet through this project,
I have begun to continue to crochet and really
get to know more about crocheting. Recently, I made
a hat, and right now, I'm working on making a neck
warmer." —Maria Staack

"As I worked, I became more conscious of the
acrylic yarns I was using and the likely
environmental impact of these ending up in
a landfill." —Kerry Flynn

"I learned that there is such a thing as hyperbolic
geometry. Through crocheting the coral patterns,
it gave me a fundamental understanding of the
concept." —Barb Oliver

"It was an opportunity that came along at just the
right time. We were already one year into the
pandemic with no end in sight. Crocheting corals
was a real blessing. It gave me something creative
to work on, and it made me feel part of a community.
It enabled me to meet wonderful crocheters from
all over. I looked forward to our Wednesday
Zoom meetings, and I truly enjoyed our Saturday
craft circles when we were finally able to meet
in person. Everyone was so generous throughout this
whole project, sharing their projects and ideas.
Seeing what others were doing would inspire me to
push my crochet limits and create new things.
I learned so much by participating. It was great to
see so many people involved, all ages—men, women,
students, experienced and first-time crocheters.
The project made it possible for everyone to
contribute." —Karen Franciola

"I'm very excited to be part of a museum exhibit,
which makes me feel like a true, valued artist."
—Lorraine Mignone

"This experience with the Tang has renewed
my interest in connecting craft with a cause."
—Kristen Hyman

"I used remnants of yarn I had squirreled away
for nearly fifty years, including yarn from
my long-deceased mother's projects. And then,
instead of purchasing new yarn, I switched
to plastic—an acknowledgment of the plastic bags
in the ocean. I had saved a couple of years of
thin plastic bags in which the daily newspapers
were delivered. These bags came in a range of pastel
colors, and I saved the ocean-blue *New York Times*
wrappers. These were supplemented by white and
beige plastic bags from curbside-meal pickups."
—Anonymous

"I learn to crochet in a more abstract way. I loved
it!" —Caroline Declercq

"I was especially excited about the intention of
the project, to bring attention to climate change
and the peril to our magnificent coral reefs
(which I have experienced on snorkel trips in the
Caribbean)." —Deborah Soffel

"I was amazed by how much I enjoyed working on
this project and how much I learned. First, I had
never crocheted anything freeform—or just for
fun—and watching my crochet hook create the curls
and whorls of the corals as I applied the various
mathematical formulae we were given felt like
sheer magic. It's one thing to be told it will work—
another to see it happening! Further, many of my
pieces were made from plastics gathered around
my home and neighborhood, things that would
normally have been thrown away or recycled. This
gave me a new and very tangible demonstration
and understanding of the extent of this particular
environmental problem confronting our world."
—Rebecca Knutson

"It was exciting to work on a project with a real
goal, not just another scarf or mitten, ho-hum.
All knitters and crocheters habitually think about
the person they are making for—this just happens.
[Crocheting for the *Reef*] made me remember the
many corals and undersea landscapes I've enjoyed."
—Betty Holck

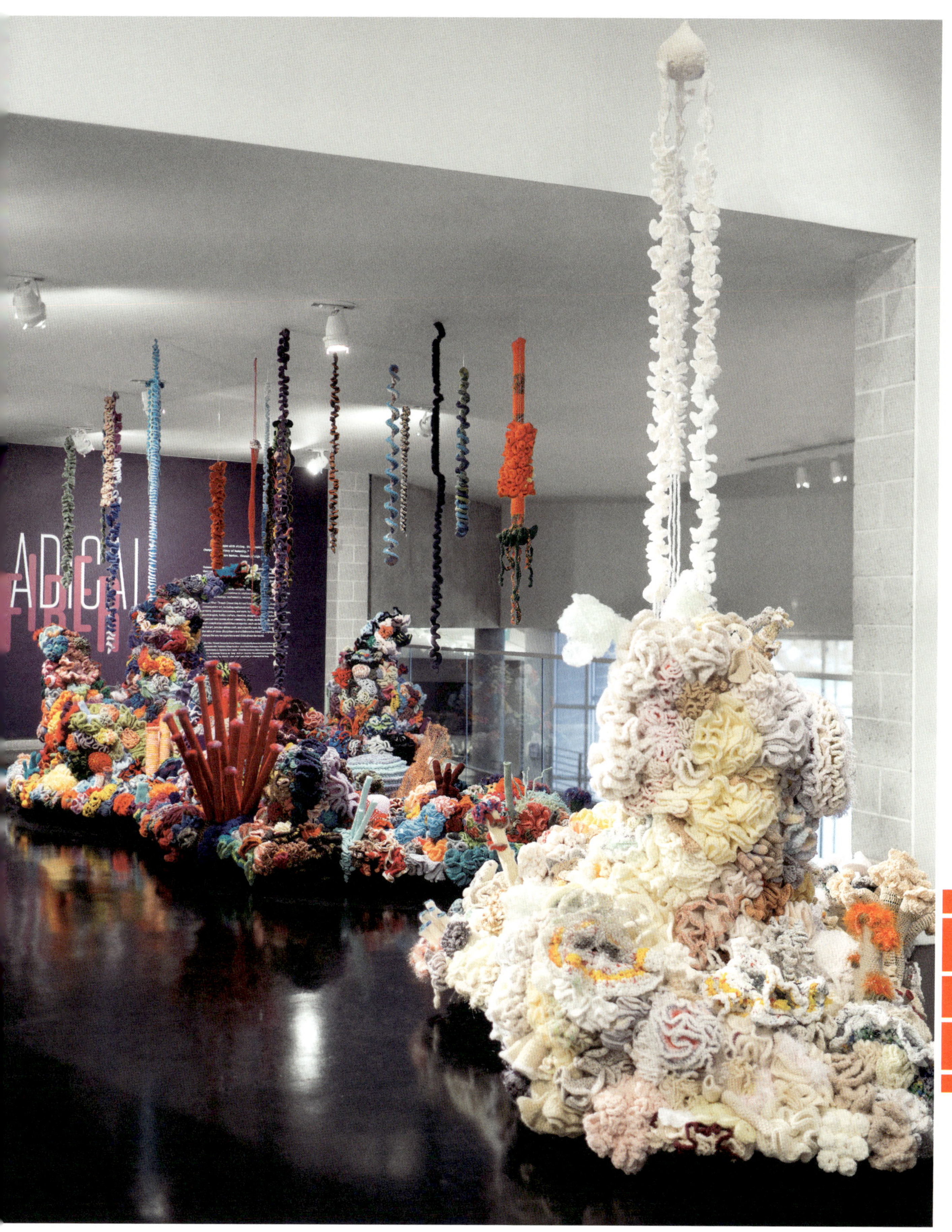

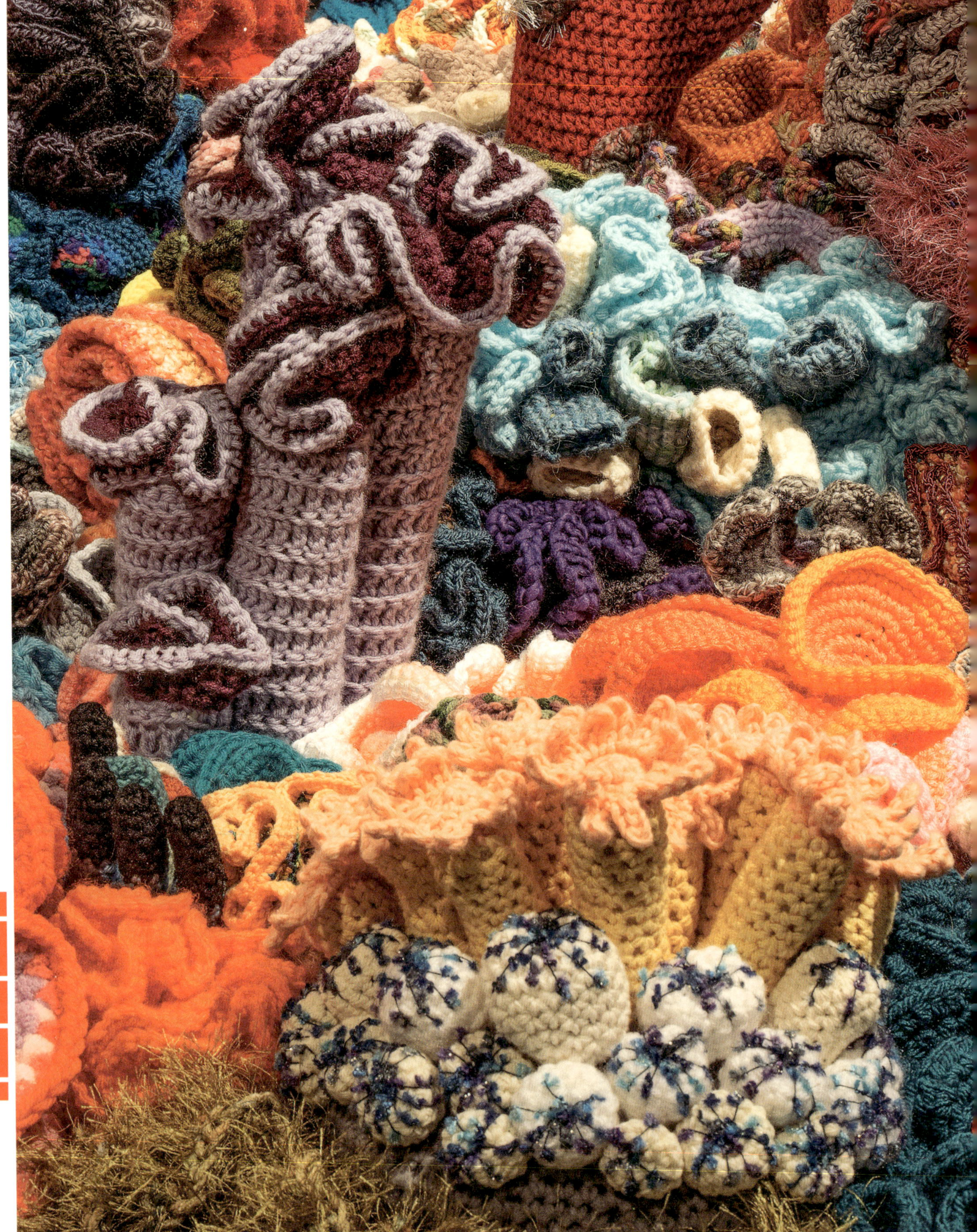

"I found the project quickly became addictive for me, especially since I was recycling grocery bags and other bits of household plastic to create my corals. I became obsessed with the idea of turning the bags into art rather than just reusing and recycling them." —*Jen Schildge*

"I made two pieces using natural fiber, being mindful of the ocean's growing burden of microplastic pollution. Both pieces incorporated ideas about today's real coral reefs and the challenges of a rapidly changing world, as well as my own family and arts & crafts background...One piece represents a sponge, a key element in coral reef ecosystems with a long evolutionary history. Sponges filter and clarify huge volumes of water using some very efficient fluid dynamic physics, which helps to keep the reef environment healthy for corals and other animals. The sponge is made with new natural fiber wool space-dyed in hues from pinks and yellows to greens and purples. With this piece, I chose to diverge mostly from the hyperbolic central structure, in order to highlight one particular knitting pattern that is very close to my heart and family history, my Great-Great Aunt Alice's slipper pattern. She would knit all year to make these slippers in different colors and sizes for the entire clan from toddlers to elders, and every household would receive a box around Christmastime. They are the most comfortable homemade slipper pattern that I have ever seen, using two yarns to knit soft cushiony tubes with a crisscrossing of interior footbed that reminds me of tubular sponges and sponge spicules...I knit this sponge piece based on Aunt Alice's handwritten slipper pattern, showcasing the tubular sponge structure that differs markedly from the interior sponge spicules. I adapted the original pattern's hyperbolic aspect to enhance the sponge shape, and finished the edges of the sponge with rows of crocheted yarn in a hyperbolic pattern with respect to the larger project."
—*Amy Frappier*

"These were the first 3-D objects I've made, and they were lots of fun. I was addicted to making them for several weeks—a great activity to experience, especially during COVID." —*Kate Dudding*

"It's such a unique opportunity to have a very personal connection to a museum exhibition...along with the fact that I got to learn something that connected me with all the other people making the *Reef*, so it was both museum connections and interpersonal connections that pushed me to want to learn this new craft...What surprised me the most about the project was how infectious it was...It's such a simple thing both to teach other people and easy enough to learn." —*Paul Seggev*

"I was surprised how satisfying is was to be part of a large collaborative project." —*Angie Bowles*

"It pushed my ability to crochet out of my normal habits of making garments. More importantly, it gave me a lot to do during the pandemic...It helped me deal with the isolation. I became addicted to it. Then I started studying, because I do care about the environment but I don't know much about the ocean. When I read that half the Great Barrier Reef was gone, that bothered me. For me, that's a justice issue...As one person, I may be able to do something, even if it's just making one person aware of work that has to be done." —*Betty Davis*

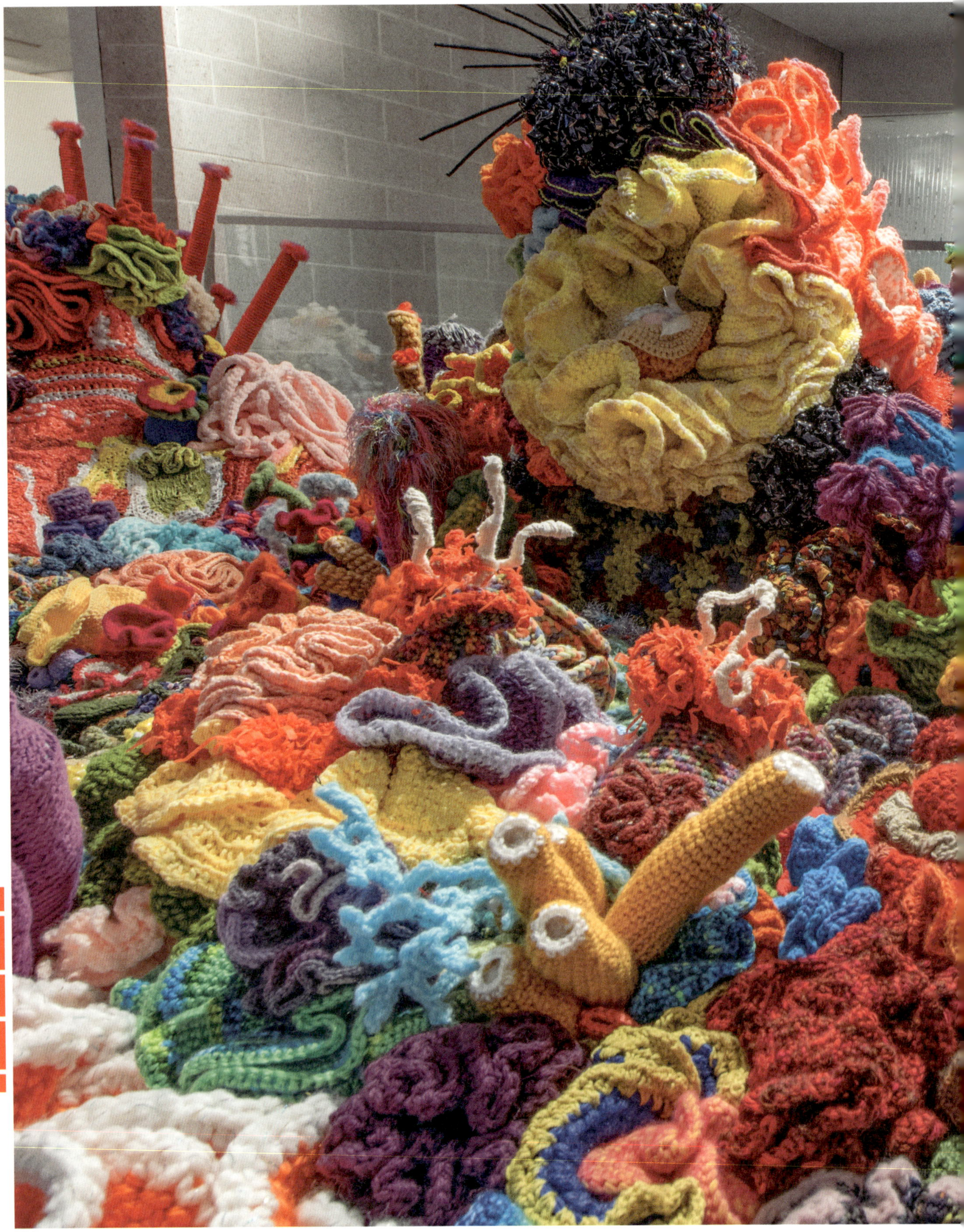

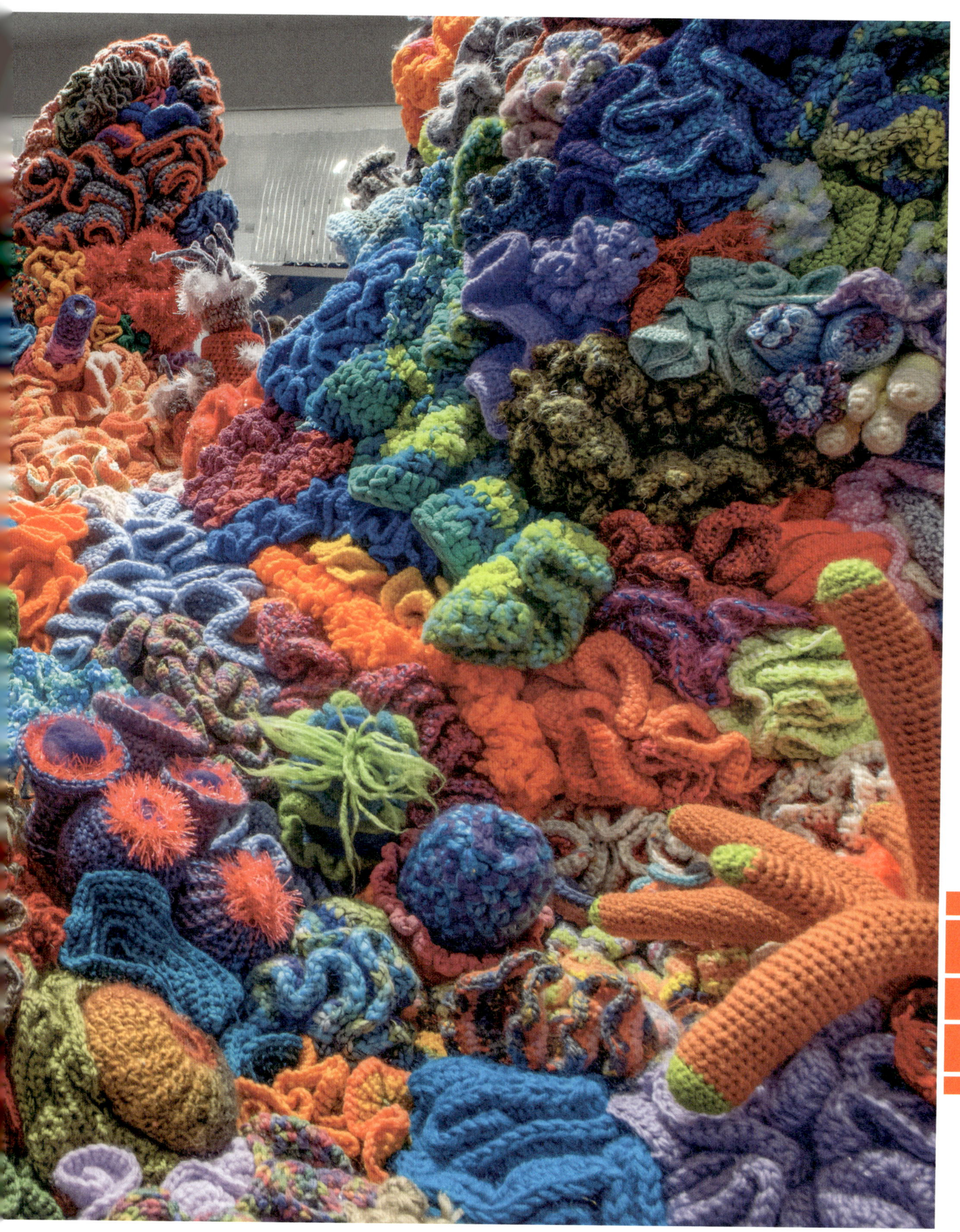

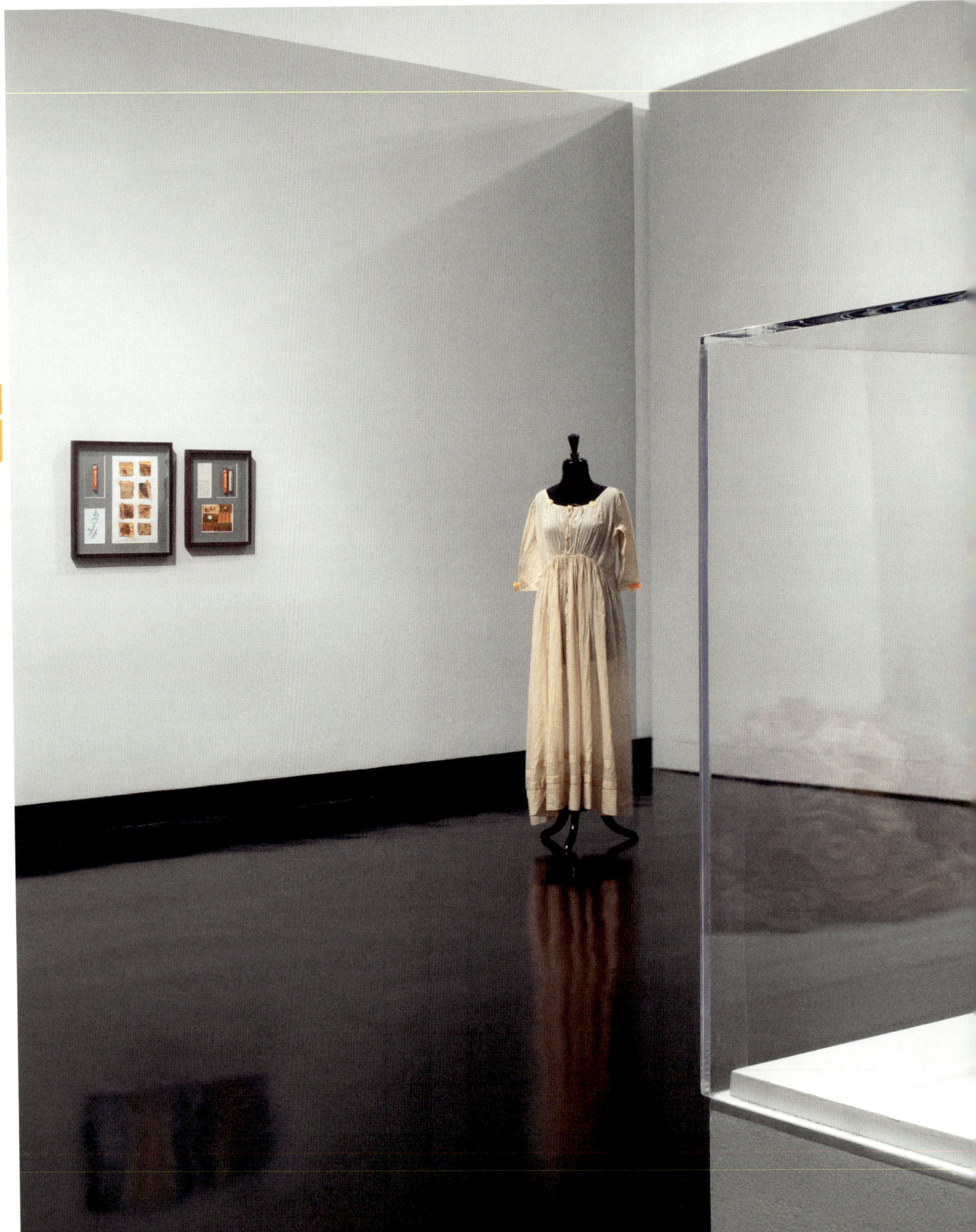

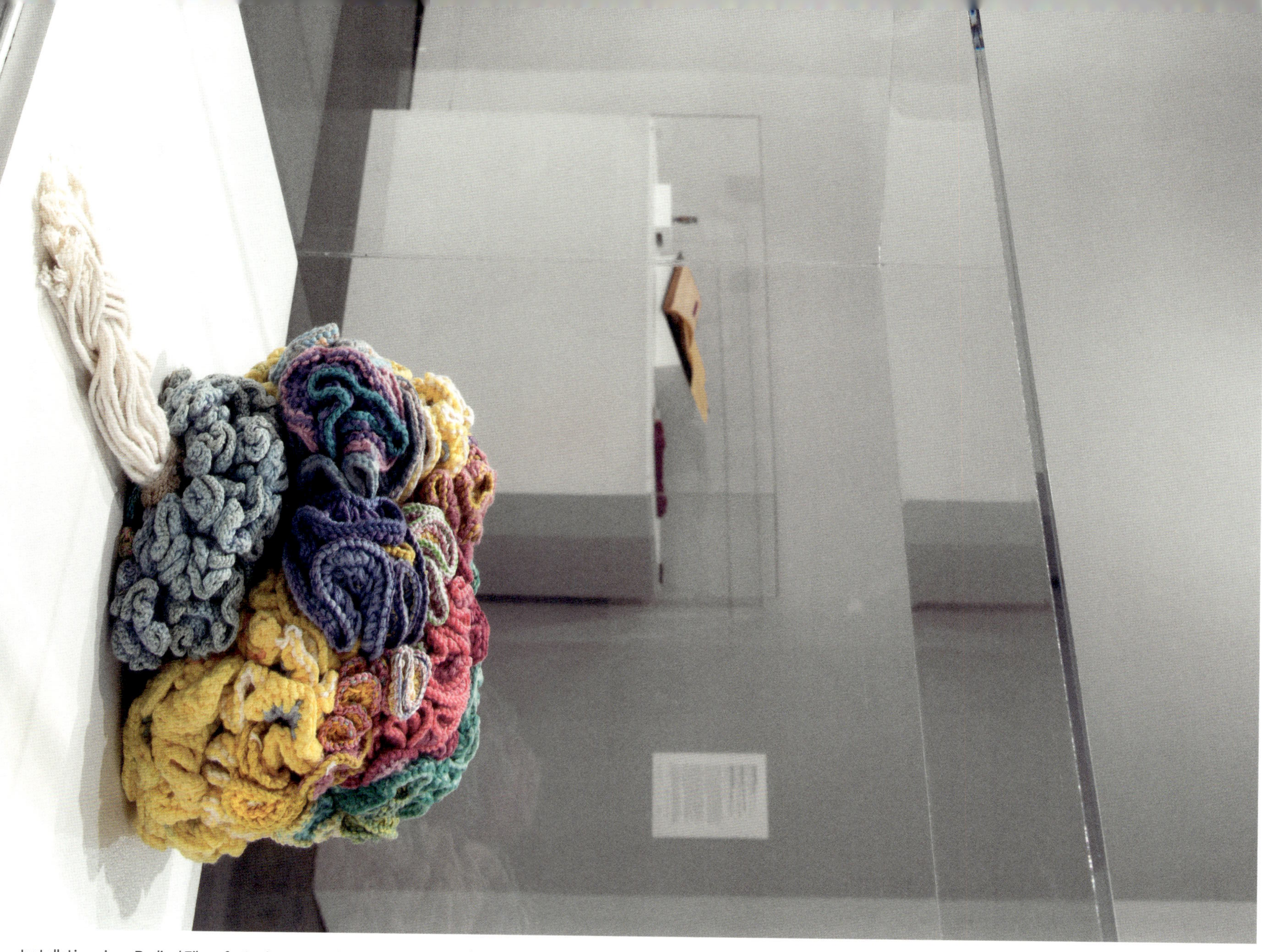

Installation view, *Radical Fiber*, featuring artwork by Anna Dumitriu (background, left), materials related to the discovery of mauveine dye (background, center), and artwork by Karen Norberg (foreground, right)

Auther, Elissa. *String, Felt, Thread: The Hierarchy of Art and Craft in American Art*. Minneapolis: University of Minnesota Press, 2009.

Barber, *Elizabeth Wayland. Women's Work: The First 20,000 Years: Women, Cloth, and Society in Early Times*. New York: W. W. Norton, 1994.

belcastro, sarah-marie and Carolyn Yackel, eds. *Crafting by Concepts: Fiber Arts and Mathematics*. New York: AK Peters, 2011.

belcastro, sarah-marie and Carolyn Yackel, eds. *Figuring Fibers*. Providence, Rhode Island: American Mathematical Society, 2018.

belcastro, sarah-marie and Carolyn Yackel, eds. *Making Mathematics with Needlework: Ten Papers and Ten Projects*. New York: AK Peters, 2013.

Bryan-Wilson, Julia. *Fray: Art and Textile Politics*. University of Chicago Press, 2017.

Buechley, Leah, and Benjamin Mako Hill. "LilyPad in the Wild: How Hardware's Long Tail is Supporting New Engineering and Design Communities." In *DIS '10 Proceedings of the Eighth ACM Conference on Designing Interactive Systems*, 199–207. New York: Association for Computing Machinery, 2010. doi. org/10.1145/1858171.1858206.

Ellison, Elaine, and Diana Venters, *More Mathematical Quilts: No Sewing Required!* Emeryville, California: Key Curriculum Press, 2003.

Garfield, Simon. *Mauve: How One Man Invented a Color That Changed the World*. New York: W. W. Norton & Company, 2000.

Hartshorne, Robin. *Geometry: Euclid and Beyond*. New York: Springer, 2000.

Hunter, Clare. *Threads of Life: A History of the World Through the Eye of a Needle*. New York: Abrams, 2019.

Kittelmann, Udo, Christine Wertheim, and Margaret Wertheim, eds., *Christine and Margaret Wertheim: Value and Transformation of Corals*. Baden-Baden, Germany: Museum Frieder Burda, 2022.

Kurbak, Ebru, ed. *Stitching Worlds: Exploring Textiles and Electronics*. Berlin: Revolver Publishing, 2018.

Le Couteur, Penny, and Jay Burreson. *Napoleon's Buttons: How 17 Molecules Changed History*. New York: TarcherPerigee, 2004.

Mathematical Art Galleries. "Exhibitions." gallery. bridgesmathart.org/exhibitions.

Murro, Juan Antonio, and Jeffrey C. Splitstoser. *Written in Knots: Undeciphered Accounts of Andean Life*. Washington, DC: Dumbarton Oaks Museum, 2019.

Ornes, Stephen. *Math Art: Truth, Beauty, and Equations*. New York: Sterling, 2019.

Osinga, Hinke M., and Bernd Krauskopf, "Crocheting the Lorenz Manifold," *The Mathematical Intelligencer* 26 (2004): 25–37. doi.rg/10.1007/ BF02985416.

Parkes, Clara. *The Knitter's Book of Yarn: The Ultimate Guide to Choosing, Using, and Enjoying Yarn*. New York: Potter Craft, 2007.

Perkov, Kayleigh. *The Computer Pays Its Debt: Women, Textiles, and Technology, 1965–1985.* Asheville, NC: Center for Craft, March 13, 2020–October 9, 2020.

Postrel, Virginia. *The Fabric of Civilization: How Textiles Made the World.* New York: Basic Books, 2020.

Rosner, Daniela K., et al. "Making Core Memory: Design Inquiry into Gendered Legacies of Engineering and Craftwork." *CHI Conference on Human Factors in Computing Systems,* (2018): 1–13. faculty.washington.edu/dkrosner/files/CHI-2018-Core-Memory.pdf.

Shorey, Samantha, and Daniela K. Rosner. "A Voice of Process: Re-Presencing the Gendered Labor of Apollo Innovation." *communication +1* 7, no. 2 (2019): Article 4. doi.org/10.7275/yen8-qn18.

Taimina, Daina. *Crocheting Adventures with Hyperbolic Planes: Tactile Mathematics, Art and Craft for All to Explore.* 2nd ed. Boca Raton: CRC Press, 2018.

Wertheim, Margaret. *A Field Guide to Hyperbolic Space.* Los Angeles: Institute For Figuring Press, 2005.

Wertheim, Christine and Margaret Wertheim. *Crochet Coral Reef: A Project.* Los Angeles: Institute For Figuring Press, 2014.

Wolz, Ursula, Michael Auschauer, and Andrea Mayr-Stalder. "Programming Embroidery with TurtleStitch." In *SIGGRAPH '19: ACM SIGGRAPH 2019 Studio* (Association for Computing Machinery, July 2019). doi.org/10.1145/3306306.3328002.

Lia Cook

Connectome, 2013
Woven cotton, rayon
72 × 51 inches
Collection of the artist
MACHINE | BRAIN

Veronica Dry

Nature Calls, 2017
Cushion pad, corduroy fabric, wool,
cord, twine raffia, fabric scraps, wire
12 × 22 × 8 inches
Collection of the artist
COMMUNITY | MACHINE | BRAIN

Anna Dumitriu

Magic Bullet Series, 2015
Wood, metal, glass, textile dyed with
natural dyes, drawing on paper, vintage
printed materials, sulfa drugs in tubes
17 ¼ × 12 ¾ × 1 ¼ inches (left panel)
20 ⅛ × 18 × 1 ¼ inches (right panel)

The Romantic Disease Dress, 2014
Fabric, walnut husks, safflower-
and madder root–dyed silk, prontosil-
dyed silk
54 ¾ × 21 ¼ × 17 ½ inches (display size)

Collection of the artist
BODY

Ellis Developments
Peter Butcher, textile designer

**Biomplantable device for
reconstructive shoulder surgery**,
developed 1997–2003, made 2021
Embroidered polyester
5 ¾ × 5 ¾ inches

**Biomplantable device for torn
rotator cuff**, 1997
Embroidered polyester
2 ⅜ × 1 ½ inches

**Biomplantable device for torn
rotator cuff**, developed 1997–2003,
made 2021
Embroidered polyester
6 ¼ × 1 ¼ inches

Tang Teaching Museum collection,
gifts of Ellis Developments, Ltd.,
2021.32.1–3
MACHINE | BODY

Elaine Krajenke Ellison

Greek Cross to Square Dissection,
2013
Pieced, appliquéd, quilted cotton
53 × 53 ¼ inches
Collection of the artist
COMMUNITY | MACHINE

Imperial Chemical Industries, Dyestuffs Division

Hank of yarn dyed with mauve,
c. 1956–68
1 ⅜ × 12 × 3 ½ inches (display size,
yarn only)
Collection of the Science Museum
Group, William Henry Perkin Papers,
lent by Helen Beaufoy and Michael
Kirkpatrick, YL1999.2/5/105
COMMUNITY | BODY

Hanne Kekkonen

Crochet samples, 2021
Crocheted cotton yarn
2 ⅝ × 3 ¼ × 3 inches, each

Klein bottle, 2020
Crocheted cotton yarn
5 × 3 ½ × 3 ¼ inches

Möbius band, 2021
Crocheted cotton yarn, metal ring
2 × 4 ¾ × 4 ¼ inches

Möbius snail, 2018–19
Crocheted cotton yarn, metal ring
2 ¾ × 4 × 3 ½ inches

Seifert surface of a trefoil knot, 2021
Crocheted cotton yarn, plastic
rod, wool
4 ¼ × 5 ¼ × 5 ¼ inches

Seifert surface of a (3,3) torus link,
2018–19
Crocheted cotton yarn, plastic
rod, wool
3 ⅝ × 4 × 4 inches

Collection of the artist
SHAPE

Kintra Fibers

**Kintra Fibers components and
samples** (wheat straw, lab vials
containing resin and fiber, spool of yarn,
dyed and undyed hand-knit fabric
samples), 2021
Dimensions variable
Collection of the artist
COMMUNITY | MACHINE

Karen Norberg

Knitted Brain (KB #1), c. 1993/2021
Knitted cotton yarn, nylon and
polyester zipper, Velcro, metal snaps
6 × 7 × 15 inches

Knitted Brain #2 (KB #2), 2020–22
Knitted cotton yarn, nylon zippers,
metal zipper pulls, metal snaps, magnets
3 ½ × 6 × 8 ¾ inches and 5 × 6 ¼ × 4 inches

Collection of the artist
SHAPE | BRAIN

Dario Robleto

The Creative Potential of Disease,
2004
A self-portrait doll made by a Union
Civil War soldier amputee while
recovering in the hospital, mended and
repaired with a modern-day surgeon's
surgical needle and thread, new
pant leg material made from a modern-
day soldier's uniform, cast leg made
from femur bone dust, and prosthetic
alginate treated with *Balm of a 1000
Foreign Fields*, vegetable ivory, collagen,
melted shrapnel and bullet lead, cold
cast steel and zinc, polyester resin, rust
12 ⅛ × 10 × 1 ⅝ inches
Collection of Barry Sloane and
Michael Duncan
COMMUNITY | BODY | BRAIN

Henry Edward Schunck

Mauveine dye, a sample chemical, late
19th century
Mauveine dye, glass bottle, paper label
3 ¾ × 1 × 1 inches
Collection of the Science Museum
Group, gift of the University of
Manchester, Y1997.7.1.271
COMMUNITY | BODY

John Sims

*The Hanging of Knots up to 8
Crossings*, 2007
A Tribute to Sol LeWitt
Knotted rope
Dimensions variable, 120 × 144 inches
(installation size)
Collection of the artist
SHAPE

Soft Monitor
(Victoria Manganiello and Julian
Goldman)

c o m p u t e r 1.0, 2018/2021
Hollow polymer tubing, natural fiber
thread, liquid, operating system
Dimensions variable
Collection of the artists
COMMUNITY | MACHINE

**Helen Remick, Daniela Rosner,
Samantha Shorey, Brock Craft**

Core Memory Quilt, 2018
Woven patches, core memory
artifacts, embroidered patches
68 × 57 ¾ × ⅛ inches
Collection of the artists
COMMUNITY | MACHINE

Daina Taimina

Hyperbolic sketch, 2006
Crocheted microfiber yarn
19 × 16 inches

Day and Night, 2007
Crocheted wool yarn
18 ⅛ × 18 ⅛ × 15 ⅞ inches

Collection of the artist
SHAPE

Unidentified Inca artist

Quipu, c. 1470–1520
Cotton, camelid fiber
29 ¾ × 21 ¼ × ⅛ inches
Collection of the Staten Island Museum,
gift of Mr. and Mrs. Harold Kaye,
A1955.93.9
COMMUNITY | MACHINE

Unidentified maker

Model of a Jacquard loom, 1825
Wood, metal, paper punch cards
15 × 23 × 7 inches
Courtesy of the Computer History
Museum, B117.80
MACHINE

Unidentified maker

Sample of Perkin's mauve
*Journal of the Society of Dyers and
Colourists*, November 1906
Published by the Society of Dyers
and Colourists, England-Bradford
(West Yorkshire)
11 × 17 × 4 inches (open book)
Collection of the Science History
Institute
COMMUNITY | BODY

***Chanccani Quipu**, 2012*
Ink on knotted cords of unspun wool
and bamboo
78 ½ × 17 × 4 ¼ inches (display size)
Collection of the artist, courtesy
of the artist and Lehmann Maupin, New
York, Hong Kong, Seoul, and London
COMMUNITY | MACHINE

Carolyn Yackel

***Abundance**, 2021*

***Interconnection**, 2021*

***Yearning**, 2021*

Embroidered thread over three
Styrofoam balls
3 ⅜ × 3 ⅜ × 3 ⅜ inches, each
Collection of the artist
SHAPE

***Saratoga Springs Satellite Reef**,*
part of the worldwide *Crochet Coral
Reef* project by **Christine and
Margaret Wertheim** and the Institute
For Figuring
Conceptualized 2005, made 2021–22
Crocheted yarn, plastic, wire
Dimensions variable
COMMUNITY | SHAPE | BRAIN

Crochet Coral Reef is a project created
by sisters Christine Wertheim and
Margaret Wertheim of the Institute For
Figuring in Los Angeles. Residing at the
intersection of mathematics, marine
biology, handicraft, and community art
practice, the project responds to the
environmental crisis of global warming
and the escalating problem of oceanic
plastic trash by highlighting not only
the damage humans do to earth's
ecology, but also our power for positive
action. The Wertheims' *Crochet Coral
Reef* collection has been exhibited
worldwide, including at the 2019 Venice
Biennale, Helsinki Biennial (Finland),
Andy Warhol Museum (Pittsburgh),
Hayward Gallery (London), Science
Gallery (Dublin), Museum Frieder
Burda (Germany), and the Smithsonian
National Museum of Natural History
(Washington, DC). The project also
encompasses a community-art program
in which more than 20,000 people
around the world have participated
in making fifty locally based *Satellite
Reefs*—in New York, London, Chicago,
Melbourne, Ireland, Latvia, Germany,
Finland, the United Arab Emirates,
and elsewhere. The *Saratoga Springs
Satellite Reef* is one of the latest
additions to this ever-evolving wooly
archipelago.

Lia Cook, Emeritus Professor of Arts at California College of the Arts, is a fiber artist who combines photography, painting, and digital technology to explore the materiality of the woven image. Her work has been exhibited and is in the permanent collections of museums worldwide, and she is the recipient of numerous awards, including, most recently, the American Craft Council's Gold Medal for Consummate Craftsmanship in Fiber. In her collaborations with neuroscientists, she investigates emotional connections to textile images.

Brock Craft is Associate Teaching Professor in Human Centered Design and Engineering at the University of Washington. He has taught classes on digital and electronic media at the Royal College of Art; Goldsmiths, University of London; and DePaul University. His work focuses on interactive design and usability in various domains, including human-computer interaction, product design, digital art, and learning. Craft has also collaborated on projects that connect the digital world to the physical through interactive technology, which he incorporates into his pedagogy design.

Veronica Dry is a severely sight-impaired retired civil servant living in Buckinghamshire, England. She has no art or crafting background but has discovered skills she did not know she had and has explored making work involving weaving and sound to produce both tactile and audio "scenes." She is now delving into close-up plant photography.

Anna Dumitriu is an award-winning, internationally renowned British artist who works with BioArt, sculpture, installation, and digital media to explore our relationship to infectious diseases, synthetic biology, and robotics. She has exhibited extensively internationally, including at ZKM, Ars Electronica Center in Linz, Bozar Centre for Fine Arts, the Picasso Museum, Künstlerhaus in Vienna, the MIT Museum, Liljevalchs, Kunsthal Charlottenborg, MOCA Taipei, HeK (Basel), Art Laboratory Berlin, Taipei Fine Arts Museum, the 6th Guangzhou Triennial, and the History of Science Museum in Oxford. Her work is held in several major collections, including ZKM, the Science Museum in London, and the Eden Project, and has been featured in publications including *Frieze*, *Artforum*, *Leonardo*, the *Art Newspaper*, *Nature*, and the *Lancet*.

Elaine Krajenke Ellison is a retired high school mathematics teacher who creates quilts to teach mathematical concepts. She conducts workshops and speaks at various institutions around the world, including the Science Museum in London, and at meetings of school, quilt, mathematical, and science groups to instruct on using mathematical visualization to aid in problem solving. She is the coauthor, with Diana Venters, of *Mathematical Quilts* and *More Mathematical Quilts*.

Ellis Developments is a research and development company involved in contract research and in-house product development in the field of textile material science, particularly for composites and surgical implants. Founder Julian Ellis has a master's degree in philosophy for research of the technology of fabrics and specializes in the technical aspects and quality control of yarn, fabric, and garment manufacturing. Ellis has earned an OBE for Services to the Technical Textiles industry and services to the welfare of prisoners. Peter Butcher is an award-winning textile designer.

Hanne Kekkonen is Assistant Professor of Applied Mathematics at the Delft Institute of Applied Mathematics. Her research focuses on Bayesian inverse problems, Bayesian statistics, and nonparametric statistics. She uses fiber craft to better understand and teach mathematical principles and to encourage engagement in the field of applied mathematics. Kekkonen obtained her PhD in applied mathematics at the University of Helsinki.

Kintra Fibers is a materials science company that has developed a proprietary bio-based and biodegradable polyester. Kintra's material is designed to address the environmental impact caused by traditional polyester at every stage, from the start- to end-of-life. By utilizing bio-based inputs and designing a biodegradable material from the outset, Kintra Fibers aims to transform the apparel industry to work in harmony with the planet.

Karen Norberg is a retired epidemiologist and child psychiatrist with more than thirty years of clinical and research experience, particularly in applications of econometric methods and of demography and public health. Throughout her professional career, she has had several occasional adventures at the intersections of medicine and visual art and craft. Her *Knitted Brain* project began in part because her family, like many others, has been deeply affected by differences in how each of our brains work (or don't work—at times). In this ongoing project, she combines her knowledge of neuroanatomy with her skills in fiber craft to construct anatomically accurate models of the human brain— with color coding, some gentle humor (she hopes) and some artistic license (she admits). She earned her MD from Harvard Medical School.

Helen Remick is a quilt artist based in Seattle, Washington. Since retiring from the University of Washington in 2005, quilting has been her primary passion. Her designs typically include mandalas, spirals, and symmetrical patterns.

Dario Robleto is an artist, researcher, writer, and teacher working at the intersections of art and science. He has been a visiting scholar and artist-in-residence at the Smithsonian National Museum of American History, the SETI Institute, the Robert Rauschenberg Foundation, and the Radcliffe Institute for Advanced Study at Harvard University, among other institutions, and was most recently Artist-at-Large at Northwestern University's McCormick School of Engineering and the Block Museum of Art. In 2016, he was appointed as the Texas State Artist.

Daniela Rosner is Associate Professor in Human-Centered Design and Engineering at the University of Washington and an associate member of the Einstein Center Digital Future in Berlin. Her work focuses on the external factors of technology development with a stress on under-recognized innovation practices, from maintenance to needlework. She has won multiple awards from the US National Science Foundation and has published articles on techno-culture that have been featured in the *Journal of New Media & Society* and the *Journal of Material Culture*.

Samantha Shorey is Assistant Professor in Communication Studies at the University of Texas at Austin. She is a design researcher whose work focuses on the process of people becoming innovators and the way women are often overlooked in their contributions to technology design. Prior to becoming a professor, she worked at the Smithsonian's Lemelson Center for the Study of Invention and Innovation, National Museum of American History, where she examined the computer hardware created by women during the Apollo moon missions.

John Sims was a multimedia math-artist, writer, and activist whose work spans topics that include mathematics, art, text, performance, and political-media activism. At the Ringling College of Art and Design, he designed a visual mathematics curriculum for artists and visual thinkers. He cocurated the 2002 exhibition *MathArt/ArtMath* and created a system of fifteen mathematical art exhibitions called *Rhythm of Structure*, including a year-long 2009 Bowery Poetry Club exhibition/film project. This work led to his own exhibition, *SquareRoots: A Quilted Manifesto*, featuring math-art quilts made in collaboration with Amish quilters, and the creation of the Pi Day Anthem.

Soft Monitor is an art and design collective focused on telling stories of materials, technology, and culture through physical experience. Founded in 2017 by Victoria Manganiello and Julian Goldman, Soft Monitor makes experiential objects driven by collaboration between people, materials, and information. Their work has been shown at Center for Craft in Asheville, North Carolina; Ars Electronica Center in Linz, Austria; the Indianapolis Museum of Contemporary Art; and the Museum of Arts and Design in New York. Manganiello is an award-winning textile artist and educator whose work focuses on the intersections between textiles, rituals, gender, and technology.

Goldman is an award-winning industrial designer focused on cutting-edge biomaterials, sustainable design, and computer-aided making for varied product-directed companies.

Daina Taimina, retired Adjunct Associate Professor of Mathematics at Cornell University, created the first crocheted hyperbolic plane for use in a non-Euclidean geometry class in 1997. Since then, she has crocheted many more, turning a geometric model into a fiber artwork. Taimina has given many lectures and participated in art exhibitions around the world. Her book *Crocheting Adventures with Hyperbolic Planes* received the Diagram Prize from *Bookseller* magazine and the Euler Book Prize from the Mathematical Association of America in 2012.

Cecilia Vicuña, a poet, artist, filmmaker, and activist, was born, raised, and subsequently exiled from Santiago de Chile after a military coup against President Salvador Allende in the early 1970s. Her work often begins as poetry—which she then transforms into film, song, sculpture, or performance—and addresses contemporary concerns of ecological destruction, human rights, and cultural homogenization. Her work has been exhibited and collected worldwide, and she is the recipient of numerous awards, including the 2022 Venice Biennale's Golden Lion for Lifetime Achievement.

Margaret Wertheim is a science writer, artist, and author of books on the cultural history of physics. She has written for *The New York Times*, *Los Angeles Times*, *Cabinet*, *Aeon*, and many other publications about the intersection of science, mathematics, culture, and the arts. **Christine Wertheim** is an experimental poet, performer, artist, and writer, and a former faculty member at the California Institute of the Arts. Margaret and Christine conduct the *Crochet Coral Reef* project through their Los Angeles–based practice, the Institute For Figuring, which is dedicated to "the poetic dimensions of science and mathematics." The IFF is at once an art endeavor and a framework for innovative public science engagement.

Carolyn Yackel is Professor of Mathematics at Mercer University, focusing on recreational mathematics and expanding mathematical pedagogical tools, such as fiber-based models. She has worked with sarah-marie belcastro to organize the Knitting Network at various national mathematics meetings, such as the annual Joint Mathematics Meetings (JMM), and has coedited books on visualizing mathematics through fiber craft, including *Figuring Fibers* and *Crafting by Concepts: Fiber Arts and Mathematics*.

Trisha Andrew is Professor of Chemistry and Chemical Engineering at the University of Massachusetts Amherst. She directs the Wearable Electronics Lab, a multidisciplinary research team that innovates methods and materials to coat fibers, fabrics, and garments and to transform them into electronically active devices. The coated fibers and fabrics can harvest solar light, store energy, capture carbon dioxide, and sense temperature, touch, motion, or physiological signals. She is founder and CTO of Soliyarn, a Boston-based start-up that is commercializing vapor-coating technology to transform off-the-shelf fabrics into smart clothing.

Preeti Arya, Assistant Professor of Textile Development and Marketing at the Fashion Institute of Technology, is a strong supporter and spokesperson of sustainable products and practices in the textile industry. Her research includes sustainability in chemical finishes, textile processing, performance textiles, natural composites, and closed-loop textile production systems.

Nurcan Atalan-Helicke is Associate Professor of Environmental Studies and Sciences at Skidmore College. An interdisciplinary social scientist, she conducts research at the intersection of food production and access to clean, healthy food and focuses on the conservation of agricultural biodiversity in Turkey, genetically modified food from an Islamic perspective, and the gender dimensions of food consumption. She cocurated the 2019 exhibition *Like Sugar* at the Tang Teaching Museum.

Elissa Auther is Deputy Director of Curatorial Affairs and William and Mildred Lasdon Chief Curator at the Museum of Arts and Design. She has published widely on the history of modernism and its relationship to craft and the decorative, and on the material culture of the American counterculture and feminist art. Her book *String, Felt, Thread: The Hierarchy of Art and Craft in American Art* focuses on the broad use of fiber in art of the 1960s and 1970s and the changing hierarchical relationship between art and craft at that time.

Alissa Baier-Lentz is the cofounder and COO of Kintra Fibers, a materials science company that provides apparel brands with performance-oriented, 100 percent bio-based and compostable fibers. A fashion industry entrepreneur, Baier-Lentz grew frustrated by the transparency issues, misinformation, and challenges of sourcing a planet-friendly material. She joined nanoengineer and Kintra CEO Billy McCall to provide brands with farm-to-fabric traceable materials that make no compromises between performance, price, and the planet.

Ian Berry is Dayton Director of the Tang Teaching Museum and Professor of Liberal Arts at Skidmore College. Berry is a leader in the field of academic museums and is a regular speaker on interdisciplinary, inventive curatorial practice and teaching in museums. He is well known for exhibitions and publications with contemporary artists, including Terry Adkins, Nancy Grossman, Corita Kent, Nicholas Krushenick, Dona Nelson, Tim Rollins and KOS, Alma Thomas, Fred Tomaselli, and Kara Walker.

Denise Evert is Associate Professor in the Psychology Department, a member of the Neuroscience Program, and currently holds the Susan Kettering Williamson '59 Endowed Chair in Neuroscience at Skidmore College. She received her PhD from Princeton University and completed postdoctoral training at Boston University School of Medicine, Harvard Medical School, and associated Boston-area Veterans Affairs Medical Centers. Her research interests include hemispheric special-ization for attentional and emotional processing, and she has taught courses on the topic of creativity entitled "Creating Minds" and "Creativity and the Brain."

Emilie Giles is a researcher, artist, and educator. Her work is situated within human-computer interaction and focuses on the linking of creative tech-nology, craft practice, and accessibility. She earned her PhD from The Open University, United Kingdom, and is Senior Lecturer of Graphic Design at the Arts University Bournemouth. Her research explores how people who are often excluded from the benefits of creative technology can create on their own terms, giving them a greater sense of agency and empowerment. Her work guides blind and visually impaired people to make their own e-textile interactive artwork and explores how personal stories or memories can be told through the work.

Juan Hinestroza is the Rebecca Q. Morgan '60 Professor of Fiber Science and Apparel Design and directs the Textiles Nanotechnology Laboratory at the College of Human Ecology at Cornell University. His research focuses on understanding, at the nanoscale, fundamental phenomena that are of relevance to fiber and polymer science. Hinestroza is an inventor of more than thirty-three granted international patents, an author of more than one hundred peer-reviewed articles and five book chapters, and an editor of a seminal book on cellulose-based green composites. His pioneering work has enabled the creation of three start-up companies, and he has served as a consultant to major Fortune 50 cor-porations and investment banks in the field of smart and interactive textiles and fibers.

Mark Huibregtse recently retired as Professor of Mathematics at Skidmore College. He received his PhD at the Massachusetts Institute of Technology and completed a graduate-level program at the Institute for Retraining in Computer Science at Clarkson University. His research focuses on algebraic geometry, and he has published numerous articles on the topic.

Sarita Lagalwar is Associate Professor in the Neuroscience Program at Skidmore College. She received her PhD in cell and molecular neuroscience at Northwestern University and completed postdoctoral research at Northwestern and the University of Minnesota. Her research focuses on molecular mechanisms in neuro-degenerative disease.

Elaine Larsen is Senior Instructor of Biology at Skidmore College. She received her PhD at Tufts University and has done research regarding cell biology. Among the introductory-level courses she teaches is the interdisciplinary "Fiber Arts Sciences." Her research interests include the science behind various fibers—the ways they're generated, harvested, and prepared—and the ways their physical properties interact with the dyeing process to create textiles.

Sang-Wook Lee is Professor of Art at Skidmore College. He received a BFA and an MFA from Dong-A University in South Korea before coming to the United States, where he earned an MFA in fabric design from the University of Georgia. His expertise on fiber arts and his Korean upbringing allow his installa-tion works to be representations of cross-cultural expressions between Eastern and Western traditions. His work has been exhibited in both the United States and Asia.

Rebecca McNamara is Associate Curator at the Tang Teaching Museum, where she frequently organizes exhibitions on themes related to identity, racial and social justice, and understanding the human condition. Her research focuses on American art and material culture from the nine-teenth century to the present, with a particular interest in fiber-based practice. She is curator of *Radical Fiber: Threads Connecting Art and Science*, among numerous other exhibitions, and coeditor of the award-winning three-year journal *Accelerate: Access and Inclusion at the Tang Teaching Museum*.

Stephen Ornes is an award-winning math and science writer who loves finding and telling stories at the intersection of science, math, art, and culture. He writes from a shed in his backyard in Nashville, Tennessee, and is Writer-in-Residence at Vanderbilt University. His book *Math Art: Truth, Beauty, and Equations* highlights the work of artists who find inspiration in the ideas, proofs, and visualizations of mathematics. He produced and hosted the award-winning podcast series *Calculated* to accompany the book.

Aarathi Prasad, Assistant Professor of Computer Science at Skidmore College, earned her PhD in computer science from Dartmouth College, where she completed a thesis that proposed techniques to protect users' privacy and allow them to share information when using mobile health applications. Her research interests lie in developing secure and usable applications for mobile and wearable technologies.

Rachel Roe-Dale is Professor of Mathematics and Statistics at Skidmore College. She received her PhD at Rensselaer Polytechnic Institute and focuses her research on mathematical biology, medicine, and modeling other physical systems. She cocurated *Sixfold Symmetry: Pattern in Art and Science* at the Tang Teaching Museum, her first foray into curating, and coteaches the Skidmore course "Math in the Museum," a quantitative reasoning course that explores the connections between mathematics and art.

Rachel Seligman is Assistant Director for Curatorial Affairs and Malloy Curator at the Tang Teaching Museum. Her curatorial projects include interdisciplinary collaborations on topics such as democracy and citizenship, social class, activism, civil rights, and art/science connectedness, among others. As Skidmore Lecturer in Mathematics and Statistics, Seligman coteaches the course "Math in the Museum." Her exhibitions include *Sixfold Symmetry: Pattern in Art and Science* and *Like Sugar*, among others.

Jeffrey C. Splitstoser, Research Professor at George Washington University and Vice President of the Boundary End Archaeology Research Center, has studied ancient Andean textiles for more than twenty years, having recently discovered (with Tom Dillehay, Jan Wouters, and Ana Claro) the world's earliest known use of indigo blue in a 6,200-year-old cotton textile from the prehistoric site of Huaca Prieta. Splitstoser specializes in Wari quipus, colored-and-knotted-string devices that ancient Andean peoples used to record information. His research includes reproducing the quipus and textile structures he encounters: processing, spinning, and dyeing the fibers as well as growing cotton and dye plants. He received his PhD in anthropology from The Catholic University of America, Washington, DC.

Rebecca Trousil is Senior Teaching Professor of Mathematics and the S3M Program Director at Skidmore College. She received her PhD in physics at Washington University in St. Louis. Her interdisciplinary research interests range from applications of physics and mathematics in medicine and biology to cryptology and pedagogy.

Ursula Wolz holds a PhD in computer science, with a focus on artificial intelligence, from Columbia University and consults on machine learning for innovative textile production and intelligent tutoring systems. Her volunteer work centers on textile handcrafts for social justice. As a computer science educator and entrepreneur, she studies and builds computer-based learning environments that include nontraditional media to teach coding. She founded RiverSound Solutions with a mission to empower computer users to become creators with, rather than consumers of, computing. Her recent work, called "code crafting," explores how computer science theory is firmly grounded in textile production, borrowing from concepts and techniques established by textile handcrafters millennia ago. She promotes learning environments modeled on ancient women-centered crafting circles, in which skills and knowledge grow organically from the interaction between mentor and student as meaningful work produces useful artifacts.

Acknowledgments

When Dario Robleto visited Tang Dayton Director Ian Berry's "The Artist Interview" course in 2017, he encouraged us to ask not how arts and science intersect, but rather: how can artists impact, or even transform, the sciences? Two years later, I learned about Daina Taimina's crocheted models of hyperbolic planes, which she started making in 1997, and I recognized exploring fiber, today and historically, as a distinct yet wide-ranging approach to answering this question. The more I looked, the more examples I found of how fiber materials, techniques, and narratives impacted scientific theory, discovery, and pedagogy. The exhibition was still an untitled, bare-bones idea when I brought it to Ian, and I am incredibly grateful for his early enthusiasm and support for what became a multifaceted, multiyear project. Thank you for seeing the possibilities in an idea that was not yet worked out and for your mentorship through this and so many other curatorial endeavors.

This type of project could not happen without the interdisciplinary expertise of Skidmore College faculty. Thank you to the curatorial advisory group, who guided on all manner of checklist decisions, wrote labels, developed and participated in programming, and contributed to this catalog: Mark Huibregtse, Rachel Roe-Dale, and Rebecca Trousil from Mathematics and Statistics; Sarita Lagalwar from Neuroscience; Elaine Larsen from Biology; Aarathi Prasad from Computer Science; and Sang-Wook Lee from Art. Additional thank you to Nurcan Atalan-Helicke from Environmental Studies and Sciences, and to Denise Evert from Neuroscience and Psychology for your contributions. And to Tang Assistant Director for Curatorial Affairs & Malloy Curator and Skidmore Lecturer in Mathematics Rachel Seligman for co-writing an essay in this catalog and for constant support.

Programming was a vital part of the *Radical Fiber* project, beginning with community crochet workshops and craft circles in February 2021.

Thank you to crochet instructors Victoria Manganiello, Lucy Beizer, and Cal Patch. To Amy Frappier and Margaret Wertheim for participating in our Dunkerley Dialogue series; the students in Elaine Larsen's biology course, "Straw into Gold: Science in the Fiber Arts," who led public tours of the exhibition; and to artists Hanne Kekkonen and Karen Norberg for meeting with Skidmore courses. The exhibition programming benefited from Skidmore's IdeaLab, and especially its manager, Darren Prodger, who collaborated on two programs.

The speakers in the two-day program *Radical Fiber: A Symposium on Art and Science*, held January 29–30, 2022, introduced Tang audiences to new ideas, and I'm thrilled to share edited versions of the four panels in this catalog. Thank you to all participants.

Thank you to the artists and institutions who loaned work to the exhibition: the Computer History Museum, Mountain View, California; Lia Cook; the collaborators Brock Craft, Helen Remick, Daniela Rosner, and Samantha Shorey; Veronica Dry; Anna Dumitriu; Julian Ellis; Elaine Krajenke Ellison; Hanne Kekkonen; Kintra Fibers (especially to Alissa Baier-Lentz); Karen Norberg; the Science History Institute, Philadelphia; the Science Museum Group, England; John Sims; Barry Sloane and Michael Duncan; Soft Monitor (Victoria Manganiello and Julian Goldman); the Staten Island Museum, New York; Daina Taimina; Cecilia Vicuña and Lehmann Maupin; and Carolyn Yackel. An additional thank you to the artists who spent time talking with me about fiber-science connections in preparation for this exhibition: Lia, Samantha, Anna, Julian, Hanne, Alissa, John, Victoria and Julian, Daina, Carolyn, Dario, and Margaret. And of course, to the more than 200 *Saratoga Springs Satellite Reef* participants who joined in online and in-person workshops and craft circles and who contributed creations to our community sculpture—as well as to the Tang staff, volunteers, and students who stitched each coral onto the *Reef* islands.

John Sims passed away in December 2022 during the production of this catalog. He was creative, energetic, and generous, with an unassailable creativity for connecting mathematics to all manner of topics in our world and was fearless in confronting racism through his art and teaching.

Thank you to copy editor Alison Hagge and proofreader Stephanie Cash for your careful attention to text. Barbara Glauber's ability to turn folders of texts and images into a beautifully designed object is insurmountable. Thank you, Barbara, for your sharp design and unending creativity.

Thanks to all the photographers whose work appears on these pages, especially Arthur Evans, Mindy McDaniel, and Shawn LaChapelle. Thank you to Tony Manzella at Echelon for your discerning and careful image editing and color separations.

Thank you to each of my Tang Museum colleagues, whose enthusiasm and dedicated work made this project possible: Olivia Cammisa-Frost, Michael Janairo, Kara Jefts, Elizabeth Karp, Annelise Kelly, Eric Kuhl, Evan Little, Kelsey Renko, Rachel Seligman, Patti Sopp, kelly ward, Tom Yoshikami, and Cynthia Zellner, as well as former Tang staff Jean Tschanz-Egger and Sophie Heath. From initial concept and through production of this catalog, Tang student interns proved instrumental. Thank you to the students who researched artists and artworks, made scale models, updated checklists, proofread texts, researched image rights, drafted biographies and other texts, made condition reports of artwork, and more, especially to Nathan Bloom '21, Helen Branch '24, Theo Carol '23, Will Scarlett '23, Silas Seno Mitchell '21, Brynnae Newman '22, Olivia Sessions '19, and Ali Vassiliou '23. Thanks to our skilled installation crew: Christian Heidsieck, Val Tran, and Nick Squadere. And to the many Tang Guides who led tours of the exhibition and engaged visitors in conversations about art and science.

Rebecca McNamara

Image Credits

Every effort has been made to secure copyright approval and ensure accuracy. Any additional rights holders, please contact us at tang@skidmore.edu for inclusion in future editions.

All *Radical Fiber* installation, exhibition artwork, and event photography is courtesy the Tang Teaching Museum, unless otherwise noted. Museum photography by: Arthur Evans, Shawn LaChapelle, Mindy McDaniel, and Megan Mumford.

16 left: Photo by Emilie Giles
20, 21: Musée Flaubert et d'histoire de la médecine, Réunion des musées Métropole Rouen, Normandie, France
22, 36, 37, 39, 45: Courtesy Dario Robleto
24 top: Courtesy the Texas Heart Institute
24 bottom: Division of Medicine and Science, National Museum of American History, Smithsonian Institution, gift of the Texas Heart Institute, 1978.1002
26, 27: Courtesy the Texas Heart Institute
33: Photo by Thomas R. DuBrock
34: Tang Teaching Museum collection, gift of Peter Norton, 2014.7.8. Photo by Arthur Evans
40: Courtesy the Australian National University
44 top: Reprinted from Martin W. King, et al., "Structural Design, Fabrication and Evaluation of Resorbable Fiber-Based Tissue Engineering Scaffolds," in Biotechnology and Bioengineering (London: IntechOpen, 2019), 61–90, with permission from Elsevier
44 bottom: Courtesy Nicolas L'Heureux
49 top: Royal Collection Trust | © His Majesty King Charles III 2023
49 bottom: Photo by Bob Jacobs, Laboratory of Quantitative Neuromorphology, Department of Psychology, Colorado College, Creative Commons Attribution–Share Alike 3.0 Unported
52: Courtesy Instituto Cajal
62: Courtesy Lia Cook

63: Photos by Sarah Condon-Meyers, Skidmore College
65: Robert Koch, The Aetiology of Tuberculosis, trans. Berna Pinner and Max Pinner (New York: National Tuberculosis Association, 1932).
68 far left: Courtesy Science History Institute
74 left top: Linden K. Allison, University of Massachusetts, Amherst
74 left bottom: S. Zohreh Homayounfar, University of Massachusetts, Amherst
74 right, 76 left and right top: Photos by Emilie Giles
76 right bottom: Photo by Rebecca McNamara
78, 79, 88: Courtesy Ursula Wolz
90: Courtesy Ruth Scheuing
91: Courtesy Robin Kang
93: Chronicle / Alamy Stock Photo
104: Photos by Annelise Kelly
106: Paul N. Hasluck, ed., Cassell's Cyclopaedia of Mechanics, Containing Receipts, Processes, and Memoranda for Workshop Use Based on Personal Experience and Expert Knowledge, volume 1 (London: Cassell and Company, 1904), 106.
110 top: Courtesy John Sims
110 bottom: 2023 The LeWitt Estate | Artists Rights Society (ARS), New York
111: Courtesy John Sims
113: Royal Danish Library, GKS 2232 kvart: Felipe Guamán Poma da Ayala, El primer nueva corónica y buen gobierno (c. 1615), page 362.
115: Photo by Ansis Starks, courtesy of Kim? Contemporary Art Centre, Riga
116: Photo by Heidi Jaansoo, courtesy of the Riga International Biennial of Contemporary Art
118: Dumbarton Oaks, Pre-Colombian Collection, Washington, DC
121: Courtesy Daina Taimina
128: Courtesy Jeffrey C. Splitstoser
140: Courtesy Daina Taimina, reproduced with permission from Springer Nature
144, 146, 147: Courtesy Hinke Osinga and Bernd Krauskopf, reproduced with permission from Springer Nature
145: Courtesy Carolyn Yackel

148, 149: National Museums Scotland
153: Courtesy Elaine Krajenke Ellison
156: Photos by Boyden Gallery, St. Mary's College of Maryland, courtesy Susan Goldstine
162 top: Photo by Mohammad Taherzadeh
162 middle: Ding Yun / Science China Press
162 bottom: AP Photo / Sakchai Lalit
186, 187, 189: Screenshots and photos by Olivia Cammisa-Frost, Star Herrera, Annelise Kelly, Rebecca McNamara

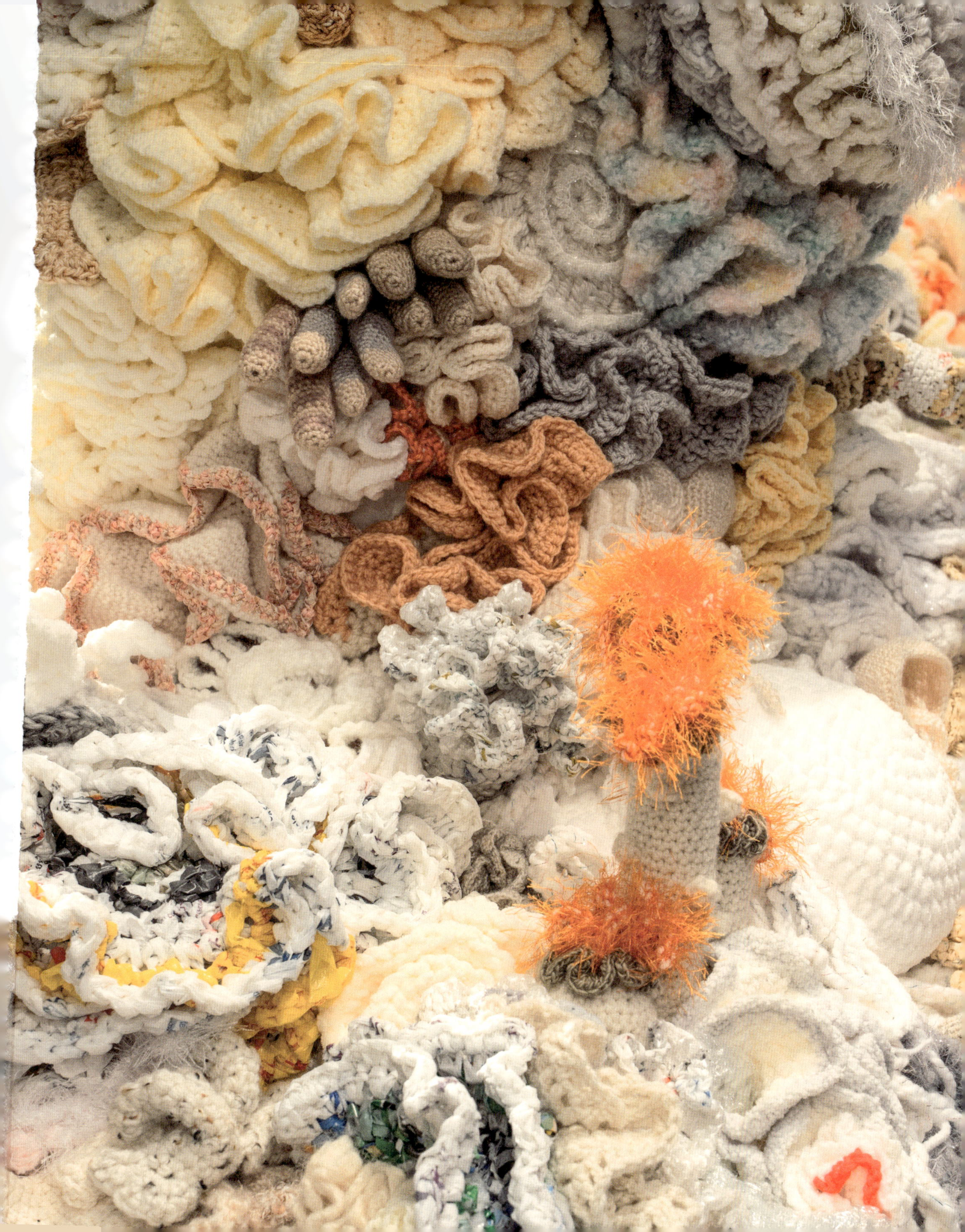

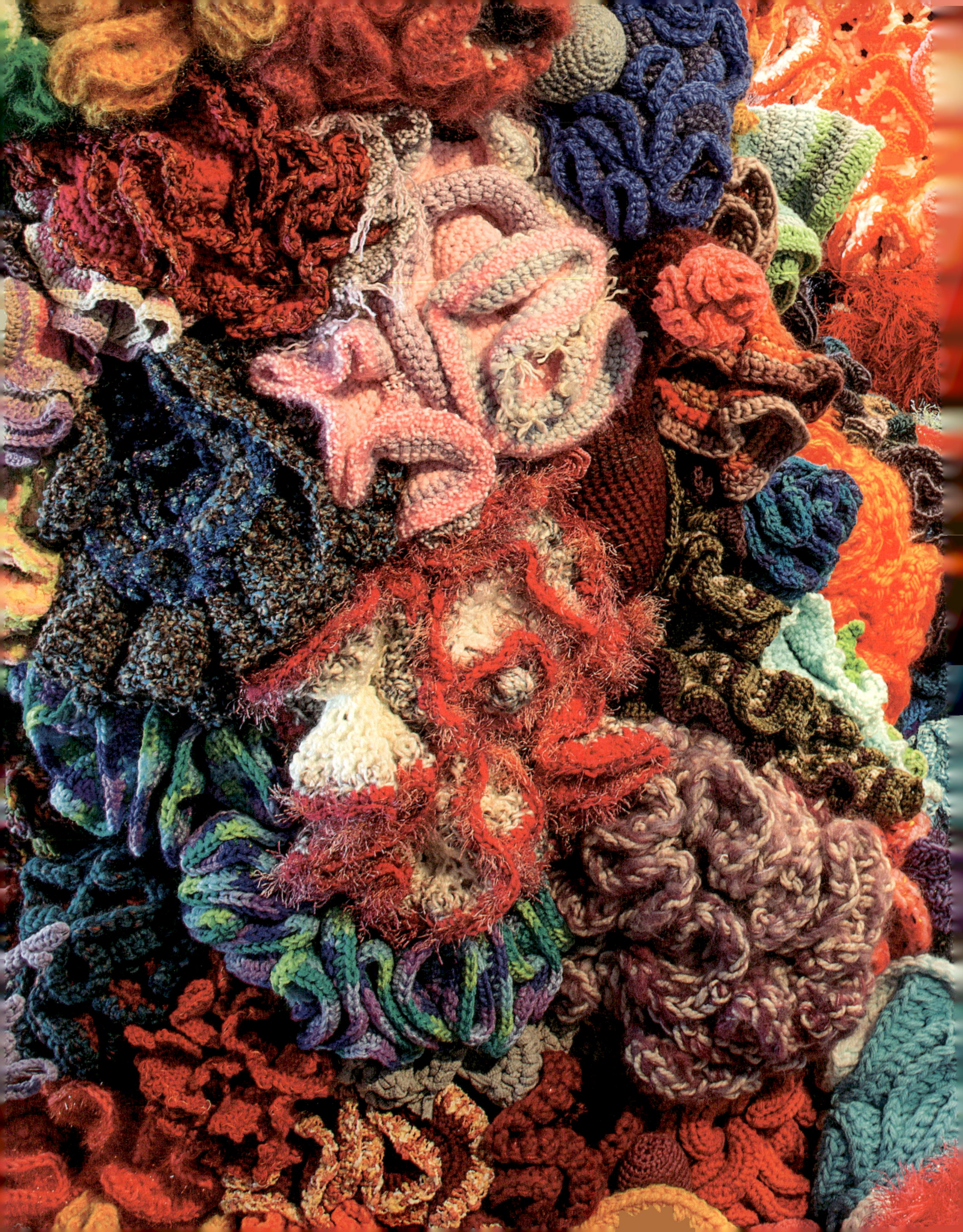